$$\iint f(x,y)\, dy\, dx \qquad \text{double integral of } f$$

$\{x \mid \quad\}$ set of all x such that

$\in$ is an element of

ϕ or $\{\ \}$ empty set

$\subset$ subset

$\cup$ union

$\cap$ intersection

$\times$ Cartesian product

A' complement of Λ

R   set of real numbers or the number line

I   set of integers

Q   set of rational numbers

Z   set of irrational numbers

$[a,b]$ closed interval from a to b

(a,b) either the point with coordinates a and b or the open interval from a to b (depending on context)

$[a,\infty), (a,\infty),$
$(-\infty,a], (-\infty,a)$ half-lines

A SHORT
CALCULUS
an applied
approach

A SHORT CALCULUS
an applied approach

DANIEL SALTZ
California State University, San Diego

REVISED EDITION

Goodyear Publishing Company, Inc.
Pacific Palisades, California

© 1974 by Goodyear Publishing Company, Inc.
Pacific Palisades, California

Current printing (last digit):
10 9 8 7 6 5 4 3 2 1
ISBN 0-87620-842-1
Library of Congress Catalog Card Number: 73-91644
Y-8421-3

Printed in the United States of America

Contents

1

FUNCTIONS AND GRAPHS

2

LIMITS THE DERIVATIVE

3 APPLICATIONS OF THE DERIVATIVE

Supplementary Topics:
Applications of the Derivative to Economics

4 THE INTEGRAL AS AN AREA

5 SEQUENCES. EXPONENTIAL AND LOGARITHMIC FUNCTIONS

6 MORE ON INTEGRATION

Supplementary Readings:
Applications of the Integral to the
Biological Sciences and Economics

7 TAYLOR POLYNOMIALS AND SERIES

8 MULTIVARIABLE CALCULUS

9   THE TRIGONOMETRIC FUNCTIONS

Appendix

Part 1. SETS

Part 2. A REVIEW OF ALGEBRA

Introduction

This book is meant to introduce some of the important concepts and methods found in calculus, and to indicate how these concepts can be used in constructing mathematical models in such areas as economics, the life sciences, and psychology. To keep on the mainstream of the general ideas of calculus, attention is restricted in the early chapters to those functions that are obtained by taking powers and roots of ratios of polynomial functions, together with exponential and logarithmic functions. Once these basic concepts are understood for these functions, it is not difficult to extend them to the trigonometric functions, and this is precisely what is done in the last chapter.

The principal audience addressed is the nonmathematician, nonphysical scientist. The philosophy guiding this book is that to fully appreciate and understand any model using calculus—be it the demand equations in economics, or the probabilistic learning models, or the numerous statistical models—it is necessary to have an appreciation of the underlying mathematical concepts. It is also the philosophy of this text that the most impelling and convincing arguments that can be given to justify mathematical facts are arguments based on the geometry of the problems; that is, an appropriate picture clarifies and makes the fact more transparent. Thus, there are no involved formal proofs. The only "proofs" that appear are those involving the algebraic properties of limits, since such proofs give additional practice in working with limits. Limit itself is defined informally and geometrically; there is no attempt to give a formal definition; and the basic properties of limits are given essentially as an axiom.

The opening chapters in the book introduce the calculus of one real variable. Every effort has been made to motivate and discuss the concepts in terms of examples from the life sciences and economics. In addition to this type of motivation accompanying the concepts as they arise, two mini-chapters called **Supplementary Topics** and **Supplementary Readings** pursue even further the use of calculus in application. Multivariable calculus is introduced toward the end of the book. The final chapter uses the concepts of calculus to study the trigonometric functions. The choice of topics throughout the book was guided by existing models and applications in the life sciences, as well as by topics that might have some future value to the reader (for example, the chapter in multivariable calculus should prove useful in the study of some probabilistic or statistical models).

Finally, a two-part appendix appears as a handy, quick reference source in elementary set theory and basic algebra. The algebra section is deliberately brief; its main purpose is to provide a quick review and examination of algebraic techniques and formulas.

A NOTE TO THE READER

Some of the examples in the book commencing in Chapter 2 where calculus begins should be completed by filling in the appropriate blanks or boxes. Such examples are placed where their practice will aid in the understanding of the technique being discussed. Answers appear at the end of the *section* in which the example is found. If there is no ambiguity, the correct answers are written one next to the other, each answer filling in the appropriate blank. If the example reads:

$$10 = 5 + \underline{\hspace{1cm}} = 3 + \underline{\hspace{1cm}}$$

the answers will read: 5, 7.

We have enclosed certain formulas, examples and results which should be emphasized in heavy horizontal lines running the length of the type column. This will help the reader pick out the important concepts.

A NOTE TO THE INSTRUCTOR

This book can be used in two different ways. A class that is unfamiliar with basic set concepts (for example, the notation for sets, and the concepts of set union and intersection), or whose algebra is weak, should first read one or both parts of the two-part appendix, the first on sets and the other on algebra. The class can then proceed to Chapter 1. A class with a stronger background in algebra (two years of high school algebra, or a basic algebra course at the college level, or the equivalent, should suffice) can start immediately in Chapter 1.

A SHORT
CALCULUS
an applied
approach

1

FUNCTIONS
AND
GRAPHS

The central concern of this chapter is the concept of function. Preliminary to this concept, we first introduce the notion of a coordinate plane in which it is possible to identify points (in a plane) with pairs of numbers. This will provide a setting in which geometrical objects can be classified in terms of algebraic equations. For example, we shall derive an algebraic "formula" for the distance between points, and later algebraic equations for geometrical objects such as lines and circles. Conversely, starting from an algebraic equation, we can ask for the associated picture, or *graph*, of the equation. Finally, we introduce functions and their graphs.

A word of caution: Section 1 uses the fact that it is possible to identify real numbers with points on a line, and it uses the concept of absolute value. If these ideas are new to you or if you wish to refresh your memory, they are discussed in sections 1 and 2 of part 2 of the appendix. Section 4 (the section on functions) uses the concepts of interval and half-line. These concepts are discussed in section 8, part 2 of the appendix.

SECTION 1

THE COORDINATE PLANE. DISTANCE

A fairly standard exercise is to show that points on a line can be identified with real numbers.* We shall now show that points in a plane can be identified with *pairs* of real numbers, called the **coordinates** of the point. This identification will allow us to see that we are able to describe many geometrical configurations in the plane by algebraic equations involving their coordinates.

To accomplish the identification, we construct a so-called **coordinate plane** as follows. Choose a horizontal real line in the plane and suppose that the real numbers are already placed on the line.

Next, construct a vertical line perpendicular to the given line, intersecting it at 0, and labeling it the same way as the horizontal line by using the same unit interval with increasing values pointing up. Call the horizontal line the **x**-axis, or the **horizontal axis** and the vertical line the **y**-axis, or the **vertical axis.** To-

---

*For a more complete discussion of this identification, see section 1, part 2 of the appendix.

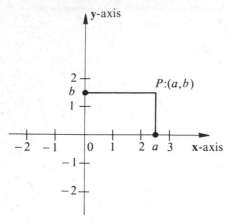

Figure 1-1

gether, these axes are called the **coordinate axes.** We are now in a position to label points in the plane. From the point P, draw a perpendicular to the **x**-axis, and suppose this perpendicular intersects the **x**-axis at $x = a$. Construct a perpendicular from P to the **y**-axis, and suppose the point at which it intersects this axis is $y = b$ (see Figure 1-1). We call the value a the **x-coordinate** of P, the value b the **y-coordinate** of P. We label P with the pair of real numbers (a,b) in that order; that is, x-coordinate first, y-coordinate second. Observe that for each point P there are unique real numbers, $x = a$, $y = b$ for which P can be labeled (a,b), and for each pair of real numbers (a,b) there is a unique point in the plane with coordinates (a,b). In this way, we have identified all points in the plane with pairs of real numbers. This scheme will be necessary when we wish to construct graphs in the next section. Example 1 illustrates some points in the plane.

EXAMPLE 1   See Figure 1-2 in which various points in the plane are plotted.

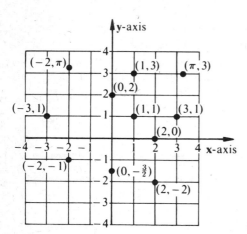

Figure 1-2

As an illustration of the usefulness of this coordinate scheme, we shall find the length of the line segment between two points, say P and Q; that is, we shall find the *distance between P and Q*. Suppose that we have a coordinate plane, and $P = (a,b)$ and $Q = (c,d)$, with $a \neq c$ and $b \neq d$ (that is, the line is neither vertical nor horizontal).

As in Figure 1-3, construct the right triangle PQR with hypotenuse the segment between P and Q. In the diagram, R is the point (c,b). From the theorem of Pythagoras

$$PQ^2 = PR^2 + RQ^2$$

where PQ is the length of the segment from P to Q, with PR and RQ defined similarly. In Figure 1-3, $c > a$, $d > b$, $PR = c - a$, and $RQ = d - b$. More precisely, since, in general, we would not know whether $c > a$ or $c < a$, we would set PR equal to the absolute value* of $c - a$ (that is, $PR = |c - a|$) to guarantee that $PR > 0$. Similarly, we would set $RQ = |d - b|$. Hence, if we denote the distance between P and Q by D, we see that

$$D^2 = |c - a|^2 + |d - b|^2$$

But $|c - a|^2 = (c - a)^2,$ $|d - b|^2 = (d - b)^2$

(Why?) Hence,

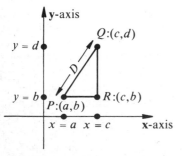

Figure 1-3

(1.1) $$D = ((c - a)^2 + (d - b)^2)^{1/2}$$

*Absolute value is discussed in section 2, part 2 of the appendix.

is the **distance between the points** (a,b) **and** (c,d), and, hence, also the distance between (c,d) and (a,b).

Formula (1.1) is also valid, though more simple, if the line segment is parallel to an axis. In fact, if the line through P and Q is parallel to the y-axis, then $a = c$, and we can see immediately that the distance between P and Q is $|d - b|$, which is exactly what we obtain if we set $a = c$ in (1.1).

EXAMPLE 2 Find the distance between $(-1, 2)$ and $(3, 5)$.

SOLUTION The required distance is

$$\sqrt{(-1 - 3)^2 + (2 - 5)^2} = \sqrt{16 + 9} = 5$$

EXAMPLE 3 Find a relationship between numbers x and y for which the point (x, y) is equidistant from the points $(1,3)$ and $(-1,2)$.

SOLUTION Note that the square of the distance from (x,y) to $(1,3)$ is $(x - 1)^2 + (y - 3)^2$, whereas the square of the distance from (x,y) to $(-1,2)$ is $(x + 1)^2 + (y - 2)^2$. Thus, (x,y) is equidistant from $(1,3)$ and $(-1,2)$ if and only if

$$(x - 1)^2 + (y - 3)^2 = (x + 1)^2 + (y - 2)^2$$

or, equivalently

$$x^2 - 2x + 1 + y^2 - 6y + 9 = x^2 + 2x + 1 + y^2 - 4y + 4$$

or, finally

$$4x + 2y - 5 = 0$$

That is, the point (x,y) is equidistant from $(1,3)$ and $(-1,2)$ if and only if the coordinates of the point, x and y, satisfy the algebraic relationship $4x + 2y - 5 = 0$. We shall see in a later section that

$$\{(x,y): \quad 4x + 2y - 5 = 0\}$$

corresponds to a straight line in the plane (in fact, the line that is the perpendicular bisector of the segment connecting $(1,3)$ to $(-1,2)$).

Problems In problems 1–6, plot the given points and then find the distance between them.

1. $(1,0)$ and $(5,0)$
2. $(-5,3)$ and $(-5,-7)$
3. $(2,3)$ and $(-2,5)$
4. $(-1,1)$ and $(3,-2)$
5. $(-2,-3)$ and $(4,1)$
6. $(-2,-1)$ and $(-1,-2)$

7. Any three points in a plane which do not lie on a common line are the vertices of a triangle. Show that each of the following sets form right triangles with vertices A, B, and C.
   (a) A: $(1,0)$, B: $(5,3)$, C: $(4,-4)$
   (b) A: $(0,1)$, B: $(2,3)$, C: $(0,5)$
   (Hint: Use the theorem of Pythagoras.)

8. Find all numbers x for which the distance between $(-1,x)$ and $(2,0)$ is 5.

9. Find a relationship between x and y for which the point (x,y) is equidistant from the points $(1,3)$ and $(-2,4)$.

10. Show that if point P with coordinates (x,y) is equidistant from the point $(0,4)$ and the x-axis, then $x^2 - 8y + 16 = 0$.

SECTION 2
GRAPHS

A **graph** *is any collection of points.* Hence, a graph in a coordinate plane is a collection of points (x,y) where x and y are numbers. The following examples illustrate graphs whose coordinates satisfy an algebraic relation.

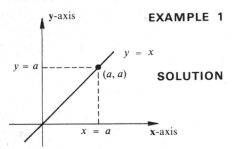

Figure 1-4

EXAMPLE 1   Graph the set of points (x,y) for which $y = x$; that is, graph the equation $y = x$.

SOLUTION   Every point (x,y) for which $y = x$ (that is, every point (x,x)) is equidistant from each of the coordinate axes. Thus, the desired graph is the straight line that makes a 45° (counterclockwise) angle with the positive horizontal axis. See Figure 1-4.

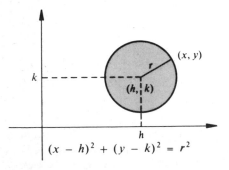

Figure 1-5

EXAMPLE 2   A **circle** is defined as the set of all points in a plane at a fixed distance, say r, from a fixed point, say (h,k) (see Figure 1-5). We can use the formula for distance to find the algebraic equation of the circle. Let (x,y) be any point on the circle; then the distance D from (x,y) to (h,k) is given by $D = \sqrt{(x-h)^2 + (y-k)^2}$, and this has the constant value of r, so that

$$r = \sqrt{(x-h)^2 + (y-k)^2}$$

or, equivalently

$$r^2 = (x-h)^2 + (y-k)^2$$

Hence, this is the equation of the circle of radius r and center

(h,k). If the center of the circle is $(0,0)$, the circle has the special form

$$r^2 = x^2 + y^2$$

EXAMPLE 3   Describe geometrically

$$\{(x,y): \quad x^2 + y^2 + 2x - 2y = 23\}$$

SOLUTION   Let x and y be numbers; completing the square (see Appendix 1, section 7), we see that (x,y) satisfies the relationship $x^2 + y^2 + 2x - 2y = 23$ if and only if it satisfies

$$x^2 + 2x + y^2 - 2y = 23$$
$$(x^2 + 2x + 1) + (y^2 - 2y + 1) = 23 + 1 + 1$$

or, finally

$$(x + 1)^2 + (y - 1)^2 = 25$$

Hence, the graph of $\{(x,y): \quad x^2 + y^2 + 2x - 2y = 23\}$ is the circle with radius 5 and with center $(-1,1)$.

EXAMPLE 4   The graph of the equation

$$(x - 1.2)^2 + (y + \sqrt{2})^2 = 3.7$$

is the circle with radius $\sqrt{3.7}$ and with center $(1.2, -\sqrt{2})$.

EXAMPLE 5   Graph the equation $y = x^2$.

SOLUTION   For the time being, we shall rely on plotting points. Later we shall see that calculus will settle our present uncertainties about the shape. If $x = 0$, then $y = x^2 = 0^2 = 0$; thus $(0,0)$ is a point on the graph. If $x = 3$, then $y = x^2 = 3^2 = 9$; and thus $(3,9)$ is a point on the graph. If $x = 5$, then $y = x^2 = 5^2 = 25$; and so $(5,25)$ is on the graph. In Figure 1-6 we construct a table, plot points, and draw a smooth curve through these points. Note that whenever (x,y) is on the graph, so is $(-x,y)$, since $y = x^2 = (-x)^2$.

x	y	(x,y)
-2	4	$(-2,4)$
-1	1	$(-1,1)$
$-\frac{1}{2}$	$\frac{1}{4}$	$(-\frac{1}{2}, \frac{1}{4})$
0	0	$(0,0)$
$\frac{1}{2}$	$\frac{1}{4}$	$(\frac{1}{2}, \frac{1}{4})$
1	1	$(1,1)$
2	4	$(2,4)$

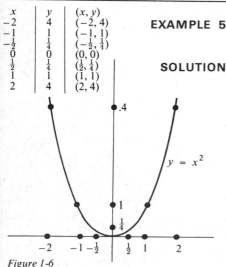

$y = x^2$

Figure 1-6

EXAMPLE 6   Graph each of the equations (a) $y = x^2 + 1$, (b) $y = x^2 - 1$, (c) $y = x^2 + 2$, (d) $y = x^3$, and (e) $y = x^3 + 1$.

SOLUTION   The graphs are sketched in Figure 1-7.   You should construct tables to convince yourself of the general shape.   Note the graph of $y = x^2 + c$ is the graph of $y = x^2$ moved up c units if $c > 0$ and down $|c|$ units if $c < 0$.

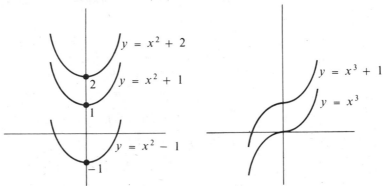

Figure 1-7

Problems   In problems 1–23, sketch a graph of the set of (x,y) which satisfies the given condition (that is, graph the given equation).

1. $y = x + 1$ 2. $y = x + 3$
3. $y = x - 1$ 4. $y = x - 3$
5. $y = -x$ 6. $y = -x + 1$
7. $y = -x - 2$ 8. $y = 3x$
9. $y = 3x + 1$ 10. $y = 3x - 1$
11. $y = x^2 + 3$ 12. $y = x^2 - 5$
13. $y = -x^2$ 14. $y = -x^2 + 1$
15. $y = -x^3 + 2$ 16. $y = -x^3$
17. $y = x^4$ 18. $y = -x^4$
19. $y^2 = x$ 20. $(x - 3)^2 + (y + \frac{3}{2})^2 = 4$
21. $(x + \sqrt{2})^2 + (y - \pi)^2 = 16$
22. $(x - \pi + 1)^2 + (y - \pi + 2)^2 = 25$
23. $(x + 3.5)^2 + (y + 1.5)^2 = \frac{9}{4}$

In problems 24–27, find the center and radius for each of given circles: the set of all points in the plane (x,y) for which

24. $x^2 + y^2 - 2x - 4y = 1$
25. $x^2 + y^2 - 4x + 4y + 3 = 0$
26. $x^2 + x + y^2 - 8y = -\frac{21}{4}$
27. $x^2 + y^2 - 4x + 6y + 11 = 0$
28. Sketch a graph of the set:

$$\{(x,y): \ |x| + |y| = 1\}$$

SECTION 3

LINES IN A PLANE

We shall now investigate lines in a (coordinate) plane. Specifically, we shall derive equations of lines.

We know from plane geometry that a line L is determined by any point on L and some measure of the inclination, or direction, of L. For example, a line is completely determined if we know the angle θ the line makes with the positive horizontal axis together with a point on the line. However, the degree (or radian) measure of θ turns out not to be useful; instead, we introduce the following measure for θ; that is, we introduce the following measure for the direction of a line.

Let L be a nonvertical line (vertical lines will be considered separately), and let (a,b) and (a',b') be two points on L (since L is not vertical, we see that $a \neq a'$). We define the **slope** of the line L, which we denote by m, as follows:

$$m = \frac{b - b'}{a - a'}$$

The slope m is a measure of the inclination of the line L (in fact, those of you who know trigonometry will recognize that $m = \tan \theta$ where again θ is the angle between L and the positive horizontal axis). By using arguments from plane geometry that involve similar triangles, we can show that each (nonvertical) line has *precisely one* slope, and that the slope is independent of the particular points used to compute it. See Figure 1-8 (in this figure, the pair of points (a,b) and (a',b') are used to find m,

Figure 1-8

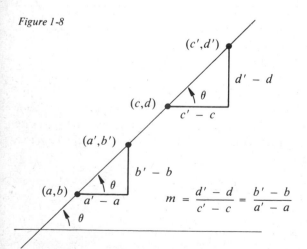

$$m = \frac{d' - d}{c' - c} = \frac{b' - b}{a' - a}$$

and then the points (c,d) and (c',d') are used). Observe also that, since slope is a measure of inclination and lines with the same inclination are parallel, *parallel lines have the same slope.* Some qualitative results on slope are given in Figure 1-9.

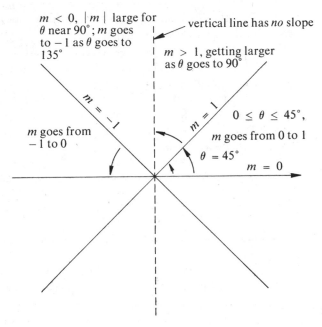

$m < 0,\ |m|$ large for θ near $90°$; m goes to -1 as θ goes to $135°$

vertical line has *no* slope

$m > 1$, getting larger as θ goes to $90°$

$m = -1$

m goes from -1 to 0

$m = 1$

$0 \le \theta \le 45°$, m goes from 0 to 1

$\theta = 45°$

$m = 0$

Figure 1-9

EXAMPLE 1 The slope of the line through the points $(-2,3)$ and $(4,-5)$ is
$$\frac{-5-3}{4-(-2)} = \frac{-8}{6} = -\frac{4}{3}.$$

We shall next derive an equation for L, *assuming that we know a point on L, say (a,b), and that the slope of L is m.* Let (x,y) be any point on L distinct from (a,b). If we use (x,y) and (a,b) to compute m, we see that

$$\frac{y-b}{x-a} = m$$

or, equivalently

$$y - b = m(x - a)$$

Notice that (a,b) also satisfies this condition, since $b - b = m(a - a)$. Thus, if (x,y) is on L, then $y - b = m(x - a)$. The converse is also true (although we shall not show it); hence, *(x,y) is a point on the line with slope m and containing (or passing through) the point (a, b) if and only if*

(1.2) $$y - b = m(x - a)$$

Equivalently, *(1.2) is an equation for the line with slope m and containing (or passing through) the point (a,b).*

EXAMPLE 2   Find an equation of the line passing through $(3, -4)$ and whose slope is 7.

SOLUTION   Using formula (1.2), (x,y) is on this line if and only if

$$y - (-4) = 7(x - 3)$$
$$y + 4 = 7x - 21$$

or, finally
$$y = 7x - 25$$

and hence, $y = 7x - 25$ is an equation of the line. Notice that $y - 7x = -25$, $2y = 14x - 50$, and $-7x + y + 25 = 0$ are also equations for this line.

Given two points on a line, say (a,b) and (a',b'), the slope of the line is $(b - b')/(a - a')$, and hence from (1.2), *an equation of the line containing (a,b) and (a',b') is*

(1.3) $$y - b = \frac{b - b'}{a - a'} (x - a)$$

EXAMPLE 3   Find an equation of the line through $(2,1)$ and $(-1,4)$.

SOLUTION   An equation of this line is

$$y - 1 = \frac{4 - 1}{-1 - 2} (x - 2) = -(x - 2)$$

or $$y - 1 = -x + 2$$
or $$y = -x + 3$$

Thus, (x,y) is on the line if and only if $y = -x + 3$. (Note that we get the same equation if we use the point $(-1,4)$ instead of $(2,1)$.)

EXAMPLE 4   Is $(4,2)$ on the line given in Example 3?

SOLUTION $(4,2)$ is on the line if and only if, with $x = 4$ and $y = 2$, we have $y = -x + 3$; that is, we have $2 = -4 + 3$. Since $2 \neq -1$, it follows that $(4,2)$ is *not* on the line.

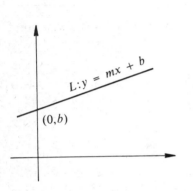

Figure 1-10

Suppose next that we have a line L with slope m that crosses the vertical axis at $(0,b)$. See Figure 1-10. Then, by (1.2) the equation of L is $y - b = m(x - 0)$, or

(1.4) $$y = mx + b$$

In (1.4) the number b is called the **y-intercept** of the line, and formula (1.4) is called the **y-intercept form for L**. Note that every nonvertical line must cross the vertical axis somewhere, and thus *every nonvertical line has the form (1.4)* for some m and some b.

We shall now consider vertical lines. Suppose we have a vertical line that passes through $(3,0)$. Then for every real number y, note that $(3,y)$ is on this line, and thus the equation of the line is $x = 3$. See Figure 1-11.

In general, *the equation of the vertical line passing through $(a,0)$ is $x = a$.*

Since we know that the graph of the equation $y = mx + b$ (where m and b are in **R**) is a straight line, all we have to do to find the line which is the graph of the equation is to find *any* two points which satisfy the equation.

Figure 1-11

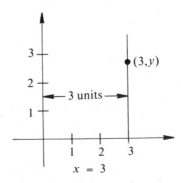

EXAMPLE 5   Graph the equation $y = -3x + 2$.

SOLUTION   The graph is a line. Thus, all we have to do is to find *any* two points on the graph. For example, if $x = 0$, then $y = 2$, and $(0,2)$ is on the line; and if $x = 1$, then $y = -1$, and $(1,-1)$ is on the line. See Figure 1-12.

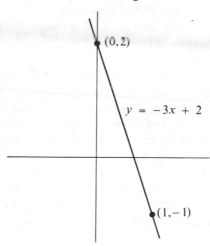

Figure 1-12

Problems   In problems 1–5, find the equation of the lines through the specified points:

1. $(1,4)$ and $(3,-2)$ 2. $(2,-4)$ and $(-6,-4)$
3. $(-\frac{1}{2},1)$ and $(1,2)$ 4. $(1,1)$ and $(1,6)$
5. (a,b) and $(0,d)$, $a \neq 0$, $b \neq 0$, $d \neq 0$.

In problems 6–11, graph the given equation. Give the slope and y-intercept of each line.

6. $y = -3x + 5$ 7. $y = 2x - 1$
8. $y = (x/2) + 5$ 9. $y = (x/2) - 1$
10. $2x - 3y = 7$ 11. $x + 2y = 1$

In problems 12–15, find the slope of the line whose equation is given, and then find where the line intersects each axis.

12. $2x + 3y - 1 = 0$ 13. $-x - y + 4 = 0$
14. $4y + 4 = 0$ 15. $2x - 3y - 1 = 0$

In problems 16–18, find the equation of the line:

16. Through $(2,-1)$ and parallel to the line given by $y = -2x + 1$.
17. Through $(-1,-4)$ and parallel to the line given by $y = 3x + 1$.
18. Through $(-2,1)$ and parallel to the line given by $y = 2x + 1$.

SECTION 4
FUNCTIONS

The function concept is central to our work in the remainder of this book. In calculus, we shall analyze numerical functions. The specific definitions of these objects will appear later. Our concern first is to analyze the general concept of function.

Loosely speaking, a **function** consists of two sets and a rule which associates (or pairs) elements in one set with elements in the other. For example, assume that we have a gas in a closed container, and we want to determine the effect of the temperature of the gas on the pressure the gas exerts on the walls of the container. The function we have in mind consists of two sets: the set of admissible temperatures and the set of possible pressures of the gas (on the walls of the container). The rule we want is the one which associates to each element in the first set of temperatures the associated pressure from the second set.

We shall say that *a function from (the set)* A to (the set) B is a rule which associates to each element of A precisely one element of B*. In Figure 1-13, where $A = \{a,b,c\}$ and $B = \{1,2,3\}$, three distinct functions f, g, and h, are indicated, each by the appropriate arrows; that is, the arrows show which element in B is being associated, or paired, with the corresponding element in A.

Thus, from Figure 1-13(a), f is the function that "sends" or

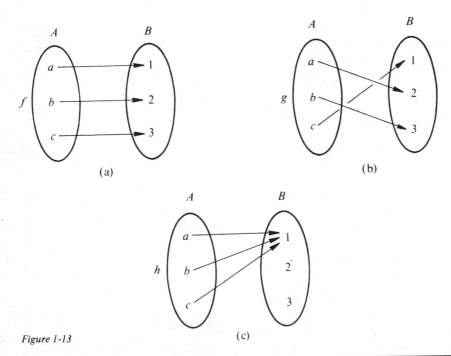

Figure 1-13

"maps" *a* to 1, *b* to 2, and *c* to 3. A common notation for this is: $f(a) = 1, f(b) = 2, f(c) = 3$. Thus, from Figure 1-13(b), $g(a) = 2, g(b) = 3, g(c) = 1$, and in Figure 1-13(c), $h(a) = h(b) = h(c) = 1$. Note that, unlike the functions *f* and *g*, the function *h* does not map *A* onto all of *B*, but only to the number 1. Also notice that, as required by our definition of function, each element in *A* is mapped into precisely one element in *B* (for the function *h*, the element happens to be the same for *a*, *b*, and *c*). These functions should be contrasted to the pairing indicated in Figure 1-14, which is *not* a function from *A* to *B*, since *a* is *not* mapped into precisely one element, but instead is mapped into 2 and 3. If the arrows are reversed, would the mapping from *B* to *A* be a function?

A bit more terminology: If *f* is a function from *A* to *B*, then *A* is called the **domain** of *f* and *B* is called the **codomain** of *f*. We say that *f is undefined at a*, or, *f*(a) *is undefined, if a is not in the domain of f*. The set of all numbers $f(x)$ (as *x* varies over *A*) is called the **range** of *f*; that is,

$$(\textbf{The range of f}) = \{f(x): x \in (domain\ of\ f)\}$$
$$= \{y: y = f(x)\ and\ x \in (domain\ of\ f)\}$$

Two functions *f* and *g* which have the same domain are said to be *equal*, denoted by $f = g$, if $f(x) = g(x)$ for all *x* in their common domain.

The concept of function is really a very natural one for many areas of application. For instance, when the physicist says that the position of a particle can be given as a function of time, he is asserting that there is some relation connecting time with the position of the particle and that, at a given time, the particle is to be found in no more than one position. When the economist says that demand is a function of price, he means that price and demand are related in such a way that to each price (in some domain of prices), there corresponds exactly one demand. Observe, however, that the same demand might result from different prices (the analogy here is that there corresponds, at most, one *y* to each *x*, but different *x*'s might correspond to the same *y*).

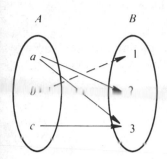

A   *B*

Figure 1-14

EXAMPLE 1   In an early learning-theory model proposed by L. L. Thurstone ("The Learning Curve Equation," *Psychological Bulletin*, 1917), the quantity *x*, the number of times one must practice an act (in order to be able to complete that act successfully) was related to *y*, the number of successful acts, by the equation

$$y = \frac{a(x + b)}{x + c}$$

(where *x* is a positive integer) where *a*, *b*, and *c* are certain empirical constants. Thus, the number of successful acts is functionally related to the number of practice acts; that is, there is a "successful act" function, call it G, given by

$$G(x) = \frac{a(x + b)}{x + c}$$

where x, a positive integer, is the number of times one must practice an act. The domain of G is some set of positive integers.

EXAMPLE 2   Assume that a gas is placed in a container and the volume of the gas is held fixed. If we denote the "pressure function" by P, the pressure (on the walls of the container) as a function of (absolute) temperature t is given by

$$P(t) = ct$$

where c is a constant that depends on such things as the volume of the gas.

EXAMPLE 3   Let $R(x)$ and $C(x)$ be the total revenue received and the total costs incurred, respectively, in the production of x units of some product. R and C are functions, since to each x, the number of units produced, we correspond exactly one total received revenue and one total cost incurred. The function P given by

$$P(x) = R(x) - C(x), \qquad \text{with} \qquad x \geq 0$$

is the **profit function**. Typically, $P(0) = 0$, since we expect no profits if no commodities are produced. The problem of finding all x (with $x > 0$) such that $P(x) = 0$ is called the **break-even problem**; that is, the number x for which $P(x) = 0$ is the point or time at which setup costs, such as developmental costs, are recovered and the product becomes profitable.

In the following examples, we shall give some **numerical functions**; that is, functions whose domain and range are each sets of real numbers (and hence subsets of **R**).

EXAMPLE 4   Let $f(x) = x$, $x \in \mathbf{R}$; that is, let f be the function which makes each real number x correspond to the same number x. Hence, $f(0) = 0$, $f(-1) = -1$, $f(7) = 7$, $f(125) = 125$, etc. It is legitimate to call this relation a function, since to every real number x we associate exactly one real number, namely, the number x itself. The domain of the function is the entire *x-axis*, and the range is the entire *y-axis*.

EXAMPLE 5 Let $g(x) = x^2 + 1$, $x \in \mathbf{R}$; some values of g are

$$g(-1) = (-1)^2 + 1 = 2, g(0) = 1, g(1) = 2, g[(2)^{1/2}] = 3, \text{etc.}$$

Observe that this is a functional relation, since to each x there corresponds exactly one number, x^2, and hence exactly one number $x^2 + 1$. The domain is the entire **x**-axis. Observe, however, that $g(x) = -2$ has no real solution, since $x^2 + 1 = -2$ implies that $x^2 = -3$, and there is no real number x whose square is -3; hence, -2 is not in the range of 9. We shall exhibit the graph of this function later. Observe that $g(1) = g(-1) = 2$, but this does not violate the definition of function.

EXAMPLE 6 Let $f(x) = (x)^{1/2}$, for $x \geq 0$. For each nonnegative number x there corresponds exactly one number $\sqrt{x}$. For example, $f(2) = (2)^{1/2}, f(4) = 2, f(16) = 4$.

EXAMPLE 7 Let $f(x) = 2$, $x \in \mathbf{R}$. This states that to each x, f associates the same number 2. Thus, $f(-1) = 2$, $f(0) = 2$, $f(1,000,000) = 2$. The domain is the set of all real x; the range is the set containing the single number 2.

EXAMPLE 8 Let $f(x) = |x|$, $x \in \mathbf{R}$. By definition, f is the **absolute value function**. Observe $f(0) = |0| = 0$, $f(-1) = |-1| = 1$, $f(1) = |1| = 1$, $f(-2) = f(2) = 2$, and for any real number a, $f(a) = f(-a) = |a|$. The domain of f is the entire real line, and the range is the nonnegative real line (that is, the closed half-line $J: y \geq 0$).*

It will be useful to establish a **domain convention** to be used henceforth throughout the book:

Unless otherwise stated, the (implicit) domain of a numerical function will always be taken to be the set of all real numbers for which the function makes sense (as a real number).

For example, if $f(x) = x$, we will understand that the (implicit) domain of f is all of $\mathbf{R}$, whereas if $f(x) = \sqrt{x}$, we will assume that the domain is (implicitly) $\{x: x \geq 0\}$.

EXAMPLE 9 Let $f(x) = 1/x$. By our domain convention, the domain of f is

*More precisely, $J = \{y: y \geq 0\}$. Half-lines and intervals are discussed in section 8, part 2 of the appendix.

the set of all numbers except 0. Observe $f(0.1) = 10$, $f(0.01) = 100$, and $f(1/10^n) = 10^n$ for any positive integer n. Hence, if x is close to 0 and positive, $f(x)$ is large and positive. Also, $f(10) = 1/10$, $f(100) = 1/100$, and $f(10^n) = 1/10^n$ for any positive integer n; in fact, if x is positive and large, $f(x)$ is positive and near 0. Similar remarks can be made for $x < 0$.

EXAMPLE 10   Let $f(x) = \dfrac{x^2 - 1}{x - 1}$. We shall analyze f.

We know that $\dfrac{x^2 - 1}{x - 1} = \dfrac{(x + 1)(x - 1)}{x - 1}$

If $x - 1 \neq 0$, that is, if $x \neq 1$, then

$$\frac{(x + 1)(x - 1)}{x - 1} = x + 1$$

(division by 0 is *not* meaningful). But

$$f(1) = 0/0$$

which is meaningless; that is, $f(1)$ is undefined (recall this means that 1 is *not* in the domain of f). For $x \neq 1$,

$$f(x) = x + 1$$

which is everywhere defined. Hence, the domain of f is all x except $x = 1$. Finally, note that $f(x) = (x^2 - 1)/(x - 1)$, with $x \neq 1$, can also be described by $f(x) = x + 1$ for $x \neq 1$.

EXAMPLE 11   Let $[x]$ denote the greatest integer that is $\leq x$. For instance, $[3/2] = 1$, $[1/2] = 0$, $[7.5] = 7$, $[5] = 5$, $[-1.5] = -2$, and $[-3.1] = -4$.

Let $f(x) = [x]$. Then, by definition, f is the **greatest integer function**. Thus, $f(3/2) = [3/2] = 1$, $f(1/2) = [1/2] = 0$, etc.

$$f(-1.5) = -2 \quad \text{and} \quad f(-3.1) = -4.$$

There is a convenient picture we can associate with numerical functions. To each x in the domain of f, associate the number $f(x)$ and plot the point $(x, f(x))$.

The set of all points $(x, f(x))$ in the plane obtained as x varies over the domain of f is called the **graph** *of f.*

Equivalently,

$$(\text{Graph of } f) = \{(x, f(x)): x \in (\text{domain } f)\}$$

In Figure 1-15, we graph the functions given in Examples 4 to

11. Note that, if $f(x) = x$, the graph of f is the graph of the equation $y = x$; and if $f(x) = x^2 + 1$, the graph of f is the graph of the equation $y = x^2 + 1$; these equations were graphed in section 2 of this chapter.

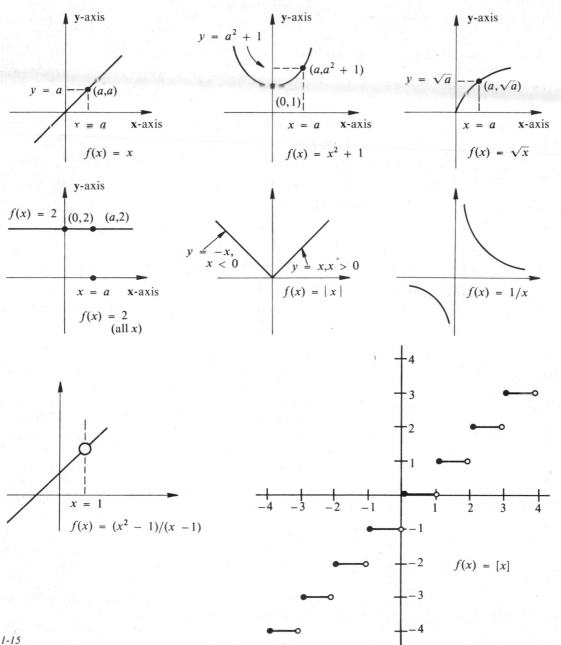

Figure 1-15

EXAMPLE 12   Let $f(x) = \sqrt{x^2 - 2x - 3}$. Find the implicit domain of f.

SOLUTION   The domain of f consists of all numbers x for which $x^2 - 2x - 3 \geq 0$. To solve this inequality,* note that for $x \in \mathbf{R}$,

$$x^2 - 2x - 3 = (x + 1)(x - 3)$$

Thus (using Figure 1-16, for example), the implicit domain of f is**

$$\{x: \ x \leq -1\} \cup \{x: \ x \geq 3\}$$

or, equivalently $(-\infty, -1] \cup [3, \infty)$

Figure 1-16

$$x + 1 \quad - - - - - - \ 0 + + + + + + + + + +$$

$$x - 3 \quad - - - - - - - - - - - \ 0 + + + + +$$

$$(x + 1)(x - 3) \quad + + + + + + \ 0 - - - - \ 0 + + + + +$$

$$\underset{-1 \qquad\qquad 3}{\rule{6cm}{0.4pt}}$$

The expression "y is a function of x" shall mean that there is a function, say f, which makes values y correspond to certain values x, and we write this expression as $y = f(x)$. Since there is nothing special about the symbol f, the equation $y = g(x)$ or $y = \theta(x)$ also designates "y is a function of x."

Such notation as "$y = y(x)$" is often used in mathematical literature. Although logically faulty, the notation is highly suggestive. The logical difficulty is that the y on the right of the equality, representing the *function*, is conceptually different from the y on the left side of the equality, representing the *real number* on the y-axis obtained by applying the function to x. The notation is suggestive because it implies that the result of applying the function to x (that is, $y(x)$) is exactly y. With this word of caution, the symbol $y = y(x)$ will appear occasionally where circumstances make it useful. In the same spirit of utility (over propriety), we shall occasionally refer to "the function $f(x) = x^2$" or "the function $g(x) = x^3$" when we mean "the function f whose value at x is x^2" or "the function g given by $g(x) = x^3$," respectively.

The first class of functions that we shall consider in the calculus will be the **polynomial functions** (in one variable). Let $a_0, \ldots, a_n$

---

*See section 8, part 2 of the appendix.
**This set notation is discussed in section 2, part 1 of the appendix.

be $n + 1$ real numbers, which we shall think of as fixed. Suppose further that $a_0 \neq 0$. The function P given by

$$P(x) = a_0x^n + a_1x^{n-1} + \cdots + a_{n-1}x + a_n$$

is called a **polynomial function in x of degree n.** It is a function of x. (Why?)

EXAMPLE 13   1. $P(x) = x^2 - 2x + 1$ is a polynomial of degree 2.
2. $P(x) = x^{15} - 3x^6 + x$ is a polynomial of degree 15.
3. $P(x) = 2$ is a polynomial of degree 0.

We conclude with a brief observation concerning when a relation is *not* a function.

EXAMPLE 14   Is the graph in Figure 1-17 the graph of a function (whose domain is a subset of the horizontal axis)?

SOLUTION   No, since there are points x for which more than one y corresponds; that is, there are points x for which there is *more* than one point (x,y) on the graph (for example, the point x in Figure 1-17).

Figure 1-17

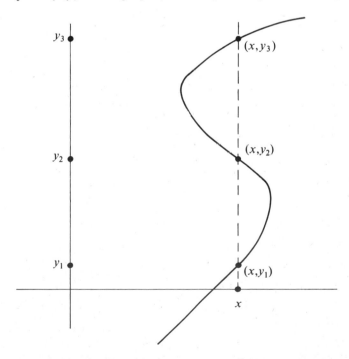

Problems   1. Let $f(x) = x^4 + 1$. Find $f(-1)$, $f(0)$, $f(1)$, $f(2)$, $f(1/2)$, $f(-1/2)$, $f(1/a)$, $a \neq 0$.

2. Let $f(x) = x^2 + (1/x) - 1$. Find $f(-2)$, $f(-1)$, $f(1)$, $f(2)$, $f(a + 1)$ where $a \neq -1$. Why is there this restriction on a?

3. Let $f(x) = x^3 - x^2 + x + 1$. Find $f(0)$, $f(1)$, $f(-1)$, $f(2)$, $f(-2)$.

In problems 4–8, find the implicit domain of the given function.

4. $f(x) = (x^2 - x - 2)^{1/2}$

5. $f(x) = (-x^2 - x + 2)^{1/2}$

6. $f(x) = (x^2 + 1)^{1/2}$

7. $f(x) = \dfrac{1}{x^2 - 1}$

8. $f(x) = \dfrac{1}{(x^2 - 1)^{1/2}}$

9. Draw the graph of the set:

$$\left\{ (x,y) : \quad y = \frac{x^2 - 9}{x - 3} \right\}$$

Is "y a function of x"?

In problems 10–22, graph the given function.

10. $f(x) = 3x - 8$

11. $f(x) = -4x + 1$

12. $f(x) = x^2 + 1$

13. $f(x) = x^2 - 3$

14. $f(x) = (1/x) + 1$

15. $f(x) = \dfrac{1}{x + 1}$

16. $f(x) = 1/x^2$

17. $f(x) = (1/x^2) - 5$

18. $f(x) = x|x|$

19. $f(x) = |x + 1|$

20. $f(x) = |x| + 1$

21. $f(x) = x/|x|$

22. $f(x) = x - [x]$

23. Let $f(x) = \begin{cases} 1 \text{ if } x \geq 0 \\ -1 \text{ if } x < 0 \end{cases}$. Find $f(-3)$, $f(-2)$, $f(0)$, $f(1)$, and $f(2)$. Graph f.

24. Let $f(x) = \begin{cases} x^2 \text{ if } x \geq 1 \\ x \text{ if } x < 1 \end{cases}$. Find $f(-5)$, $f(-4)$, $f(-3)$, $f(-2)$, $f(-1)$, $f(0)$, $f(1)$, $f(2)$, $f(3)$, $f(4)$, and $f(5)$. Graph f.

25. Let x and y satisfy the relationship $x^2 + y^2 = 1$. Can y be expressed as a function of x? (That is, is there a function f for which $y = f(x)$?)

26. Let $f(x) = \sqrt{x}$, $g(x) = 1/x$. Let c and h be real numbers, $h \neq 0$. Evaluate

$$\frac{f(c + h) - f(c)}{h} , \frac{g(c + h) - g(c)}{h}$$

27. In the learning theory model of Example 1, suppose that $a = 5, b = 2$, and $c = 10$; therefore

$$G(x) = \frac{5(x + 2)}{x + 10}$$

How many times must an act be practiced in order that there will be
a. One successful act?
h. Two successful acts?
c. Or n successful acts, where n is a positive integer?
Assume for this problem that an act can be practiced a fractional number of times.

28. Let $A = \{a,b,c\}$ and $B = \{1,2,3\}$. How many functions are there with domain A and range B if we require that distinct values of B are associated with distinct elements of A?

SECTION 5

THE ALGEBRA OF FUNCTIONS

Starting from numerical functions f and g, we can construct "new" functions, the **sum** and **product** functions, in a rather natural way. For example, if $f(x) = x^2$, $x \in \mathbf{R}$, and $g(x) = 3x$, $x \in \mathbf{R}$, then by $f + g$, we mean the function that associates the number $x^2 + 3x$ to each number x; that is, if $f(x) = x^2$, and $g(x) = 3x$, then $(f + g)(x) = x^2 + 3x$. Similarly (still with f the function given by $f(x) = x^2$, $x \in \mathbf{R}$), the *function* $7f$ is the function which associates the number $7x^2$ to each number x; that is, if

$$f(x) = x^2$$

then $$(7f)(x) = 7x^2$$

Note that $f + g$ and $7f$ are again functions!

EXAMPLE 1

Let f, g, and h be functions given by $f(x) = x^2$, $g(x) = 3x$, and $h(x) = 1$ (in each case, for $x \in \mathbf{R}$). Then

1. $(f - g)(x) = x^2 - 3x$
2. $(2f + 3g)(x) = 2x^2 + 9x$
3. $(-2f + 2g - 5h)(x) = -2x^2 + 6x - 5$

In general, then, if f and g are functions and c is any number, the *sum function* $\mathbf{f} + \mathbf{g}$ is given by

$$(f + g)(x) = f(x) + g(x)$$

and the *product of f by c*, **cf**, is given by

$$(cf)(x) = c \cdot f(x)$$

The **product** of two functions is defined in a similar way. For example, if $f(x) = x^2$ and $g(x) = 3x$, then $(f \cdot g)(x) = 3x^3$. If $u(x) = x^5$ and $v(x) = 3x^{10}$, then $(uv)(x) = 3x^{15}$. Some care must be taken with the domain of a product function. For example, if $f(x) = 1/x$, $x \neq 0$, and $g(x) = x^2$, all x, then $(fg)(x) = (1/x) \cdot x^2 = x$, but the domain of fg is all x except 0 (because $(1/x) \cdot x^2 = x$ is true only for nonzero x).

EXAMPLE 2 We shall compute some product functions.

1. If $f(x) = 5x$, all x,
    and $g(x) = x + 1$, all x,
    then $(fg)(x) = 5x(x + 1) = 5x^2 + 5x$, all x.
2. If $f(x) = x + 4$, all x,
    and $g(x) = x - 4$, all x,
    then $(fg)(x) = x^2 - 16$, all x.
3. If $f(x) = 1/(x^2 - 16)$, $x \neq 4$, $x \neq -4$,
    and if $g(x) = x^2 + 4x$, all x,
    then $(fg)(x) = \{1/[(x + 4)(x - 4)]\} \cdot x(x + 4) = x/(x - 4)$,
    with $x \neq 4$, $x \neq -4$.
4. If $f(x) = 3x^4$,
    then $f^2(x) = (f \cdot f)(x) = 9x^8$
    $f^3(x) = (f^2 \cdot f)(x) = 27x^{12}$
    and $(f)^{1/2}(x) = x^2 \cdot (3)^{1/2}$.

In general, if f and g are functions, then the *product function*, **fg**, is given by

$$(fg)(x) = f(x) \cdot g(x)$$

and the *power function*, **f^n**, is given by

$$f^n(x) = [f(x)]^n$$

(whenever $[f(x)]^n$ is meaningful).

An important reason for introducing such functions is that it is often possible to predict the nature of the sum or product function by knowing the behavior of each of the constituent functions. This will become more apparent as the calculus is developed.

Another important operation between functions is their so-called *composition*. An easy way to explain the composition of functions is by illustration.

EXAMPLE 3   Let $f(x) = u^2 + 3$. We now consider the case that u is also a function. Assume that $u(x) = 2x - 1$. Then, substituting

$$f[u(x)] = [u(x)]^2 + 3$$
$$= (2x - 1)^2 + 3$$
$$= 4(x^2 - x + 1)$$

If we let $w(x) = 4(x^2 - x + 1)$ then

$$f[u(x)] = w(x)$$

Thus, from the functions f and u, we have constructed a function w, which we call the **composition of f with u**. In particular, for example

$$f[u(0)] = w(0) = 4$$
$$f[u(2)] = w(2) = 12$$

and $$f[u(-1)] = w(-1) = 12$$

Equivalently

$$u(0) = -1 \quad \text{implies} \quad f[u(0)] = f(-1) = 4$$
$$u(2) = 3 \quad \text{implies} \quad f[u(2)] = f(3) = 12$$

and $$u(-1) = -3 \quad \text{implies} \quad f[u(-1)] = f(-3) = 12$$

In general, the composition of f by u is denoted by $f \circ u$. *Thus, $f \circ u$ is the function whose value at x is given by*

$$(f \circ u)(x) = f[u(x)]$$

and in this example, $(f \circ u)(x) = w(x) = 4(x^2 - x + 1)$.

EXAMPLE 4   Assume that $d(p)$ is the demand for a commodity if it is sold at a price p. However, the price at which the commodity is sold depends, in turn, on the manufacturing cost, c, of the commodity.

Thus, demand is a function of price, and price is a function of cost, and, therefore, we have a "function of a function," the composition $d \circ p$, with

$$d \circ p(c) = d[p(c)]$$

Thus, $d \circ p$ is the function whose value at *cost c* is found by evaluating the demand at the value of the price that must be charged if the manufacturing cost is c.

Problems Let $f(x) = x^2$, and let $g(x) = 1$. In problems 1–4, graph the given function.

1. $2f + g$ 2. $3f - 2g$
3. $f^2 + g$ 4. $f^2 + f$
5. Let $f(x) = 1/(x - 1)$ and $g(x) = x/(x^2 - 1)$. Find all points where the graph of $f + g$ crosses (or touches) the x-axis.
6. Let $f(x) = 1/(x^2 - 1)$ and $g(x) = x/(x^2 - x - 2)$. Find all x for which $(f + g)(x) = 0$.

In Problems 7–11, find the required composition, $f \circ u$.

7. $f(u) = u^2 + 1, u(x) = 2x$
8. $f(u) = u^2 + u, u(x) = 2x - 1$
9. $f(u) = (u - 1)^3, u(x) = 3x + 1$
10. $f(u) = \sqrt{u}, u(x) = x^2$
11. $f(u) = u^4, u(x) = x + 1$ (*Hint:* Use the binomial theorem.)

2

LIMITS.
THE
DERIVATIVE.

An important question in applications is "How fast?" How fast is a population growing? Is money flowing in an economy? Is a process taking place? Is a car moving? The rate-of-change problem (or, equivalently, the velocity problem) is one of the major considerations of this chapter. To characterize such problems, a concept called **limit** is needed. Therefore, the first discussion in this chapter is that of LIMIT—one of the most profound and significant concepts in all of mathematics.

Although rate of change is one of the most important uses for the limit, we will encounter others; for example, limit will be used as an aid in analyzing graphs of functions.

The rate-of-change problem soon gets subsumed under another heading, the derivative, of which it is an example. Another example of the use of the derivative is the geometrically significant concept of tangent line to the graph of a function.

A slight problem then arises. Although the derivative is needed in order to find the velocity, it is often difficult to compute from "scratch" (that is, from its basic definition). The chapter ends with a discussion of the algebra of derivatives (the formulas for computing derivatives as painlessly as possible).

SECTION 1
LIMITS

The concept of **limit** is the concept that distinguishes calculus from the other mathematical disciplines. In a real sense, calculus is the study of limits. Even though the concept plays a central role in calculus, it is one of those concepts which eludes immediate application to a nonmathematical area. With apologies, all we can do in this section is to introduce the concept with a "wait and see how important it is" attitude. As we shall see in subsequent chapters, ideas such as rate of change of processes, maximum and minimum values of quantities, and elasticity of demand (in economics) are formed from the limit concept. Rather than attempting to formulate a precise definition of limit, the direction will be instead to develop an intuitive understanding of the concept.

The idea of limit is centered about the notion of "closeness."

Until now we have asked for the value of a function *at* a point, say c (that is, for the number $f(c)$). We now ask (and this in essence is the limit question): What number is $f(x)$ "near" when x is "near" c? Let us consider some elementary examples.

EXAMPLE 1   Let $f(x) = x^2 + 1$, x in **R** (that is, for all real numbers x). Let us arbitrarily pick the number 2 and ask: What number does $f(x)$ stay arbitrarily near when x is near 2? That is, as numbers x get "closer and closer" to 2, is there a number to which the corresponding numbers $f(x)$ get (and stay) "closer and closer"? We might check some values near 2. For example,

If $x = 1.9$,   then $f(x) = f(1.9) = (1.9)^2 + 1 = 4.61$
If $x = 1.99$,   then $f(x) = f(1.99) = (1.99)^2 + 1 = 4.9601$
If $x = 1.999$, then $f(x) = f(1.999) = (1.999)^2 + 1 = 4.996001$

We could continue in this way (in fact, we should also consider the possibility that $x = 2.01$, $x = 2.001$, $x = 2.0001$, etc.) and increase our intuitive feeling that as x gets closer and closer to 2, x^2 gets closer and closer to 4, and thus, $f(x) = x^2 + 1$ gets closer and closer to 5. We abbreviate these statements about closeness by saying: the limit of $f(x)$ as x approaches (or, tends to) 2 is 5, and we write

$$\lim_{x \to 2} f(x) = 5$$

or

$$\lim_{x \to 2} (x^2 + 1) = 5$$

(See Figure 2-1.)

In a completely analogous way, choosing, say, $x = 1$, we see that as x gets closer and closer to 1, x^2 gets closer and closer to 1; hence, $f(x) = x^2 + 1$ gets closer and closer to 2; and thus,

$$\lim_{x \to 1} f(x) = \lim_{x \to 1} (x^2 + 1) = 2$$

Similarly, choosing $x = -2$,

$$\lim_{x \to -2} f(x) = \lim_{x \to -2} (x^2 + 1) = 5$$

(that is, as x gets closer and closer to -2, x^2 gets closer and closer to 4, and thus, $x^2 + 1$ gets closer and closer to 5). In fact, in general, for any number c,

$$\lim_{x \to c} f(x) = \lim_{x \to c} (x^2 + 1) = c^2 + 1$$

Figure 2-1

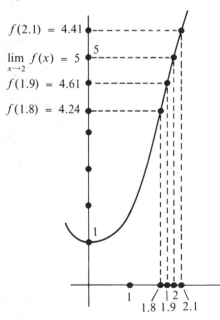

$f(2.1) = 4.41$

$\lim_{x \to 2} f(x) = 5$

$f(1.9) = 4.61$

$f(1.8) = 4.24$

1

1

1.8 1.9 2.1

As x gets closer and closer to 2, $f(x)$ gets closer and closer to 5.

We shall not give any formal definition of the limit of a function. An informal "definition," however, is suggested by the examples we have just worked:

lim$_{x \to c}$ *f(x)* = *L means that we can always arrange it so that if we take numbers x that are sufficiently close to (but not equal to) c, then the corresponding numbers f(x) will stay as close to the number L as we please.*

A more formal treatment of limit would, of course, be based on our previously stated informal definition; *we now want to consider more precisely the meaning for*

$$\lim_{x \to c} f(x) = L$$

In a more formal treatment, we would need to verify that, no matter how close we insist that $f(x)$ is to L (this is the "as close as we please" part of our informal definition), there is a corresponding "sufficiently close" distance between x and c which will insure that the distance between $f(x)$ and L is within the prescribed "degree of closeness." We shall not pursue this. The next examples are designed to increase our intuition about our informal meaning of limit.

EXAMPLE 2   Find $\lim\limits_{x \to 3} (2x^2 + 3x - 1)$.

SOLUTION   We break the problem into parts. First, as x gets closer and closer to 3 (that is, as x "approaches" 3), we see that x^2 gets closer and closer to, or approaches, 9, and thus $2x^2$ gets closer and closer to 18. Next, as x approaches 3, we see that $3x$ approaches 9. Finally, since -1 is a fixed number, as x approaches 3, we see that -1 remains unaffected as it is still simply -1. Accumulating these observations, we see that as x approaches 3, $f(x)$ approaches $18 + 9 - 1$, or 26. Thus,

$$\lim_{x \to 3} (2x^2 + 3x - 1) = \lim_{x \to 3} (2x^2) + \lim_{x \to 3} (3x) + \lim_{x \to 3} (-1)$$
$$= 18 + 9 - 1$$
$$= 26$$

"Just a moment," you might say, "all this seems like much ado about nothing. In Example 1, $\lim_{x \to 2} f(x) = \lim_{x \to 2} (x^2 + 1) = 5$, but $f(2)$ *is* 5. If you want to find $\lim_{x \to 2} f(x)$, simply compute $f(2)$. In fact, to find $\lim_{x \to c} f(x) = \lim_{x \to c} (x^2 + 1)$, simply compute $f(c) = c^2 + 1$, and you get the answer without all these considerations of 'nearness.' Example 2 isn't any different."

And for the examples we have given so far, you would be right! The problem is that we have chosen very elementary examples to illustrate the limit concept as a concept of nearness, so as not to confuse the issue at first exposure. We shall see later that functions with the property that $f(c)$ is the value that $f(x)$ approaches as x approaches c are, in some sense, very nice functions and are called **continuous at c**. Let us now turn to a slightly more sophisticated example. The type of limit we are about to consider (in Example 3) is more representative of the limit process that we will soon consider for the limit process called the **derivative**. But we had better not get too much ahead of the game.

EXAMPLE 3   Let $f(x) = (x^2 - 1)/(x - 1)$, $x \neq 1$. Find $\lim_{x \to 1} f(x)$.

SOLUTION   Let us first observe a fundamental difference between this problem and the problems in Examples 1 and 2. In the earlier examples, because the domain of the function considered was in each case all of **R**, we necessarily took a limit as x approached a point in the domain. In this example, we are asking for the limit as x approaches 1, but 1 is *not* in the domain of f. Thus, in this example if there is a limit, *it cannot be $f(1)$ since there is no number $f(1)$*. (In fact, if we set $x = 1$, we get $(1^2 - 1)/(1 - 1) = 0/0$, which is meaningless.) The question then is whether or not *$f(x)$ approaches some number as x approaches 1*.

Let us view the problem geometrically. We are interested in what happens to f as x *approaches* 1; thus, we are concerned with *x near* 1, and *not equal* to 1. To graph f, note that for $x \neq 1$, we have

$$f(x) = \frac{x^2 - 1}{x - 1} = \frac{(x + 1)(x - 1)}{x - 1} = x + 1$$

Hence, a completely *equivalent description* of f is

$$f(x) = x + 1 \text{ if } x \neq 1 \quad \text{and} \quad f \text{ is not defined at } 1$$

Consequently, the graph of f is a straight line with a hole above 1. See Figure 2-2. Thus, since $f(x) = x + 1$ for $x \neq 1$, we see that $f(x)$ approaches 2 as x approaches 1; that is,

$$\lim_{x \to 1} f(x) = \lim_{x \to 1} \frac{x^2 - 1}{x - 1}$$
$$= \lim_{x \to 1} (x + 1)$$
$$= 2$$

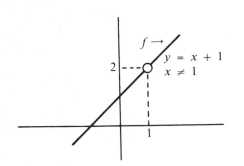

Figure 2-2

but $2 \neq f(1)$.

Let us look more closely at the graph in Figure 2-2, and the related function. Notice that, since there is a hole in the graph above $x = 1$, the graph is not a "continuous" curve at $x = 1$. Here, $\lim_{x \to 1} f(x)$ (that is, the number 2) is the number of units in the y direction that we would have to go above the number 1 (on the **x**-axis) so as to make the graph of f a continuous curve; the point $(1, \lim_{x\to 1} f(x)) = (1,2)$ is exactly the point which, when inserted into the graph of f, will make the graph a continuous curve.

Thus, we have come to our first application of limit. Given a function f and a number c:

If $\lim_{x\to c} f(x)$ is precisely what you would get by plugging c into f (that is, if $\lim_{x\to c} f(x) = f(c)$) then the graph of f is a continuous curve at c (as in Examples 1 and 2; see, for instance, Figure 2-1), and we say that the function f is **continuous** *at c.*

If $\lim_{x\to c} f(x)$ exists but is $\neq f(c)$ (as in Example 3, where, with $c = 1$, there *was no* number $f(c) = f(1)$), then the graph of f is discontinuous at c. Moreover, in this case, $\lim_{x\to c} f(x)$ gives us the information necessary to patch the graph so that it *will* be a continuous curve; specifically, $\lim_{x\to c} f(x)$ is the number of units above (or below if the limit is negative) the point c at which we must insert the point which will make the graph continuous at c. We shall pursue these ideas in section 4.

Before going further, we shall use the next example to dispel the idea that limits always "exist."

EXAMPLE 4   Discuss $\lim_{x\to 0} 1/x$.

SOLUTION   The question is this: as x approaches 0, what, if anything, does $(1/x)$ approach? If $(1/x)$ does *not* approach a number, we say that $\lim_{x\to 0} (1/x)$ does not exist. We can try some numbers x that are near 0. For $x > 0$ and "near" 0,

If $x = 0.01$,   then $\dfrac{1}{x} = \dfrac{1}{0.01} = \dfrac{1}{10^{-2}} = 100$

If $x = 0.001$,   then $\dfrac{1}{x} = \dfrac{1}{0.001} = \dfrac{1}{10^{-3}} = 1000$

If $x = 10^{-8}$,   then $\dfrac{1}{x} = \dfrac{1}{10^{-8}} = 10^8$

For $x < 0$ and "near" 0,

If $x = -0.01$,   then $\dfrac{1}{x} = \dfrac{1}{-0.01} = -100$

If $x = -0.001$,   then $\dfrac{1}{x} = \dfrac{1}{-0.001} = -1000$

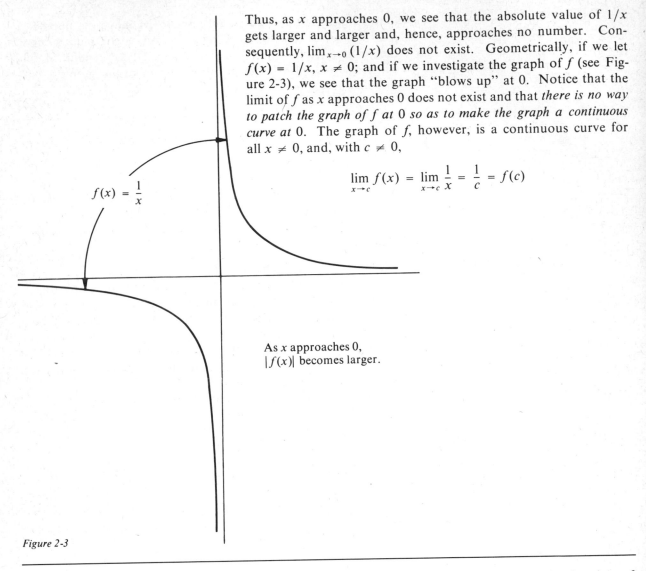

Thus, as x approaches 0, we see that the absolute value of $1/x$ gets larger and larger and, hence, approaches no number. Consequently, $\lim_{x \to 0} (1/x)$ does not exist. Geometrically, if we let $f(x) = 1/x$, $x \neq 0$; and if we investigate the graph of f (see Figure 2-3), we see that the graph "blows up" at 0. Notice that the limit of f as x approaches 0 does not exist and that *there is no way to patch the graph of f at 0 so as to make the graph a continuous curve at* 0. The graph of f, however, is a continuous curve for all $x \neq 0$, and, with $c \neq 0$,

$$\lim_{x \to c} f(x) = \lim_{x \to c} \frac{1}{x} = \frac{1}{c} = f(c)$$

$f(x) = \dfrac{1}{x}$

As x approaches 0, $|f(x)|$ becomes larger.

Figure 2-3

A word on terminology: Example 3 is an example of a class of problems in which we are asked to find $\lim_{x \to c} f(x)/g(x)$ and in which, when we let $x \to c$, we obtain a meaningless "0/0" form; that is, when $\lim_{x \to c} f(x) = 0$ and $\lim_{x \to c} g(x) = 0$. Such problems are usually handled by an appropriate factoring, as in Example 3. When the 0/0 form arises in a limit, it is referred to as an **indeterminate form**. As the previous example and the next two illustrate, when an indeterminate form arises, there may or may not be a limit; in all cases, further investigation (usually depending on algebra) is necessary.

EXAMPLE 5   Find $\lim_{x\to 0}(x^2 + x)/x$.

SOLUTION   Because $\lim_{x\to 0}(x^2 + x) = 0$ and $\lim_{x\to 0} x = 0$, we have an indeterminate form. By factoring, we see that

$$\lim_{x\to 0}\frac{x^2 + x}{x} = \lim_{x\to 0}\frac{x(\quad\quad)}{x} = \underline{\quad\quad\quad}$$

and thus there is a limit.

EXAMPLE 6   Find $\lim_{x\to 0}(x^2 + x)/x^2$.

SOLUTION   This again is an indeterminate form. But this time

$$\lim_{x\to 0}\frac{x^2 + x}{x^2} = \lim_{x\to 0}\frac{x(x + 1)}{x^2}$$

$$= \lim_{x\to 0}\frac{x + 1}{x}$$

$$= \lim_{x\to 0}\left(1 + \frac{1}{x}\right)$$

which *does not exist*. (Why?)

Answers: Example 5: $x + 1$, 1

Problems   In problems 1–8, find the limit.

1. $\lim_{x\to -1}\dfrac{x^2 + 3x}{x + 4}$

2. $\lim_{x\to 1}\dfrac{4x^2 + 3}{x + 1}$

3. $\lim_{x\to 3}(x^2 + 3x - 7)$

4. $\lim_{x\to 0}(x^{10} - 15x^8 - 3x^2 + 10)$

5. $\lim_{x\to 1/2}\dfrac{4x + 1}{6x - 7}$

6. $\lim_{x\to 0.1}\dfrac{[x^2 + (1/100)]}{[x + (1/10)]}$

7. $\lim_{x\to -3}(|x| + x)$

8. $\lim_{x\to -4}(|x| - x^2 + 2x)$

9. Let $f(x) = x^2/x$. Find $\lim_{x\to 0} f(x)$. Graph f.

10. Let $f(x) = x^2/x^2$. Find $\lim_{x\to 0} f(x)$. Graph f.

11. Let $f(x) = \dfrac{(x - 1)^2}{x - 1}$. Find $\lim_{x\to 1} f(x)$. Graph f.

12. Let $f(x) = \dfrac{(x - 1)^2}{(x - 1)^2}$. Find $\lim_{x\to 1} f(x)$. Graph f.

13. Let $f(x) = \dfrac{x^2 - x}{x}$. Find $\lim_{x\to 0} f(x)$. Graph f.

14. Let $f(x) = \dfrac{x^2 - x - 2}{x^2 - 2x}$. Find $\lim\limits_{x \to 2} f(x)$. Graph f.

15. Let $f(x) = \dfrac{x^2 + 2x + 1}{x^2 + x}$. Find $\lim\limits_{x \to -1} f(x)$. Graph f.

In problems 16–29, find the limit.

16. $\lim\limits_{x \to 2} \dfrac{x^4 - 16}{x - 2}$
    17. $\lim\limits_{x \to 1} \dfrac{x^4 - 1}{x^2 - 1}$

18. $\lim\limits_{x \to 5} \dfrac{x^2 - 3x - 10}{x^2 - 6x + 5}$
    19. $\lim\limits_{x \to 1} \dfrac{x^2 - 3x - 10}{x^2 - 6x + 5}$

20. $\lim\limits_{x \to -2} \dfrac{x^2 - 3x - 10}{x^2 - 6x + 5}$

21. Let $f(x) = \dfrac{x^2 + 2x - 8}{x^2 + 6x + 8}$. Find

    a. $\lim\limits_{x \to 2} f(x)$
        b. $\lim\limits_{x \to 0} f(x)$

    c. $\lim\limits_{x \to -4} f(x)$
        d. $\lim\limits_{x \to -2} f(x)$

22. $\lim\limits_{x \to c} \dfrac{(1/x^2) - (1/c^2)}{x - c}$, $c \neq 0$

23. $\lim\limits_{h \to 0} \dfrac{(5 + h)^2 - 25}{h}$

24. $\lim\limits_{x \to 1} \dfrac{\sqrt{x} - 1}{x - 1}$. (*Hint:* One way to find the limit is to rationalize. Another is to note that $x - 1 = (\sqrt{x} + 1)(\sqrt{x} - 1)$.)

25. $\lim\limits_{x \to 1/2} [x]$

26. $\lim\limits_{x \to 3/4} [x]$

(In problems 27–29, n is an integer)

27. $\lim\limits_{x \to n} [x]$

28. $\lim\limits_{x \to n} |[x]|$

29. $\lim\limits_{x \to n} [|x|]$

SECTION 2

MORE ON LIMITS

We continue our investigation in this section of the basic limit concept by focusing attention this time on the "algebra" of limits (that is, the relationship between the sum, difference, product, and quotient of limits and, respectively, the limit of the sum, difference, product, and quotient). Finally, we investigate the meaning of limit for functions whose domains are intervals or half-lines.

In Example 2, section 1, we found that $\lim_{x \to 3} (2x^2 + 3x - 1)$ was simply the sum of the limits $\lim_{x \to 3} 2x^2$, $\lim_{x \to 3} (3x)$, and

$\lim_{x \to 3}(-1)$. In fact, in section 1 we saw that the limit of a sum of functions is simply the sum of their limits, with similar results for differences, products, and quotients. For completeness, we state this result formally.

Assume that $\lim_{x \to c} f(x)$ and $\lim_{x \to c} g(x)$ exist. Let A be any real number. Then the following are true:

(2.1)   If $h(x) = A$ for all x,   then $\lim_{x \to c} h(x) = A$

and this holds for any real number c; that is, the limit of a constant function is a constant. This is usually abbreviated by writing

$$\lim A = A$$

(2.2) $\lim_{x \to c} [f(x) \pm g(x)] = \lim_{x \to c} f(x) \pm \lim_{x \to c} g(x)$

That is, *the limit of a sum (or difference) of functions is equal to the sum (or difference) of the limits,*

(2.3) $\lim_{x \to c} [f(x) \cdot g(x)] = \lim_{x \to c} f(x) \cdot \lim_{x \to c} g(x)$

That is, *the limit of a product is equal to the product of limits,*

(2.4) $\lim_{x \to c} \dfrac{f(x)}{g(x)} = \dfrac{\lim_{x \to c} f(x)}{\lim_{x \to c} g(x)}$ provided $\lim_{x \to c} g(x) \neq 0$

That is, *the limit of a quotient is equal to the quotient of the limits, provided the limit in the denominator is not zero.*

(2.5) $\lim_{x \to c} [f(x)^P] = [\lim_{x \to c} f(x)]^P$ provided each limit exists

We shall observe in the following examples that this result gives no information if one or both of the limits fail to exist.

EXAMPLE 1   Let $f(x) = x, g(x) = 1/x$. What is $\lim_{x \to 0} f(x) g(x)$?

SOLUTION   We know that $\lim_{x \to 0} g(x)$ does not exist. Hence (2.3) does not apply. But

$$\lim_{x \to 0} f(x) g(x) = \lim_{x \to 0} x \cdot (1/x)$$

and since the limit $x \to 0$ automatically insures $x \neq 0$ (x is *near* 0, and thus $\neq 0$), we have $x \cdot (1/x) = 1$. Therefore,

$$\lim_{x \to 0} f(x) g(x) = \lim_{x \to 0} 1 = 1$$

Consequently, although one function in a product may fail to have a limit, the product itself may have a limit, but the previous result cannot be used to find that limit.

EXAMPLE 2   Let $f(x) = 1$, and $g(x) = 1/x$. What is $\lim_{x \to 0} f(x) g(x)$?

SOLUTION   Here, $\lim_{x \to 0} f(x)$ exists and is equal to 1, but $\lim_{x \to 0} g(x)$ does not exist.   Thus, (2.3) does not apply.   We see here that $\lim_{x \to 0} f(x) g(x) = \lim_{x \to 0} 1 \cdot (1/x)$ does not exist.

EXAMPLE 3   Let $f(x) = 1/x$, $g(x) = -1/x$.

SOLUTION   Clearly neither $\lim_{x \to 0} f(x)$ nor $\lim_{x \to 0} g(x)$ exists.   Thus, (2.2) does not apply. However,

$$\lim_{x \to 0} \{f(x) + g(x)\} = \lim_{x \to 0} \{1/x - 1/x\}$$

and, since $x \to 0$ implies $x \neq 0$, $(1/x) - (1/x) = 0$, therefore,

$$\lim_{x \to 0} \{f(x) + g(x)\} = \lim_{x \to 0} 0 = 0$$

Hence, although each of two functions may fail to have a limit, their sum may have a limit.

EXAMPLE 4   Evaluate $\lim_{x \to 2} [x^2(x + 3) + (1/x)]$.

SOLUTION   Notice that

$$\lim_{x \to 2} [x^2(x + 3) + (1/x)] = \lim_{x \to 2} [x^2(x + 3)] + \lim_{x \to 2} (1/x)$$
$$= \left[\lim_{x \to 2} x^2\right]\left[\lim_{x \to 2} (x + 3)\right] + \left[\lim_{x \to 2} (1/x)\right]$$

provided that each limit in the decomposition exists and that the limit that appears in any denominator is not zero. This is certainly the case here; hence,

$$\lim_{x \to 2} [x^2(x + 3) + (1/x)] = (4 \cdot 5) + 1/2 = 41/2$$

EXAMPLE 5   Let $f(x) = x^2 + 3$, x in $[1,3]$. Find $\lim_{x \to 1} f(x)$, $\lim_{x \to 2} f(x)$ and $\lim_{x \to 3} f(x)$.

SOLUTION   Notice that the domain of f is the *closed interval* from 1 to 3; that is, $f(x)$ has meaning if and only if $1 \leq x \leq 3$. Thus, to find $\lim_{x \to 1} f(x)$, it makes no sense to investigate "$\lim_{x \to 1}$" for both the cases $x < 1$ and $x > 1$, since f has no meaning for $x < 1$! To get around this, when we investigate "$\lim_{x \to 1}$," we shall simply require that x be *in the domain of f and near 1* (that is, that x is near 1 with $x > 1$). Thus, by $\lim_{x \to 1} f(x)$, we mean $\lim_{x \to 1 \text{ and } x > 1} f(x)$; hence,

$$\lim_{x \to 1} f(x) = \lim_{x \to 1, x > 1} (x^2 + 3) = 4$$

Similarly, since 3 is the right endpoint of the domain of f,

$$\lim_{x \to 3} f(x) = \lim_{x \to 3, x < 3} f(x) = 3^2 + 3 = 12$$

Since 2 is "inside" the domain of f, to investigate "$\lim_{x \to 2}$," we investigate numbers x that are near 2, with $x > 2$ *and* $x < 2$. We obtain

$$\lim_{x \to 2} f(x) = 7$$

This problem differs from our previous ones, since here we are given a function whose domain is an *interval*, and we are asked to find the limit of $f(x)$ as x tends to an *endpoint* of that interval.

Problems

In problems 1–8, let f and g be functions for which $\lim_{x \to 5} f(x) = 3$ and $\lim_{x \to 5} g(x) = -2$. Compute the given limit.

1. $\lim_{x \to 5} 2f(x) + 3g(x)$

2. $\lim_{x \to 5} \dfrac{2f(x)}{3g(x)}$

3. $\lim_{x \to 5} \dfrac{f(x) - 3}{g(x)}$

4. $\lim_{x \to 5} \dfrac{f(x)}{g(x) - 2}$

5. $\lim_{x \to 5} \dfrac{f(x)}{g(x) + 2}$

6. $\lim_{x \to 5} \dfrac{f(x) + g(x)}{f(x) - g(x)}$

7. $\lim_{x \to 5} \left(\dfrac{1}{f(x)} + \dfrac{2}{g(x)} \right)$

8. $\lim_{x \to 5} \left(\dfrac{f(x)}{g(x)} \right)^3$

In problems 9–12, assume that $\lim_{x \to -1} f(x) = 16$. Use equation (2.5) in the list of properties of limits to find the given limit.

9. $\lim_{x \to -1} \sqrt{f(x)}$

10. $\lim_{x \to -1} \sqrt[4]{f(x)}$

11. $\lim_{x \to -1} \sqrt[4]{(f(x))^3}$

12. $\lim_{x \to -1} \sqrt{f(x) + 9}$

13. Let f be a function with the property that

$$\lim_{x \to 1} f(x) = 8. \text{ Discuss: } \lim_{x \to 1} \frac{f(x)}{x - 1}$$

14. Find: $\lim_{x \to 0} \left(\dfrac{1}{x^2} - \dfrac{1}{x^2} \right)$.

15. Show that $\lim_{x\to 0} [(1/x^2) - (1/x)]$ does not exist by completing the following:

$$\frac{1}{x^2} - \frac{1}{x} = \frac{1-x}{x^2} \qquad \text{(Why?)}$$

As $x \to 0$, $1 - x \to$ _____ and $x^2 \to$ _____; hence, $(1 - x)/x^2$ becomes large. (Why?) A not uncommon, but fallacious, line of reasoning to handle this problem is

$$\lim_{x\to 0} \frac{1}{x^2} = \infty \qquad \text{and} \qquad \lim_{x\to 0} \frac{1}{x} = \infty$$

and thus,

$$\lim_{x\to 0} \frac{1}{x^2} - \frac{1}{x} = \infty - \infty = 0$$

What is wrong with this argument?

16. Let $f(x) = (x^2)^{1/2}$, with $0 \leq x \leq 4$. Find $\lim_{x\to 0} f(x)$ and $\lim_{x\to 4} f(x)$.

17. Let $f(x) = (x^2)^{1/2}$ with $-4 \leq x \leq 0$. Find $\lim_{x\to -4} f(x)$ and $\lim_{x\to 0} f(x)$.

SECTION 3
ASYMPTOTES

When graphing a function whose domain is the entire line, it is valuable to know what happens to the graph when "$|x|$ is very large," with either $x > 0$ or $x < 0$ (or both). Sometimes this information is obvious. For example, if $f(x) = 2x + 1$, it is clear from the graph, which is a straight line, that "$f(x) \to \infty$ as $x \to \infty$" and "$f(x) \to -\infty$ as $x \to -\infty$"; that is, $f(x)$ gets large as x gets large, and with $x < 0$ and $f(x) < 0$, $|f(x)|$ gets large as $|x|$ gets large. But what about the graph of the function g given by $g(x) = [(x + 1)/x]$?

Since

$$g(x) = \frac{x + 1}{x} = 1 + \frac{1}{x}$$

and $1/x$ tends to 0 as x gets large, we see that $g(x)$ tends to 1 as x gets large, and we denote this by

$$\lim_{x\to\infty} g(x) = \lim_{x\to\infty} \left(1 + \frac{1}{x}\right) = 1$$

it should also be clear that

$$\lim_{x\to -\infty} g(x) = \lim_{x\to -\infty} \left(1 + \frac{1}{x}\right) = 1$$

Observe that g is undefined where the denominator is 0; that is, when $x = 0$. To see what happens to the graph of g near 0, note that as $x \to 0$ with $x > 0$, $1/x$ (which is positive) gets large and, hence, so does $g(x)$; whereas, as $x \to 0$ with $x < 0$, we see that

$g(x) < 0$ (for example, $1 + (1/-0.01) = 1 - 100 = -99$) and $|g(x)|$ gets large. We designate this by:

$$\lim_{x \to 0, x > 0} g(x) = \infty, \quad \lim_{x \to 0, x < 0} g(x) = -\infty$$

To further aid in graphing g, note that $g(x) = 0$ if $(x + 1)/x = 0$; that is, if $x = -1$. The graph is given in Figure 2-4. The lines $y = 1$ and $x = 0$ are called, respectively, **horizontal** and **vertical asymptotes** for g. In general, $lim_{x \to \infty} f(x) = L$ means that $f(x)$ can be made to stay arbitrarily close to L by taking x sufficiently large.

Note that, geometrically, $\lim_{x \to \infty} f(x) = L$ means that, as x gets large through positive values, the graph of f "approaches" the horizontal line given by $y = L$, with a similar interpretation if $\lim_{x \to -\infty} f(x) = L'$, in which case the lines $y = L$ and $y = L'$ are horizontal asymptotes for f. See Figure 2-5.

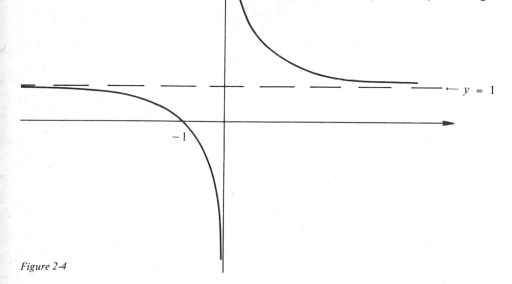

$\leftarrow y = 1$

-1

Figure 2-4

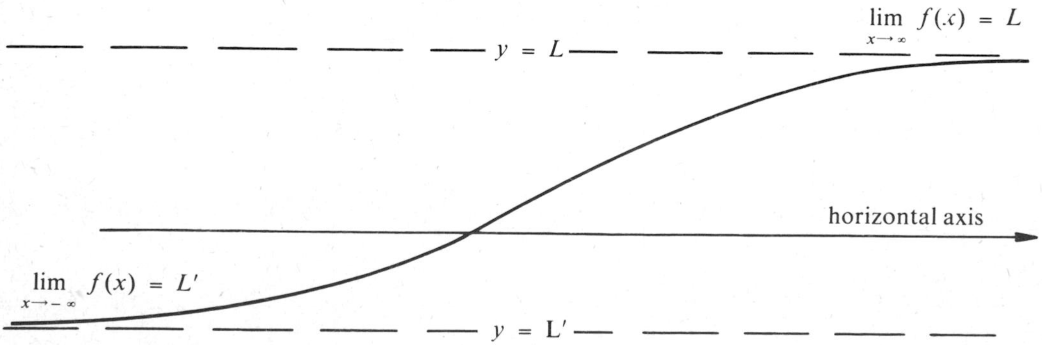

$$\lim_{x \to \infty} f(x) = L$$

$y = L$

horizontal axis

$$\lim_{x \to -\infty} f(x) = L'$$

$y = L'$

Figure 2-5

EXAMPLE 1   Find $\lim_{x \to \infty} (3x + 8)/(5x - 10)$.

SOLUTION
$$\lim_{x \to \infty} \frac{3x + 8}{5x - 10} = \lim_{x \to \infty} \frac{x[3 + (8/x)]}{x[5 - (10/x)]}$$

$$= \lim_{x \to \infty} \frac{3 + (8/x)}{5 - (10/x)}$$

$$= \frac{3}{5}$$

since $8/x$ and $10/x$ approach 0 as x gets large.

EXAMPLE 2
$$\lim_{x \to \infty} \frac{2x^2 - x + 1}{3x^2 + 2x + 5} = \lim_{x \to \infty} \frac{x^2 \left(2 - \dfrac{1}{x} + \dfrac{1}{x^2} \right)}{x^2 \left(3 + \dfrac{2}{x} + \dfrac{5}{x^2} \right)}$$

$$= \lim_{x \to \infty} \frac{2 - \dfrac{1}{x} + \dfrac{1}{x^2}}{3 + \dfrac{2}{x} + \dfrac{5}{x^2}}$$

$$= \frac{2 - 0 + 0}{3 + 0 + 0}$$

$$= 2/3$$

We have used the fact that

$$\lim_{x \to \infty} \frac{1}{x} = 0, \quad \lim_{x \to \infty} \frac{1}{x^2} = 0, \quad \lim_{x \to \infty} \frac{2}{x} = 0, \quad \text{and} \quad \lim_{x \to \infty} \frac{5}{x^2} = 0$$

Another way to approach this problem is to notice that

$$\frac{2x^2 - x + 1}{3x^2 + 2x + 5} = \frac{(2x^2 - x + 1) \cdot (1/x^2)}{3x^2 + 2x + 5 \cdot (1/x^2)}$$

$$= \frac{2 - \dfrac{1}{x} + \dfrac{1}{x^2}}{3 + \dfrac{2}{x} + \dfrac{5}{x^2}} \to \frac{2}{3} \quad \text{as} \quad x \to \infty$$

EXAMPLE 3 $\text{Lim}_{x \to \infty}(\sqrt{x^2 + x} - x) = 1/2$. To see this, observe that, for $x > 0$, we can rationalize to obtain

$$\sqrt{x^2 + x} - x = \frac{\sqrt{x^2 + x} - x}{1} \cdot \frac{\sqrt{x^2 + x} + x}{\sqrt{x^2 + x} + x}$$

$$= \frac{(x^2 + x) - x^2}{\sqrt{x^2 + x} + x}$$

$$= \frac{x}{\sqrt{x^2 + x} + x}$$

$$= \frac{x}{\sqrt{x^2(1 + 1/x)} + x}$$

$$= \frac{x}{x(\sqrt{1 + 1/x} + 1)}$$

$$= \frac{1}{\sqrt{1 + 1/x} + 1} \to \frac{1}{2} \quad \text{as} \quad x \to \infty$$

Problems In problems 1–10, find all the horizontal and vertical asymptotes.

1. $f(x) = \dfrac{x - 1}{2x^2 + 3x + 1}$

2. $f(x) = \dfrac{3x - 5}{x^2 - 2x - 8}$

3. $f(x) = \dfrac{2x^2 - 5x - 3}{x^2 - x - 20}$

4. $f(x) = \dfrac{2x^2 + 6x + 4}{-x^2 + x + 12}$

5. $f(x) = \dfrac{x^2 - 4}{2x^2 - 5x - 3}$

6. $f(x) = \dfrac{3x^2 - 2x}{x^2 - 9}$

7. $f(x) = \dfrac{x^3 - 27}{x^2 - 4}$

8. $f(x) = \dfrac{x + 1}{x^3 - 4x}$

9. $f(x) = \dfrac{x^2 + x}{x^3 - x^2 - 2x}$

10. $f(x) = \dfrac{x^3 + 3x^2 + 2x}{x^3 - x^2 - 2x}$

In problems 11–16, find the limit.

11. $\lim\limits_{x \to \infty} \dfrac{-x^4 + 2x^2 + 3}{2x^4}$

12. $\lim\limits_{x \to \infty} \dfrac{3x^{10} - 5x^2 + 10{,}000}{2x^{11} + 1}$

13. $\lim\limits_{x \to \infty} \dfrac{x^3 - x + 1}{2x^3 + x + 10}$

14. $\lim\limits_{x \to \infty} \dfrac{x^3 + 100x^2 + 10^{10}}{2x^3 - 5{,}000}$

15. $\lim\limits_{x \to \infty} \left(\dfrac{x^{101} + 1}{x^{101} - 1}\right)$

16. $\lim\limits_{x \to \infty} \left(\dfrac{x^3}{x^2 - 2} - \dfrac{x^3}{x^2 + 2}\right)$

17. $\lim\limits_{x \to \infty} [(x^2 + 1)^{1/2} - x]$

18. $\lim\limits_{x \to \infty} [(x^2 - x)^{1/2} - x]$

19. $\lim\limits_{x \to \infty} [(x^4 + x)^{1/2} - x^2]$

20. $\lim\limits_{x \to \infty} [(x^4 + x^2)^{1/2} - x^2]$

21. $\lim\limits_{x \to \infty} [(x^4 + x^3)^{1/2} - x^2]$ **22.** $\lim\limits_{x \to \infty} \{[x(x + b)]^{1/2} - x\}$

23. Assume that the market price $p(x)$ asked for x gallons of oil is given by

$$p(x) = \frac{6x^2 + 7x}{5x^2 + 1} \qquad \text{(dollars)}$$

Show that the asking price stabilizes with increasing demand; that is, the price levels off even though the demand may get larger and larger. What is the price for "infinite" demand?

SECTION 4

CONTINUITY

We have already seen how limit can be used to determine if the graph of a function is a continuous curve at a point (that is, has no holes, breaks, or jumps at that point). We found that if $f(x)$ approaches $f(a)$ as x approaches a, then the graph of f is a continuous curve at a. Thus, for a function whose graph is a continuous curve, all we have to do to compute $\lim_{x \to a} f(x)$ is to "plug" a into f. Putting the emphasis on the function rather than its graph, we shall say that

the function f is **continuous** at a if $\lim\limits_{x \to a} f(x) = f(a)$

(that is, f is continuous at a if its graph is a continuous curve at a).

EXAMPLE 1   Let $f(x) = ax^2 + bx + c$; f is a **quadratic function**, and its graph, a parabola, is certainly a continuous curve. As we expect, for any number w,

$$\lim\limits_{x \to w} f(x) = aw^2 + bw + c = f(w)$$

Remark   The quadratic function discussed in Example 1 is a special case of a polynomial function. It turns out that the graph of any polynomial function is a continuous curve; or, stated in our new terminology, polynomial functions are continuous everywhere.

A function which is not continuous at c is said to be **discontinuous at** c.

EXAMPLE 2   The total cost $c(x)$ of producing and marketing x units of a commodity is assumed to be a function of x alone, and is independent of time and other commodities. It is reasonable to assume that c is continuous everywhere in its domain; that is,

$$\lim\limits_{x \to a} c(x) = c(a)$$

Heuristically, this means that, as the number of commodities produced and marketed gets closer and closer to a (units), the total cost of production approaches the cost of producing a units. Another interpretation is that a small change in the number of items produced is accompanied by a small change in cost. Note that we are implicitly assuming that the commodities in question are fluid like milk or gasoline, and are *not* discrete like shoes or shirts.

EXAMPLE 3   Let $s(t)$ be the position of an object at time t (think of $s(t)$ as the number of miles from "home" that a car has driven in t hours). It is a reasonable physical assumption then that s is continuous; that is, small changes in time result in small changes in distance.

EXAMPLE 4   Let $f(x) = 1/x$, $x \neq 0$. As we have already seen, $\lim_{x \to 0} f(x)$ does not exist, whereas if $c \neq 0$,

$$\lim_{x \to c} f(x) = \lim_{x \to c} \frac{1}{x} = \frac{1}{c} = f(c)$$

Thus, f is discontinuous at 0 (where the graph of f "blows up"; that is, has a vertical asymptote) but f is continuous everywhere else (and the graph of f is a continuous curve except at 0).

We have already seen that polynomial functions are continuous everywhere. Certainly, since positive integer powers of polynomial functions are again polynomial functions, they are also continuous everywhere. For example, if $f(x) = (x^3 + 2x^2 + x - 1)^{20}$, then f is continuous everywhere. Moreover, if $\sqrt[n]{f(c)}$ makes sense (that is, is a real number), then

$$\lim_{x \to c} \sqrt[n]{f(x)} = \sqrt[n]{f(c)}$$

Thus, the "nth root of a polynomial function" is continuous whenever it is defined. When we get into ratios of polynomial functions, however, we encounter difficulty at those points at which the denominator is zero. Such difficulty can lead to a discontinuity at such points; an example of this is found in Example 3 in section 1.

The following is an example of a function whose graph has a "jump" at a point, and thus fails to be continuous at that point. It also provides another case in which a limit fails to exist.

EXAMPLE 5   Let $f(x) = x/|x|$. Graph f, and show that f is discontinuous at 0.

SOLUTION   We can graph f as follows. Note that $|x|$ changes values at 0. For $x < 0$, we have $|x| = -x$, and hence for $x < 0$,

$$f(x) = \frac{x}{|x|} = \frac{x}{-x} = -1$$

For $x > 0$, we have $|x| = x$; hence, for $x > 0$,

$$f(x) = \frac{x}{|x|} = \frac{x}{x} = 1$$

f is not defined at 0. Hence, a completely equivalent description of f by cases (according to the sign of x) is: $f(x) = 1$ if $x > 0$, $f(x) = -1$ if $x < 0$, and f is undefined at 0. (See Figure 2-6.) Thus, the graph of f has a jump at 0.

To investigate $\lim_{x \to 0} f(x)$, notice that, as x tends to 0 with $x < 0$, $f(x)$ tends to -1; whereas, as x tends to 0 with $x > 0$, $f(x)$ tends to 1. If there is a limit, it must be the *same number* from both directions. Hence, $\lim_{x \to 0} f(x)$ does not exist, and so f is discontinuous at 0.

Another way to think of this is as follows: as x get arbitrarily close to 0, there is no value to which $f(x)$ gets arbitrarily close. Furthermore, *there is no way to patch the graph at 0 so as to make the curve continuous there* (that is, there is no way to add a point with first coordinate 0 that will make the graph of f a continuous curve at 0). Thus, the fact that $\lim_{x \to 0} f(x)$ fails to exist totally prevents us from smoothing out the graph at 0.

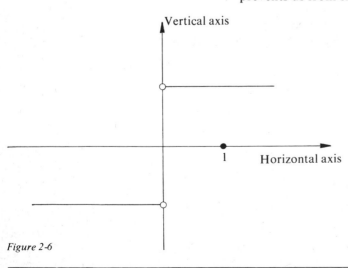

Figure 2-6

EXAMPLE 6   Let

$$f(x) = \frac{x^2 + x - 6}{x - 2}$$

for $x \neq 2$. Define f at 2 so that it will be continuous there.

SOLUTION   Fill in the missing information:

$$\lim_{x \to 2} f(x) = \lim_{x \to 2} \frac{x^2 + x - 6}{x \quad 2}$$

$$= \lim_{x \to 2} \frac{(x - 2)(\underline{\quad\quad})}{x - 2}$$

$$= \underline{\quad\quad\quad\quad}$$

Thus, f will be continuous at 2 if $f(2) = \underline{\quad\quad\quad\quad}$.

Answer   Example 6: $x + 3$; 5; 5.

Problems   In problems 1–6, determine whether the given function is continuous at the indicated points.

1. $f(x) = \dfrac{x - 1}{x}$ at $x = -1, x = 1, x = 0$.

2. $g(x) = x^2 + 2$ at $x = -1, x = 1, x = 0$.

3. $F(x) = \begin{cases} \sqrt{x} & \text{if} & x \geq 0 \\ 0 & \text{if} & -1 \leq x < 0 \end{cases}$

   at $x = -1, x = 1, x = 0$.

4. $G(x) = \begin{cases} 1 & \text{if} & x \geq 1/10 \\ -1 & \text{if} & 0 < x < 1/10 \end{cases}$

   at $x = 0, x = 1, x = 1/10$.

5. $W(x) = \dfrac{3x - 2}{x - 4}$ at $x = -3$.

6. $u(x) = \begin{cases} \dfrac{x^2 - 16}{x - 4}, & x \neq 4 \\ \\ 8, & x = 4 \end{cases}$ at $x = 1$ and $x = 4$.

7. Determine whether the function

$$f(x) = \begin{cases} \dfrac{x^2 + 2x - 3}{x + 3}, & x \neq -3 \\ \\ -4, & x = -3 \end{cases}$$

is continuous at $x = -3$.

8. Let $f(x) = [x]$.* Is f continuous at 0? Is f continuous at 1? Is f continuous at $1/2$? For what x is f continuous and for what x is f discontinuous?

9. Let $G(x) = \dfrac{x^2 - 9}{x - 3}$ for $x \neq 3$. Define $G(x)$ at $x = 3$ so that it will be continuous at $x = 3$.

10. Let $H(x) = \dfrac{x^3 - 8}{x - 2}$, $x \neq 2$. Define H at 2 so that it will be continuous there. (*Hint:* Explicitly divide $x^3 - 8$ by $x - 2$.)

11. Let

$$f(x) = \begin{cases} x^2, & x \leq 0 \\ x, & x > 0 \end{cases}$$

Observe $f(0) = 0$. Graph f. Is f continuous at 0?

SECTION 5

VELOCITY. RATE OF CHANGE

Suppose an automobile is driving on a straight road in such a way that at any (positive) time t, the distance of the automobile from a fixed starting point is, say, $f(t)$. (Naturally in any real problem, the units of time and distance would be included.) Question: What is the *velocity* of the automobile at some given time t? This problem is one in an important category of problems in application, **rate of change**. When we ask for the velocity of the automobile, we are actually asking for the time rate of change of the distance of the automobile.

Many analogous problems exist. The biologist might be interested in measuring the time rate of change of the size of a population of bacteria; the economist, in measuring the rate of change with respect to demand of total revenue (this rate of change is called **marginal revenue**), or the rate of change with respect to the total number of units produced of the total cost of production (and this is called **marginal cost**). Since all these "rate-of-change" problems are variations of a common theme, the study of any one of them will lead to a model from which all the rest can be studied. For convenience, we shall study the velocity problem.

The clue as to how velocity should be defined is found in the notion of **average velocity**. If it takes an automobile $\frac{1}{2}$ hour to go 30 miles the average velocity of the automobile is $30/(\frac{1}{2}) = 60$ miles per hour. That is, average velocity is the ratio of the distance traveled to the corresponding time elapsed; thus,

$$\text{Average velocity} = \frac{\text{change in distance}}{\text{change in time}}$$

Now at time t the automobile is $f(t)$ units from some fixed

---

*Recall that $[x]$ is the greatest integer that is $\leq x$.

reference point, and h units of time later (that is, with $h > 0$), at time $t + h$) the object will be at position $f(t + h)$. The change in distance corresponding to the time change h is, therefore, $f(t + h) - f(t)$. Consequently,

(Average velocity from time t to time $t + h$)

$$= \frac{f(t + h) - f(t)}{h}$$

Notice that

$$
\begin{aligned}
\text{(Change in time)} &= \text{(final time)} - \text{(initial time)} \\
&= (t + h) - t \\
&= h
\end{aligned}
$$

EXAMPLE 1 An object is dropped from the top of a tower. Suppose the distance $f(t)$ (in feet) of the object from the top of the tower at the end of t (seconds) is given by

$$f(t) = 16t^2, t \geq 0$$

This distance function f describes so-called "free-fall" and assumes the object is simply dropped and not pushed or projected downward, and thus the only force acting on the object is the force due to gravity. (a) What is the average velocity of the object over the first 10 seconds of motion? (b) What is the average velocity of the object from time 2 to time 4?

Of course, we have replaced our automobile by an object in free-fall, but the idea of average velocity is the same.

SOLUTION (a) For the first 10 seconds of motion, we see that the average velocity is

$$\frac{f(10) - f(0)}{10} = \frac{16(10^2) - 16 \cdot 0^2}{10} = \frac{1600}{10} = 160 \text{ (feet per second)}$$

(b) The average velocity from time 2 to time 4

$$\frac{16 \cdot 4^2 - 16 \cdot 2^2}{4 - 2} = \frac{16(4^2 - 2^2)}{2} = 96 \text{ (feet per second)}$$

Let us return to the problem of determining (instantaneous) velocity of our automobile at time t. As before, f is the distance function. It seems reasonable to expect that the average velocity from time t to a time *close* to t, say $t + h$ for some small time change h, should approximate the velocity; for h close to 0,

$$\frac{f(t + h) - f(t)}{h}$$

should approximate velocity at time t. Furthermore, the smaller h (that is, the smaller the change in time from t, or, equivalently, the closer we are to time t), the better should be the approximation. We are led to *define the (instantaneous) velocity at time t, which we shall denote by $v(t)$, as the limit of the average velocity as the change in time, h, tends to* 0; that is,

$$v(t) = \lim_{h \to 0} \frac{f(t + h) - f(t)}{h}$$

Thus, to compute $v(t)$, we must
1. Compute $f(t + h)$ (for $h \neq 0$).
2. Compute $f(t)$, and then subtract $f(t)$ from $f(t + h)$, thus obtaining the *difference* $f(t + h) - f(t)$.
3. Divide the quantity found in (2) above by h, thus obtaining the *difference quotient*

$$\frac{f(t + h) - f(t)}{h}$$

4. And finally, *take the limit as h tends to 0 of the difference quotient found in* (3); that is, compute

$$\lim_{h \to 0} \frac{f(t + h) - f(t)}{h}$$

This number is $v(t)$, the velocity of the automobile at time t. This limit, of course, is the velocity of *any* object moving in a straight-line path in such a way that its distance (from some fixed point) at time t is given by $f(t)$.

In taking the limit, it is important to recall that we must determine what number $[f(t + h) - f(t)]/h$ approaches as h *approaches* 0, and that we *cannot* "plug in" $h = 0$ in this indeterminate form. In fact, if $h = 0$, we then get

$$\frac{f(t + h) - f(t)}{h} = \frac{f(t) - f(t)}{0} = \frac{0}{0}$$

which is meaningless.

EXAMPLE 2   Assume that our automobile drives on a straight road in such a way that at t seconds the automobile is t^2 feet from our starting position. What is the velocity of the automobile (a) at time 0? (b) at the end of 2 seconds? (c) at any time t seconds?

SOLUTION   Let f be the distance function for this problem; thus, $f(t) = t^2$.

To answer (a), we set $t = 0$ and perform the following calculations:

1. For $h \neq 0, f(t + h) = f(0 + h) = f(h) = h^2$
2. $f(0) = 0^2 = 0$; thus, $f(0 + h) - f(0) = h^2 - 0 = h^2$
3. $\dfrac{f(0 + h) - f(0)}{h} = \dfrac{h^2}{h} = h$
4. $\lim\limits_{h \to 0} \dfrac{f(0 + h) - f(0)}{h} = \lim\limits_{h \to 0} h = 0$

Thus, $v(0)$, the velocity of the automobile at time 0, is equal to 0. All this says is that our automobile is starting from rest (there is no motion at time (0)).

To answer (b), the reader is invited to fill in each of the steps below.

$$f(2 + h) = \text{_____} \quad \text{and} \quad f(2) = \text{_____}$$

Hence, $f(2 + h) - f(2) = \text{_____}$

and $\dfrac{f(2 + h) - f(2)}{h} = \text{_____}$

Hence, $v(2) = \lim\limits_{h \to 0} \dfrac{f(2 + h) - f(2)}{h} = \text{_____}$

To answer (c), we have to find the velocity at any time t, and this is

$$v(t) = \lim_{h \to 0} \frac{f(t + h) - f(t)}{h}$$

$$= \lim_{h \to 0} \frac{(t + h)^2 - t^2}{h}$$

$$= \lim_{h \to 0} \frac{(t^2 + 2th + h^2) - t^2}{h}$$

$$= \lim_{h \to 0} (2t + h) \qquad (\text{since } h \neq 0)$$

$$= 2t$$

That is, if $f(t) = t^2$, then $v(t) = 2t$. Note that $v(0) = 2 \cdot 0 = 0$ and $v(2) = 2 \cdot 2 = 4$, which we obtained directly when answering (a) and (b).

The example of velocity of a moving object is one example of the limit that describes the rate-of-change process. We shall give one more example now; and in the next section, we shall connect the rate-of-change concept to a concept we will then define, the derivative.

EXAMPLE 3   A sense organ receives a stimulus which causes action potentials to be produced. Let $N(t)$ be the total number of action potentials produced t seconds after the start of the stimulus. Discuss the rate of production of the action potentials at any time t.

SOLUTION   Fix a time t. The total number of action potentials produced after a total time $t + h$ is $N(t + h)$, and thus the average (or, mean) rate of production of action potentials over the h second time interval is

$$\frac{N(t + h) - N(t)}{(t + h) - t}$$

that is,

$$\frac{N(t + h) - N(t)}{h}$$

Notice that this is like the "average velocity" of the production of action potentials. As the time interval h gets smaller and smaller, the average rate of production becomes a better and better approximation to the rate of production of action potentials at time t, and thus we are led to define

The rate of production of
action potentials at time t

$$= \lim_{h \to 0} \frac{N(t + h) - N(t)}{h}$$

Answer   Example 2b: $(2 + h)^2 = 4 + 2h + h^2$; 4; $2h + h^2$; $2 + h$; 2.

Problems
1. A ball is dropped from the roof of a house. Its distance above the ground at time t seconds is $16t^2$ feet, $t \geq 0$.
   a. Find the average velocity over the time interval $t = 0$ to $t = 1$(second)(that is, over the first second of flight).
   b. Find the average velocity over the time interval $t = a$ to $t = a + h$.
   c. Find the velocity at time a.
2. An automobile drives on a straight road in such a way that the distance (in feet) $f(t)$ it has traveled after t seconds of motion is given by $f(t) = 3t - 2$. Find the velocity of the automobile (a) at 3 seconds, (b) at any time t.
3. Repeat problem 2 if now $f(t) = 2t^2 - 1$.
4. Assume that t seconds after an organism receives a stimulus, the total number of action potentials $N(t)$ is given by $N(t) = t^3$. Find the rate of production of action potentials at (a) 2 seconds, (b) any time t.

5. Repeat problem 4, if now $N(t) = 3t^2 + 2$.
6. Assume that water is flowing from a tank in such a way that at the end of t minutes, the volume of water in the tank is $(10 - t)^2$ gallons, where $0 \leq t \leq 10$. Note that initially there are 10 gallons of water in the tank, and at the end of 10 minutes, the tank is empty. Find the rate at which the tank is emptying at any time t.
7. Assume that a population of organisms is growing so that at time t its population is $P(t)$.
    a. In the spirit of this section, define the average rate at which the population is increasing, starting at any (positive) time t, and ending at $t + h$. Explain your reasoning. Then,
    b. Define the (instantaneous) rate of growth of the population at any time.
    c. Assume that $P(t) = 4t^2 + 2t$. Find the rate of growth of the population at any time t.

SECTION 6

THE DERIVATIVE. MORE ON RATE OF CHANGE

In this section, we shall continue to investigate the limit that gives the velocity. If f is a function and x is a number, we define the function f' by

$$f'(x) = \lim_{h \to 0} \frac{f(x + h) - f(x)}{h}$$

This limit plays an important role in mathematics, and gets a special name, the **derivative of f**. Owing to historical accident, there are many notations for the derivation of f. Some of the more common ones are:

$$f', \quad Df, \quad \frac{d}{dx} f, \quad \text{and} \quad \frac{df}{dx}$$

Thus, the **derivative** *of f at x is f'(x) where*

$$f'(x) = \lim_{h \to 0} \frac{f(x + h) - f(x)}{h}$$

---

or, it is
$$\frac{d}{dx} f(x)$$

where $$\frac{d}{dx} f(x) = \lim_{h \to 0} \frac{f(x + h) - f(x)}{h}, \text{etc.}$$

We should at once recognize that $f'(x)$, that is, the derivative of f at x, has already been used to *define* velocity. In our new notation, we see that if an object moves in a straight-line path in such a way that at time t it has gone $f(t)$ units, then the velocity v at time t is given by

$$v(t) = f'(t) \qquad \text{or, in other notation,} \qquad v(t) = \frac{df(t)}{dt}$$

that is, *the velocity function is the derivative of the distance function.*

Using the result of Example 2, section 5, we see that if $f(x) = x^2$, then $f'(x) = 2x$; or, equivalently (and again, this is simply because of the different notations for the derivative):

$$\frac{d}{dx}(x^2) = 2x \qquad \text{or} \qquad \frac{dx^2}{dx} = 2x \qquad \text{or} \qquad D x^2 = 2x$$

(Note that t has been systematically replaced by x.) What we have shown is that v, the velocity function which is nothing more than the rate of change of the distance function f with respect to time, is the derivative of f. In general,

Given a function G which "depends" on a quantity u, the rate of change of G with respect to u is defined to be $G'(u)$

For example, since acceleration (call it a) is the rate of change of the velocity v with respect to time, we have

$$a(t) = v'(t)$$

Summarizing, we see that

$v(t) = f'(t)$; that is, velocity is the (time)
derivative of distance.
$a(t) = v'(t)$; that is, acceleration is the (time)
derivative of velocity.

EXAMPLE 1   If $P(t)$ is the size of a population of bacteria (at time t), then the rate of change of the size of the population at time t is $P'(t)$.

EXAMPLE 2   If $N(t)$ is the total number of action potentials produced t seconds after the start of a stimulus, then $N'(t)$ (or, in other notation, $DN(t)$), the derivative of N with respect to t, is the rate of production of action potentials at time t.

EXAMPLE 3   The total cost of producing and marketing x units of a commodity is assumed to be a function of x alone, that is, independent of time and other commodities. Designate this function by $c(x)$. It is natural to assume that if no commodities are produced, there is no cost; that is, $c(0) = 0$. The average cost to produce x units is given by

$$\frac{c(x) - c(0)}{x - 0} = \frac{c(x)}{x}$$

The rate of change of total cost with respect to the number of units produced is called marginal cost, and thus is given by $c'(x)$, the derivative of $c(x)$ with respect to x, the number of units produced.

We next consider the problem of *computing* derivatives. Our objective is to find rules, or formulas, that will allow us to evaluate derivatives without explicitly computing the required limit.

Using the definition of derivative (as a limit), it is possible to show that

$$D\,x = 1$$
$$D\,x^2 = 2x \quad \text{(previously shown)}$$
$$D\,x^3 = 3x^2$$
$$D\,x^4 = 4x^3$$
$$D\,x^5 = 5x^4$$

This pattern suggests the general formula

$$D\,x^n = nx^{n-1}$$

(where n is a positive integer), which in fact is correct. What is even more interesting is that it is correct for *any real number n*. For example,

$$D\,x^{10} = 10x^9$$
$$D\,x^{935} = 935x^{934}$$

moreover, for $x > 0$

$$
\begin{aligned}
D\,\sqrt{x} &= D\,x^{1/2} \\
&= \tfrac{1}{2}\,x^{1/2-1} \\
&= \tfrac{1}{2}\,x^{-1/2} \\
&= \frac{1}{2} \cdot \frac{1}{x^{1/2}} \\
&= \frac{1}{2\sqrt{x}}
\end{aligned}
$$

that is, for $x > 0$,

$$D\sqrt{x} = \frac{1}{2\sqrt{x}}$$

Also, for $x \neq 0$

$$D\frac{1}{x^2} = Dx^{-2}$$
$$= -2x^{-2-1}$$
$$= -2x^{-3}$$
$$= -\frac{2}{x^3}$$

The formula even works to compute Dx^π (that is, $Dx^\pi = \pi x^{\pi - 1}$); however, for the moment, it leaves unresolved the more fundamental question of the meaning of numbers x^π (for example, 2^π, or 3^π). These problems are resolved in the section on exponential functions (see page 153).

A slight generalization of this result is that for any c and n,

$$D\,cx^n = cnx^{n-1}$$

For example,

$$D\,10x^2 = 10 \cdot 2x = 20x, \qquad D(-7)x^5 = -35x^4,$$
$$D\,4\sqrt{x} = 2/\sqrt{x}, \text{ etc.}$$

To verify the formula for general n is for the time being too much. However, using the binomial theorem, it is possible to show that it works for positive integers n. To see this, note that by definition of derivative,

$$D\,cx^n = \lim_{h \to 0} \frac{c(x+h)^n - cx^n}{h}$$
$$= c \lim_{h \to 0} \frac{(x+h)^n - x^n}{h} \qquad \text{(Why?)}$$

Using the binomial theorem,*

$$(x+h)^n - x^n = -x^n + (x+h)^n$$
$$= -x^n + x^n + nx^{n-1}h + \binom{n}{2}x^{n-2}h^2$$
$$+ \binom{n}{3}x^{n-3}h^3 + \cdots + h^n$$

---

*See section 9, part 2, of the appendix.

$= nx^{n-1}h + h^2$ multiplied by an expression involving binomial coefficients, powers of x, and powers of h.

Hence,

$$\lim_{h \to 0} \frac{(x + h)^n - x^n}{h}$$

$$= \lim_{h \to 0} \frac{nx^{n-1}h + h^2 \cdot \left\{ \begin{array}{l} \text{expression involving binomial coef-} \\ \text{ficients, powers of } x, \text{ and powers of } h \end{array} \right\}}{h}$$

$$= \lim_{h \to 0} \left\{ nx^{n-1} + h \cdot \left(\begin{array}{l} \text{expression involving binomial coeffi-} \\ \text{cients, powers of } x, \text{ and powers of } h \end{array} \right) \right\}$$

$$= nx^{n-1} + 0 = nx^{n-1}.$$

Multiplying by c gives the desired result.

The proof for negative integers n is not much harder; an outline of the proof appears in the exercise set. The reasoning that is used to show that the formula holds for rational numbers n appears in the section on implicit differentiation. It is not until the chapter on exponential functions that it can be verified for arbitrary real numbers n. We shall anticipate all these results, and assume the formula holds for all real numbers n.

A word on notation: We have seen, for example, that $D\, x^5 = (dx^5)/dx = 5x^4$. But there is nothing sacred about using the letter x to characterize numbers in some domain. Thus,

$$\frac{dt^5}{dt} = 5t^4, \quad \text{and} \quad \frac{du^5}{du} = 5u^4, \quad \text{and} \quad \frac{d\Omega^5}{d\Omega} = 5\Omega^4, \text{ etc.}$$

If $f'(a)$ exists, we say that f is **differentiable** at a; similarly, if $f'(a)$ does not exist (that is, if a is *not* in the domain of f'), we say that f is not differentiable at a. For example, if $f(x) = (x)^{1/2}$, then f is not differentiable at 0. To see this, note that $D\,(x)^{1/2} = 1/(2x)^{1/2}$ which is not defined at 0.

Problems In problems 1–10, compute the given derivative:

1. $D\, 8x^{100}$ 2. $D\, \sqrt[3]{x}$ 3. $D\, 4\sqrt[4]{x}$

4. $D\, \left(\dfrac{1}{3}\right) \sqrt[6]{x}$ 5. $D - \sqrt[5]{x}$ 6. $D\, \sqrt[5]{x^3}$

7. $D\, \dfrac{5}{x}$ 8. $D\, \dfrac{2}{5x^3}$ 9. $D\, \dfrac{-3}{\sqrt[3]{x}}$ 10. $D\, \dfrac{1}{\sqrt[10]{x^3}}$

11. Assume that the distance in feet that a particle has traveled in t seconds is given by $s(t) = \sqrt{t^3}$. Find the velocity of the particle and the acceleration of the particle at an arbitrary time t.

12. Let A be the area of a square of side x. Find the rate of change of A with respect to x when $x = 10$.

13. Let A be the area of a circle of radius r. Show that the rate of change of area with respect to the radius is given by the circumference of the circle.

14. A spherical balloon has air pumped into it so that at time t the radius r is given by $r(t) = \frac{1}{2}t$. How fast is the balloon expanding; that is, how fast is the volume increasing? (Recall: the volume v of a sphere with radius r is $v = \frac{4}{3}\pi r^3$.)

15. Water is being pumped from a tank so that the volume of water V in gallons t minutes after the pumping has begun is given by $V(t) = 100(10 - t)^2$. Find the average rate at which the water is drained for the first 5 minutes, then for the first 10 minutes. Find the rate at which the water is being drained at $t = 5$, and then at $t = 10$.

16. A particle moves so that its distance s at time $s(t) = at^2$ (a, any number). Find the velocity and the acceleration of the particle.

17. Assume that the total cost of producing x units of a commodity is given by $s(x) = 2x^3$. Find the average and marginal costs of producing 100 units of the commodity.

18. Let $P(t)$ be the size of a population at time t. Suppose experiments have shown that the rate of change of the population at time t is precisely twice the size of the population at that time. Write a mathematical formula (using the derivative) to describe this.

19. The following is an outline of the verification that for any positive integer, n, $D x^{-n} = (-n)x^{(-n)-1}$. Fill in the missing steps, justifying each calculation.

$$D x^{-n} = D\left(\frac{1}{x^n}\right) = \lim_{h \to 0} \frac{\dfrac{1}{(x + h)^n} - \dfrac{1}{x^n}}{h}$$

$$= -\lim_{h \to 0} \frac{(x + h)^n - x^n}{h} \cdot \frac{1}{x^n(x + h)^n}$$

$$= -(D x^n) \cdot \frac{1}{x^{2n}}$$

$$= \frac{-nx^{n-1}}{x^{2n}}$$

$$= (-n)x^{(-n)-1}$$

SECTION 7

TANGENT LINES

In this section we shall apply the derivative to the geometrical problem of determining the line T tangent to the graph of a function f at some point on the graph, say $(a, f(a))$ (see Figure 2-7). We shall see in subsequent chapters that such information about tangent lines is useful in determining the shape of the graph of a function, which, in turn, will give information when a graph is rising or falling (useful in such problems as when a profit curve, or population curve, is rising or falling), and about the high and low points on a curve (useful in such problems as when profit, or population, is maximum or minimum).

To determine T, we can concentrate on determining the *slope* of the line T, since we already know a point on T must be $(a, f(a))$. Our initial problem is to decide what the slope of the tangent line should mean; that is, *to define the slope of the line T.* The clue is that a line drawn through $(a, f(a))$ and a point *very near* $(a, f(a))$ should have a slope that is approximately the slope of T. (See Figure 2-8.) We shall now pursue this clue in detail.

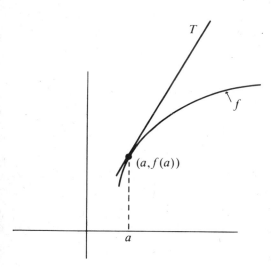

Figure 2-7

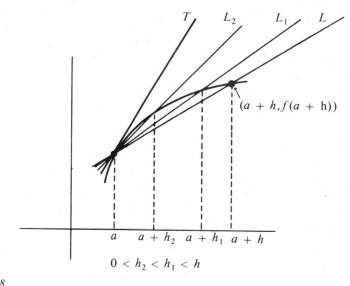

Figure 2-8

Let h be a number that is close to 0. Then $a + h$ is close to a, and the point $(a + h, f(a + h))$ is close to the point $(a, f(a))$ (see Figure 2-8). Notice that the line L through $(a, f(a))$ and $(a + h, f(a + h))$ is an approximation to the tangent line T (see Figure 2-8), and hence the slope of L,

$$\frac{f(a + h) - f(a)}{(a + h) - a} = \frac{f(a + h) - f(a)}{h}$$

should be approximately the slope of T. If we make h even smaller (for instance, replacing h by h_1 or h_2, as in Figure 2-8), then the corresponding lines (that is, L_1 and L_2 in Figure 2-8) become better and better approximations to T; and hence we would expect their slopes to be better and better approximations to the slope of T. But as h gets smaller and smaller, the slopes

$$\frac{f(a + h) - f(a)}{h}$$

get closer and closer to $f'(a)$, the *derivative* of f at a. Thus, we *define the slope of the line T tangent to the graph of f at $(a, f(a))$ to be $f'(a)$, the derivative of f at a.*

Recall that an equation for the line passing through the point (x_1, y_1) with slope m is $y - y_1 = m(x - x_1)$. Thus, since the tangent line passes through $(a, f(a))$, we see that an equation for T is

$$y - f(a) = f'(a)(x - a)$$

EXAMPLE 1   Find an equation of the line T tangent to the graph of $f(x) = \sqrt[3]{x}$ at the point $(8, 2)$.

SOLUTION   The slope of the desired line T is $f'(8)$. We see that

$$D \sqrt[3]{x} = D x^{1/3}$$

$$= \frac{1}{3} x^{-2/3}$$

$$= \frac{1}{3 \sqrt[3]{x^2}}$$

Hence, $f'(8) = \frac{1}{12}$, and thus an equation for T is

$$y - 2 = \tfrac{1}{12}(x - 8)$$

or, simplified,

$$y = \tfrac{1}{12} x + \tfrac{4}{3}$$

EXAMPLE 2   Let $R(x)$ be the total profit accrued by a firm from marketing a quantity x of some commodity. Note that $R(x + h) - R(x)$ is the total profit accrued from an increase in output of h units. The **marginal profitability** from an increase in output of h units (that is, an increase in output from x to $x + h$ units) is defined to be

$$\frac{R(x + h) - R(x)}{h}$$

In Figure 2-8, assuming that the quantity x of the commodity produced is plotted the **x**-axis, and the resulting total profit on the **y**-axis, with the graph representing the graph of R, we see that marginal profitability is precisely the slope of the line L. The **marginal profitability from the production of x units** is defined to be the limit as the increase in output (that is, h) tends to 0 of the marginal profitability in increasing output from x to $x + h$ units; that is, the marginal profitability from the production of x units is

$$\lim_{h \to 0} \frac{R(x + h) - R(x)}{h}$$

which is $R'(x)$, the slope of the line tangent to the profit curve at $(x, R(x))$.

Remark   Note that if for some function f, and some point $a \in$ (domain f), $f'(a)$ does not exist, then either

1. The tangent line T has no slope; that is, T is a vertical tangent line.
2. Or there is *no* tangent line to the graph of T at $(a, f(a))$.

We shall not pursue this in detail, but the reader might observe that if the graph of f has a corner or cusp at $(a, f(a))$, then there is no tangent line to the graph at the point (or, possibly, there is a vertical tangent line) (see Figure 2-9).

Figure 2-9

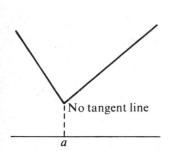

No tangent line

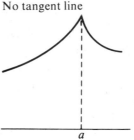

No tangent line

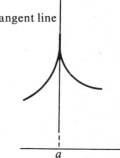

Vertical tangent line

There is an interesting relationship between the differentiability of a function at a point and the continuity of the function at that same point. If it is possible to construct a tangent line to the graph of a function at some point $(a, f(a))$, the graph of the function is necessarily continuous (no holes, breaks, gaps, or other types of discontinuities) at $(a, f(a))$; that is,* *if $f'(a)$ exists, then f is continuous at a.*

In subsequent sections, we shall derive some very useful prop-

---

*This is proved in almost any standard calculus text.

erties of derivatives. One sense in which they are useful is that they will greatly simplify the computation of derivatives. For example, we will be able to read off by inspection, using formulas, the derivative of polynomial functions.

Problems

1. Find an equation of the line tangent to the graph of $f(x) = (x)^{1/2}$ corresponding to (a) $x = 1$, (b) $x = 2$, (c) $x = 4$, and (d) $x = 9$. Sketch the graph of f. In terms of the sketch of the graph of f, what can be said about the construction of a tangent line at $(0, 0)$?

2. Find an equation of the line tangent to the graph of $f(x) = 1/x$ at (a) $x = -3$, (b) $x = 2$, and (c) $x = -2$. Sketch the graph of f. What can be said about the construction of a tangent line for $x = 0$?

3. Let $f(x) = x^{2/3}$. Find all points on the graph of f where the tangent line is parallel to the line given by $4y - 9x = 1$.

★4. Let $f(x) = \sqrt[5]{2x}$. Find an equation for the line tangent to the graph of f at $(16, f(16))$.

5. Let $f(x) = x^{3/2}$. Find an equation for the line tangent to the graph of f', the *derivative* of f, at the point $(4, f'(4))$.

6. Let $f(x) = x^2 + 1$ and $g(x) = x^2 + 3$. Graph f and g. Find f' and g'. Give a geometrical agreement to explain their similarity.

7. Let f be the function graphed in Fig. 2-10. Let g be defined by:

$$g(x) = f(x) + 1, x \text{ in } \mathbf{R}$$

Graph g. What is the relationship between f' and g'? Explain geometrically.

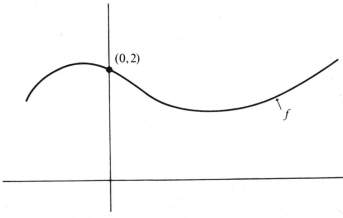

Figure 2-10

SECTION 8

TECHNIQUES FOR DIFFER-ENTIATION: AN INTRO-DUCTION. HIGHER ORDER DERIVATIVES

If $f(x) = \sqrt[5]{x^{15} + 3\sqrt{x}}$, then finding $f'(x)$ would involve the evaluation of

$$\lim_{h \to 0} \frac{\sqrt[5]{(x+h)^{15} + 3\sqrt{(x+h)}} - \sqrt{x^{15} + 3\sqrt{x}}}{h}$$

To put it mildly, the calculations would be unpleasant. We want ways to differentiate which do not explicitly make use of the definition of derivative. A step in this direction is the fact that $Dx^n = nx^{n-1}$. In this and the next few sections, we shall show that, if we know f' and g', we can then find $(f \pm g)'$, $(fg)'$, and $(f/g)'$. A useful preliminary result is:

If c is any real number, and $f(x) = c$, all x, then $f'(x) = 0$, all x.

This is often denoted by:

$$Dc = 0$$

To see that this is true, notice that if $f(x) = c$, all x, then, in particular, $f(x + h) = c$; hence,

$$Dc = \lim_{h \to 0} \frac{c - c}{h} = \lim_{h \to 0} 0 = 0$$

Notice that if $f(x) = c$, all x, then the graph of f is a horizontal line c units from the **x**-axis; the fact that $f'(x) = 0$ for all x is simply the fact that a horizontal line has slope equal to zero.

We know, for example, that $Dx^5 = 5x^4$. But what is $D[7x^5]$? That is, if $f(x) = 7x^5$, what is $f'(x)$? As we have seen, it turns out that

$$D7x^5 = 7(5x^4) = 35x^4$$

and in general,

If A is any number and f' exists, then

$$D[Af(x)] = A \cdot Df(x)$$

Equivalently, this fact in some of the other notation appears as:

$$(Af)' = Af' \quad \text{or} \quad \frac{d(Af)}{dx} = A\frac{df}{dx}$$

Thus, *the derivative of a number times a function is the number times the derivative of the function.*

To verify this, let $G(x) = Af(x)$. We then have

$$D\,Af(x) = D\,G(x) = \lim_{h \to 0} \frac{G(x + h) - G(x)}{h}$$

$$= \lim_{h \to 0} \frac{Af(x + h) - Af(x)}{h}$$

$$= \lim_{h \to 0} A \cdot \frac{f(x + h) - f(x)}{h}$$

$$= A \cdot \lim_{h \to 0} \frac{f(x + h) - f(x)}{h}$$

(Since the limit of a number times a function is the number times the limit of the function)

$$= A \cdot Df(x)$$

EXAMPLE 1   (2.6) $D\left(\dfrac{1}{3} x^3\right) = \dfrac{1}{3} D\,x^3 = \dfrac{1}{3} (3x^2) = x^2$

(2.7) $D(4\sqrt{x}) = 4D\sqrt{x} = 4D\,x^{1/2} = 4\left(\dfrac{1}{2}\right)x^{-1/2} = 2x^{-1/2} = \dfrac{2}{\sqrt{x}}$

(2.8) $\dfrac{d}{dr}(\pi r^2) = 2\pi r$

(2.9) $D(25x^{10}) = \underline{} x\underline{}$

(2.10) $D(-f(x)) = D([-1] \cdot f(x)) = (-1)\,Df(x) = -Df(x)$

We shall next show that *the derivative of a sum of functions is the sum of their derivatives; that is, if $Df(x)$ and $Dg(x)$ exist, then so does $D[f(x) + g(x)]$, and*

$$D\{f(x) + g(x)\} = \{Df(x)\} + \{Dg(x)\}$$

To see this, let $F(x) = f(x) + g(x)$. We then want to find $DF(x)$. Observe that $F(x + h) - F(x) = \{f(x + h) + g(x + h)\} - \{f(x) + g(x)\}$. Hence, we have

$$DF(x) = \lim_{h \to 0} \frac{F(x + h) - F(x)}{h}$$

$$= \lim_{h \to 0} \frac{\{f(x + h) + g(x + h)\} - \{f(x) + g(x)\}}{h}$$

$$= \lim_{h \to 0} \frac{\{f(x + h) - f(x)\} + \{g(x + h) - g(x)\}}{h}$$

$$= \lim_{h \to 0} \left\{ \frac{f(x + h) - f(x)}{h} + \frac{g(x + h) - g(x)}{h} \right\}$$

$$= \lim_{h \to 0} \frac{f(x + h) - f(x)}{h} + \lim_{h \to 0} \frac{g(x + h) - g(x)}{h}$$

$$= \{Df(x)\} + \{Dg(x)\}$$

Similarly, *the derivative of a difference is the difference of the derivatives*; that is,

$$D[f(x) - g(x)] = [Df(x)] - [Dg(x)]$$

This follows from the result on sums as follows:

$$D[f(x) - g(x)] = D\{f(x) + [-g(x)]\}$$
$$= Df(x) + D[(-1)g(x)]$$
$$= Df(x) - Dg(x)$$

EXAMPLE 2   For all x in $\mathbf{R}$,

(2.11) $D(x^3 - 2x^2 + 4x + 1) = Dx^3 - D2x^2 + D4x + D1$
$$= Dx^3 - 2Dx^2 + 4Dx + D1$$
$$= 3x^2 - 4x + 4$$

(2.12) $D(4x^8 - 7x^3 + 8x) = \underline{\hspace{1cm}} x^7 - 21x^- + \underline{\hspace{1cm}}$

(2.13) $D(ax^2 + bx + c) = 2ax + b \quad (a, b, c \in \mathbf{R})$

(2.14) $D(a_0 x^n + a_1 x^{n-1} + \cdots + a_{n-1} x + a_n)$
$$= n a_0 x^{n-1} + (n - 1) a_1 x^{n-2} + \cdots + a_{n-1}$$

We see that a polynomial of degree n has a derivative everywhere, and the derivative is a polynomial of degree $n - 1$.

EXAMPLE 3   (2.15) For $x \neq 0$, $D\left(\frac{x^2 + 1}{x^2} \right) = D\left(\frac{x^2}{x^2} + \frac{1}{x^2} \right)$
$$= D(1 + x^{-2})$$
$$= -2x^{-3}$$
$$= -\frac{2}{x^3}$$

(2.16)   For $x > 0$,

$$D[\sqrt{x}(1 + 2x + 4x^2)] = D[x^{1/2}(1 + 2x + 4x^2)]$$
$$= D[x^{1/2} + 2x^{3/2} + 4x^{5/2}]$$
$$= \frac{1}{2\sqrt{x}} + 2\left(\frac{3}{2}x^{1/2}\right) + 4\left(\frac{5}{2}x^{3/2}\right)$$
$$= \frac{1}{2\sqrt{x}} + 3\sqrt{x} + 10\sqrt{x^3}$$

Some further algebraic simplification is possible (but note that we are now talking only about algebraic simplification; the derivative has been computed). For $x > 0$, $\sqrt{x^3} = \sqrt{x^2 x} = x\sqrt{x}$. Hence, for $x > 0$,

$$D[\sqrt{x}(1 + 2x + 4x^2)] = \frac{1}{2\sqrt{x}} + 3\sqrt{x} + 10x\sqrt{x}$$
$$= \frac{1 + 6x + 20x^2}{2\sqrt{x}}$$

We have already observed (see Example 2) that if $f(x) = x^3 - 2x^2 + 4x + 1$, then $f'(x) = 3x^2 - 4x + 4$. But f' is just another function (in fact, a polynomial function), hence, f' also has a derivative, $(f')'$, or, more simply, f'', or $f^{(2)}$ given by

$$f''(x) = 6x - 4$$

Similarly,*

$$f'''(x) = 6, \qquad f^{(4)}(x) = 0, \qquad f^{(5)}(x) = 0, \qquad \text{etc.}$$

The function f'', f''', $f^{(4)}$, and so on, are called, respectively, the second, third, and fourth derivatives of f, and so on.

EXAMPLE 4   If $f(x) = x^5$, then $f'(x) = 5x^4$, $f''(x) = 20x^3$, $f'''(x) = 60x^2$, $f^{(4)}(x) = 120x$, $f^{(5)}(x) = 120$, $f^{(6)}(x) = 0$ (and, in fact, $f^{(n)}(x) = 0$ for $n \geq 6$). Notice that f is a polynomial function of degree 5, and each successive derivative is a polynomial function of degree one less than its predecessor.

$f^{(n)}$ is called the nth derivative of f

Using the other notations, this is also denoted by

$$d^n f(x)/dx^n \qquad \text{or} \qquad D^n f(x) \qquad \text{or} \qquad (d^n/dx^n)\, f(x)$$

---

*It is customary to write f'' instead of $f^{(2)}$ and f''' instead of $f^{(3)}$, but beyond this, to write $f^{(4)}$, $f^{(5)}$, etc. Of course, it is completely correct to write $f^{(2)}$ and $f^{(3)}$ for the second and third derivatives.

EXAMPLE 5   Let $f(x) = x^5 - 3x^2$. Then
$$f'(x) = 5x^4 - 6x$$
$$d^2f/dx^2 = D^2f = f''(x) = 20x^3 - 6$$
$$d^3f/dx^3 = D^3f = f'''(x) = 60x^2$$
$$d^4f/dx^4 = D^4f = f^{(4)}(x) = 120x$$
$$d^5f/dx^5 = D^5f = f^{(5)}(x) = 120$$
$$d^6f/dx^6 = D^6f = f^{(6)}(x) = 0$$
$$d^nf/dx^n = D^nf = f^{(n)}(x) = 0 \qquad (n \geq 6)$$

EXAMPLE 6   Let the position of a particle at time t be given by $f(t)$. Then, as we saw in section 7, its velocity is given by $f'(t)$ and its acceleration by $f''(t)$, the second derivative of $s(t)$.

Answers   Example 1, equation (2.9): 250; 9
Example 2, equation (2.12): 32; 2; 8

Problems

1. Find the slope of the line tangent to the graph of
$$f(x) = x^5 - 2x^4 + 3x^2 + x - 5 \qquad \text{at} \qquad x = 1$$

2. Let $f(x) = 2x^6 + x^4 + 1$. Find all the higher derivatives of f.

In problems 3–6, find f' and f''.

3. $f(x) = x^{10} - 7x^6 + 5$

4. $f(x) = \dfrac{x^5}{5} + \dfrac{x^4}{4} - \dfrac{x^3}{3} - \dfrac{x^2}{2} + x$

5. $f(x) = (x + 5)(x - 1)$

6. $f(x) = x(x + 5)(x - 1)$

7. Let $f(x) = (x + 3)(x - 1)$. Find Df, D^2f and D^3f.

8. Let $f(x) = \dfrac{x + 1}{\sqrt{x}}$. Find Df, D^2f, and D^3f.

9. Find $D[\sqrt{x}(1 + x)^2]$

10. Let $f(x) = \dfrac{2x^2 + 3x - 1}{x^2}$. Find Df and D^2f.

11. Find the points on the graph of $y = 2x^3 - 3x^2 - 12x + 20$ where the tangent line is parallel to the x-axis.

12. Find $D(x + 1)^3$.

13. Find all points on the graph of $f(x) = (\frac{1}{3})x^3 + (\frac{1}{2})x^2 - 3x + 1$ where the tangent line is parallel to the line $y = -x$.

14. Let $f(x) = x^4 + x^2 - 1$. Find the slope of the line tangent to the graph of $f'(x)$ at $x = 1$.

15. Find the number c if the line $y = x$ is to be tangent to the graph of $f(x) = x^4 + c$.

16. If the total cost of producing x units of a commodity is given by $S(x) = x^3 - \frac{1}{2}x^2 + 7x$, find the marginal cost when 10 units have been produced.

17. At the beginning of an experiment, a culture of bacteria is observed to have 3×10^5 objects. The experimenter then observed that at time t (in hours), the number of bacteria, $B(t)$, is given by

$$B(t) = 10^5(3 + 8\sqrt{t} + 4t + 3t^2)$$

How fast is the population growing at the end of (a) 15 minutes? (b) 2 hours?

18. An object is projected vertically from a horizontal platform. Its initial velocity is 128 feet per second. It is this initial velocity that pushes the object upward, while the force of gravity tends to pull the object back down. At time t (in seconds), its distance $f(t)$ (in feet) from the platform is given by

$$f(t) = 128t - 16t^2$$

a. Find the velocity and acceleration of the object.

b. When does the object reach its highest point of flight, and what is the maximum height attained? (*Hint:* Consider what the velocity must be when it reaches its maximum height and is about to descend.)

19. A tank of water is filled in such a way that at the end of t minutes there are $(t^3/3) - t^2$ gallons of water in the tank. The person filling the tank is instructed to turn off the water when the water is entering the tank at the rate of 15 gallons per minute. When should he turn off the water?

20. An object moves in such a way that in t hours its distance from its starting point is $t^3 + 2t^2 + 3t + 1$ miles. Find all times t (if there are any) at which the object is moving with equal velocity and acceleration.

SECTION 9

THE PRODUCT AND QUOTIENT RULES. THE POWER RULE

So far we have found out how to compute derivatives of sums, differences, and numerical multiples of functions. In this section, we find the derivative of products and quotients. The formulas for sums and differences were relatively simple. In contrast, the formulas for products and quotients may look strange, but such is life. First, if Df and Dg exist, then so does Dfg, and

---

(Product rule) $Df(x)g(x) = f(x)Dg(x) + [Df(x)] \cdot g(x)$

---

or, equivalently,

$$(fg)' = fg' + f'g$$

This result can be verified by what has become a standard trick for this type problem: the number $f(x)g(x + h)$ is added to and then subtracted from the numerator of the difference quotient for the derivative. Thus, the difference quotient is left unchanged in value, but the new terms aid in the evaluation of the derivative. Specifically,

$$Df(x)g(x) = \lim_{h \to 0} \frac{f(x + h)g(x + h) - f(x)g(x)}{h} \qquad \text{(By definition of derivative)}$$

$$= \lim_{h \to 0} \frac{f(x + h)g(x + h) - f(x)g(x + h) + f(x)g(x + h) - f(x)g(x)}{h}$$

$$= \lim_{h \to 0} \frac{g(x + h)[f(x + h) - f(x)] + f(x)[g(x + h) - g(x)]}{h}$$

$$= \lim_{h \to 0} \left[g(x + h) \frac{f(x + h) - f(x)}{h} + f(x) \frac{g(x + h) - g(x)}{h} \right]$$

$$= \lim_{h \to 0} g(x + h) \lim_{h \to 0} \frac{f(x + h) - f(x)}{h} + f(x) \lim_{h \to 0} \frac{g(x + h) - g(x)}{h}$$

$$= g(x)Df(x) + f(x)Dg(x)$$

Recall that, since $Dg(x)$ exists, g is continuous at x, and hence $\lim_{h \to 0} g(x + h) = g(x)$.

EXAMPLE 1

$$Dx^2(x^2 - x + 1) = x^2 D(x^2 - x + 1) + (Dx^2)(x^2 - x + 1)$$
$$= x^2(2x - 1) + (2x)(x^2 - x + 1)$$
$$= 4x^3 - 3x^2 + 2x$$

Clearly, another way to do this problem is to observe that

$$x^2(x^2 - x + 1) = x^4 - x^3 + x^2$$

and $$D(x^4 - x^3 + x^2) = 4x^3 - 3x^2 + 2x$$

In this example, it is probably better not to use the product rule. This will not always be the case.

EXAMPLE 2   Let x represent the quantity of some commodity demanded by the market and $\mathbf{p}(x)$ the price per unit of that quantity. (The reader should convince himself that functional notation is justified.) The classical definitions from economics are that **total revenue $\mathbf{R}(x)$** is equal to $x \cdot \mathbf{p}(x)$, and that **marginal revenue with respect to demand** is the rate of change of the total revenue with respect to demand. Hence, marginal revenue is simply $\mathbf{R}'(x)$. By the product rule, we see that marginal revenue is given by

$$\mathbf{R}'(x) = x\mathbf{p}'(x) + \mathbf{p}(x)$$

Under some conditions, it is better to think of demand x as a function of price p; we will write this as $\mathbf{x}(p)$. We can interpret the function $\mathbf{x}(p)$ as the number of people that will buy the commodity if it were offered at price p. For example, if 100 people will buy the commodity priced at \$2, we write $\mathbf{x}(2) = 100$ (people); if 76 people will buy the commodity priced at $\$(\frac{3}{2}) = \$1.50$, we write $\mathbf{x}(\frac{3}{2}) = 76$ (people), and so on. Total revenue $\mathbf{S}(p)$ now as a function of p is given by $\mathbf{S}(p) = p \cdot \mathbf{x}(p)$, and marginal revenue with respect to price is the rate of change of total revenue with respect to price, $\mathbf{S}'(p) = d\mathbf{S}/dp$. Again by the product rule,

$$\mathbf{S}'(p) = p(dx/dp) + \mathbf{x}(p)$$

The product rule can be used to find the derivatives of products of more than two functions. In fact, if f, g, h, u, v, and w have derivatives, then

(2.18) $$(fg)' = f'g + fg'$$

(2.19) $$(fgh)' = f'gh + fg'h + fgh'$$

(2.20) $$(fghu)' = f'ghu + fg'hu + fgh'u + fghu'$$

(2.21) $$(fghuv)' = f'ghuv + fg'huv + fgh'uv + fghu'v + fghuv'$$

(2.22) $(fghuvw)' = $ _____

To verify the first of these, note that

$$
\begin{aligned}
(fgh)' = (f[gh])' &= f'[gh] + f[gh]' \\
&= f'gh + f[g'h + gh'] \\
&= f'gh + fg'h + fgh'
\end{aligned}
$$

EXAMPLE 3
$$D(x^2 + 1)^3 = D\{(x^2 + 1)\cdot(x^2 + 1)\cdot(x^2 + 1)\}$$
$$= [D(x^2 + 1)]\cdot(x^2 + 1)\cdot(x^2 + 1) + (x^2 + 1)$$
$$\cdot[D(x^2 + 1)]\cdot(x^2 + 1) + (x^2 + 1)(x^2 + 1)$$
$$\cdot D(x^2 + 1)$$
$$= (2x)(x^2 + 1)^2 + 2x(x^2 + 1)^2 + 2x(x^2 + 1)^2$$
$$= 6x(x^2 + 1)^2$$

We can now derive a useful result on powers of functions. If we let $f = g$ in (2.18), we see that

$$D[f(x)]^2 = f'(x) f(x) + f(x) f'(x) = 2f(x) f'(x)$$

Similarly, if we set $f = g = h$ in (2.19); and $f = g = h = u$ in (2.20); and $f = g = h = u = v$ in (2.21), we get (including the case $f = g$)

$$D[f(x)]^2 = 2f(x)\cdot f'(x)$$
$$D[f(x)]^3 = 3[f(x)]^2 f'(x)$$
$$D[f(x)]^4 = 4[f(x)]^3 f'(x)$$
$$D[f(x)]^5 = 5[f(x)]^4 f'(x)$$

Continuing, we see that for any positive integer, n,

$$D[f(x)]^n = n[f(x)]^{n-1} f'(x)$$

This formula turns out to be valid for *any real number n*.

EXAMPLE 4

(2.23)
$$D(x^2 + 1)^3 = 3(x^2 + 1)^2\cdot D(x^2 + 1)$$
$$= 6x(x^2 + 1)^2$$

(2.29)
$$D\sqrt{3x + 1} = D(3x + 1)^{1/2}$$
$$= \frac{1}{2}(3x + 1)^{1/2-1}\cdot D(3x + 1)$$
$$= \frac{3}{2}\frac{1}{\sqrt{3x + 1}}$$

(2.25)
$$D\sqrt{x^4 + x^2 + 2} = D(x^4 + x^2 + 2)^{1/2}$$
$$= \frac{1}{2}(x^4 + x^2 + 2)^{1/2-1}$$
$$\times D(x^4 + x^2 + 2)$$
$$= \frac{1}{2}(x^4 + x^2 + 2)^{-1/2}(4x^3 + 2x)$$
$$= \frac{2x^3 + x}{\sqrt{x^4 + x^2 + 2}}$$

$$(2.26) \qquad D(x^3 + 5)^{10} = 10(x^3 + 5)^9 \cdot D(x^3 + 5)$$
$$= 30x^2(x^3 + 5)^9$$

Fill in the blanks:

$$(2.27) \quad D(\sqrt{x} + 1)^{3/2} = \underline{\qquad} (\sqrt{x} + 1)^{1/2} \cdot \frac{1}{2 \underline{\qquad}}$$

EXAMPLE 5 Combining the product and power rules, we see that

$$
\begin{aligned}
D(x^3 + 1)^{10}(x^2 + x - 1)^8 &= (x^3 + 1)^{10} \cdot D(x^2 + x - 1)^8 \\
&\quad + (x^2 + x - 1)^8 \cdot D(x^3 + 1)^{10} \\
&= (x^3 + 1)^{10} \cdot 8(x^2 + x - 1)^7 \\
&\quad \cdot D(x^2 + x - 1) + (x^2 + x - 1)^8 \\
&\quad \cdot 10(x^3 + 1)^9 \cdot D(x^3 + 1) \\
&= 8(x^3 + 1)^{10}(x^2 + x - 1)^7(2x + 1) \\
&\quad + 30x^2(x^2 + x - 1)^8(x^3 + 1)^9 \\
&= (x^3 + 1)^9(x^2 + x - 1)^7 \\
&\quad \cdot [8(x^3 + 1)(2x + 1) \\
&\quad + 30x^2(x^2 + x - 1)] \\
&= (x^3 + 1)^9(x^2 + x - 1)^7 \\
&\quad \cdot (46x^4 + 8x^3 + 46x - 22)
\end{aligned}
$$

We can now derive the quotient rule: *if f' and g' exist and g is not zero, then*

$$\left(\frac{f}{g}\right)' = \frac{gf' - g'f}{g^2}$$

To see this, note that

$$
\begin{aligned}
D(f/g) &= D(f \cdot g^{-1}) &\text{(Since } 1/g = g^{-1}) \\
&= f(Dg^{-1}) + (Df) \cdot g^{-1} &\text{(Product rule)} \\
&= f(-g^{-2}Dg) + (Df)g^{-1} \\
&= -\frac{f}{g^2} Dg + \frac{(Df)}{g} \\
&= \frac{(Df)g - f(Dg)}{g^2}
\end{aligned}
$$

EXAMPLE 6

$$D\left(\frac{3x + 1}{4x - 2}\right) = \frac{(4x - 2)\,D\,(3x + 1) - (3x + 1)\,D\,(4x - 2)}{(4x - 2)^2}$$

$$= \frac{3(4x - 2) - 4(3x + 1)}{(4x - 2)^2}$$

$$= -\frac{10}{(4x - 2)^2}$$

EXAMPLE 7 $D\ \dfrac{x^2 + 2x + 3}{x + 1}$

$$= \frac{(x + 1)\,D\,(x^2 + 2x + 3) - [D\,(x + 1)][x^2 + 2x + 3]}{(x + 1)^2}$$

$$= \frac{(x + 1)(2x + 2) - (x^2 + 2x + 3)}{(x + 1)^2}$$

$$= \frac{x^2 + 2x - 1}{(x + 1)^2}$$

Answers   Equation (2.22):

$$f'ghuvw + fg'huvw + fgh'uvw + fghu'vw + fghuv'w + fghuvw'$$

Example 4, equation (2.27): $\frac{3}{2}$; $\sqrt{x}$

Problems   In problems 1–30, find $f'(x)$.

1. $f(x) = \dfrac{1}{2x + 3}$ 2. $f(x) = \dfrac{1}{3x - 1}$

3. $f(x) = (x + 1)^{1/2}$ 4. $f(x) = (x + 2)^{1/2}$

5. $f(x) = (2x)^{1/2}$ 6. $f(x) = 1/[(x - 2)^{1/2}]$

7. $f(x) = \dfrac{2x + 1}{3x + 1}$ 8. $f(x) = D\ \dfrac{1}{(x^2 + 1)^{10}}$

9. $f(x) = \dfrac{x^2}{x^2 + x + 1}$ 10. $f(x) = D\ \dfrac{1}{(x^2 + 1)^{1/2}}$

11. $f(x) = \dfrac{2x^2 - 5}{3x^2 + 1}$ 12. $f(x) = D\ \dfrac{x}{(x - 1)^{1/2}}$

13. $f(x) = (-x^3 + 5)^{10}$ 14. $f(x) = (x^4 + 3x^2 - x + 1)^6$

15. $f(x) = (4x^6 + 2x^3)^7$ 16. $f(x) = [(x + 3)/(x + 2)]^2$

17. $f(x) = (x + 1)(x)^{1/2}$ 18. $f(x) = (x^5 - x^4 + 1)^{10}$

19. $f(x) = (x - 1)^3/(x + 1)^4$

20. $f(x) = x^2(x + 1)^{-1}$

21. $f(x) = (x^3 - 1)^5 \cdot \sqrt[3]{x^2 + 2}$

22. $f(x) = x^3(1 - x^3)^3$

23. $f(x) = (x + 2)(x - 5)^8$

24. $f(x) = (x^4 - 12x^2)/(x^2 - 4)^2$
25. $f(x) = (x)^{1/2}(x + 1)^{\pi + 1}$
26. $f(x) = (x^2 + 5x - 7)/(x^2 + 1)^2$
27. $f(x) = (x - 1)^3(x + 2)^5$
28. $f(x) = (x^2 + 1)^4(x^2 - 2)^3$
29. $f(x) = (x^3 - 1)^5 \cdot \sqrt{x^2 + 2}$
30. $f(x) = \sqrt[3]{x^2 + 1} \cdot \sqrt{x^3 + 1}$

In problems 31–34 find the indicated derivative.

31. $D\sqrt{x + \sqrt{x}}$ 32. $D\sqrt[3]{x^2 + \sqrt[3]{x}}$
33. $D(\sqrt{x} + x)^3$ 34. $D(\sqrt[3]{x} - \sqrt[4]{x})^8$

35. Suppose the price of a commodity with demand x is $\mathbf{p}(x) = 8/(2 + x)$. Find the total and marginal revenues.

36. The demand for a commodity as a function of price is given by $\mathbf{x}(p) = (8 - 2p)/p$.   Find the total and marginal revenues. Compare your result with the result of problem 35.

37. A sense organ receives a stimulus in such a way that at time t (seconds) after the start of the stimulus, the total number of action potentials $N(t)$ is given by

$$N(t) = 6t + \frac{1}{t^2 + 1} - 1$$

Find the rate of production of action potentials at any time t.

38. Suppose that at time t, an object has moved $(t^2 - 1)^{10} - 1$ units. Find the velocity and acceleration of the object. Find an equation whose solution is the set of all times t at which the velocity and acceleration of the object are equal.

SECTION 10

IMPLICIT DIFFER-ENTIATION RATES

Until now we have considered exclusively "functions of x" that were explicitly in terms of x. For example, in the relationship $y = x^2 + 1$, we are obviously dealing with a function of x, say G, and we might write

$$y = G(x) = x^2 + 1$$

The same relation could have been written as

$$y - x^2 - 1 = 0$$

Then this y is the same as the one in $y = G(x) = x^2 + 1$; thus, we see the relation $y - x^2 - 1 = 0$ *implicitly* gives y as a function of x. In fact, by directly substituting $G(x)$ for y, we can write this second relation as

$$[G(x)] - x^2 - 1 = 0$$

Sometimes we must deal with relationships between numbers

x and y in which it is inconvenient, or perhaps difficult, or perhaps even impossible to express y as a function (or functions) of x, and yet we need to find dy/dx. Although it is easy to solve $y - x^2 - 1 = 0$ to obtain $y = x^2 + 1$, and to see that $(dy/dx) = 2x$, we will illustrate the technique of **implicit differentiation** with this example. The only preliminary observation that we need is that if two differentiable functions are equal on some interval, then, on that interval, they have the same derivative. Thus, if in the relation $y - x^2 - 1 = 0$, we *assume* that we can solve for y to get y as a function of x, say for simplicity $y(x)$, we obtain

$$y(x) - x^2 - 1 = 0$$

and thus

$$D[y(x) - x^2 - 1] = D\,0$$

Hence,

$$(D\,y(x)) - D\,x^2 - D\,1 = D\,0 = 0$$

or,

$$y'(x) - 2x = 0$$

and, finally,

$$y'(x) = 2x$$

which is, of course, the desired derivative.

EXAMPLE 1 Consider the graph of the relation $xy - x = 1$. Find the slope of the tangent line to the graph at the point on the graph whose first coordinate is 1 (that is, corresponding to $x = 1$).

SOLUTION What we must find is (dy/dx) at $x = 1$. Assume y is a function of x, and, for simplicity, let $y = y(x)$. The relation now is

$$xy(x) - x = 1$$

Hence,

$$D[x \cdot y(x) - x] = D\,1$$

$$[D\,x \cdot y(x)] - [D\,x] = 0$$

$$xy'(x) + y(x) - 1 = 0 \qquad \text{(Using the product rule to find } D\,x \cdot y(x))$$

After some algebra, we finally obtain

$$y'(x) = \frac{1 - y(x)}{x}$$

that is, at each point $(x, y(x))$ on the graph, the slope of the tangent line is $(1 - y(x))/x$. To find $y(1)$, note that if $x = 1$, then

$$1 \cdot y(1) - 1 = 1$$

and $y(1) = 2$. Hence, the desired slope is $(1 - 2)/1 = -1$.

Remark 1   Notice in the previous example that $y'(x)$ depends on x *and* $y(x)$. In general, y' often depends on y; if solving for y leads (as it might) to more than one function, then each time a particular function is used for y, the corresponding y' will result.

Remark 2   If, in the preceding example, we would have explicitly solved for y, we would obtain

$$y = 1 + \frac{1}{x}$$

and thus

$$\frac{dy}{dx} = -\frac{1}{x^2}$$

To see that this is exactly the same result as we obtain by implicit methods, note that, substituting,

$$y'(x) = \frac{1 - y(x)}{x} = \frac{1 - [1 + (1/x)]}{x} = -\frac{1}{x^2}$$

The following example illustrates how implicit functions can be used to justify the fact that $D x^n = nx^{n-1}$ is valid when n is a rational number.

EXAMPLE 2   Let $f(x) = x^{2/3}$. Use implicit differentiation techniques to show that $f'(x) = \frac{2}{3}x^{(2/3)-1} = \frac{2}{3}x^{-(1/3)}$.

SOLUTION   Since $f(x) = x^{2/3}$, we obtain, by cubing,

$$[f(x)]^3 = x^2$$

and

$$D[f(x)]^3 = D x^2$$

$$3f(x)^2 f'(x) = 2x$$

Hence,

$$f'(x) = \frac{(2/3)x}{[f(x)]^2} = \frac{(2/3)x}{x^{4/3}} = \frac{2}{3}x^{1-(4/3)} = \frac{2}{3}x^{-(1/3)}$$

The following examples illustrate how the technique of implicit functions can be applied to certain problems involving rates.

EXAMPLE 3   A pebble is dropped into a pond causing a circular ripple. A measuring device indicates that at the time the radius is 5 inches, the radius is changing at the rate of 2 inches/second. How fast is the area changing at this instant of time?

SOLUTION Clearly, both the area and the radius are changing with time, and hence both depend upon time. Therefore, we shall write the area as $A(t)$ for the area A at time t. Similarly, we shall write $R(t)$ for the radius at time t. It is well known that $A = \pi R^2$; substituting the functions, this becomes $A(t) = \pi[R(t)]^2$. The problem tells us that at the time that $R = 5$, the radius is changing at the rate of 2 inches/second; that is,

$$\frac{dR(t)}{dt}\bigg]_{R=5} = 2 \text{ inches/second}$$

We are asked to find $dA(t)/dt$ when $R = 5$. Observe that:

$$A(t) = \pi[R(t)]^2$$

implies $\qquad \dfrac{dA(t)}{dt} = \dfrac{d}{dt}\,\pi[R(t)]^2 = \pi\,\dfrac{d[R(t)]^2}{dt}$

By the power rule,

$$\frac{d[R(t)]^2}{dt} = 2R(t)\,\frac{dR(t)}{dt}$$

Hence, $\qquad \dfrac{dA(t)}{dt} = 2\pi R(t)\,\dfrac{dR(t)}{dt}$

is the relationship at *any* time t between

$$\frac{dA(t)}{dt} \qquad \text{and} \qquad \frac{dR(t)}{dt}$$

To solve our particular problem, we see that

$$\frac{d}{dt}\,A(t)\bigg]_{R=5} = 2\pi \cdot 5 \cdot 2 = 20\pi \text{ square inches/second}$$

Once we know $A(t) = \pi[R(t)]^2$, the relationship between dA/dt and dR/dt follows and is completely determined.

EXAMPLE 4 Suppose we have a tank the shape of which is a right circular cone whose height is 10 feet and whose circular top has a radius of 2 feet. Water is poured into the tank in such a way that the volume of water is increasing at the rate of 3 cubic feet/minute at the instant that the water level is 8 feet. How fast is the water level rising at this instant?

SOLUTION The cone is drawn in Figure 2-11; $h(t)$ is the height of the water

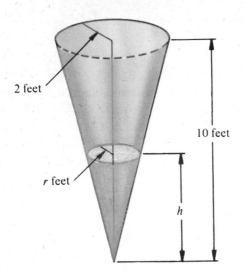

2 feet

10 feet

r feet

h

Figure 2-11

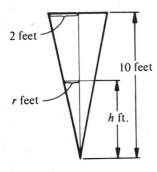

2 feet

10 feet

r feet

h ft.

Figure 2-12

at time t, and $r(t)$ is the corresponding radius (of the top of the cone of water). Let $V(t)$ be the volume of the cone of water. Using the interpretation of derivative as a measure of rate of change, we have $h'(t)$, $r'(t)$ and $V'(t)$ as the rates of change of the height, the radius, and the volume, respectively, of the cone of water at time t.

Thus, the problem can be rephrased as follows: if t_0 is the time at which $h = 8$ (that is, if $h(t_0) = 8$), find $h'(t_0)$, given that at the time t_0, $V'(t_0) = 3$.

To relate V' to h', we must first relate V to h; but this is simply

$$V(t) = \pi[r(t)]^2 h(t)$$

This relationship involves $r(t)$, for which we have no explicit information. However, by similar triangles (see Figure 2-12), we see that

$$\frac{r}{h} = \frac{2}{10} = \frac{1}{5}$$

that is,

$$r = \frac{h}{5}$$

Thus, substituting,

$$V(t) = \pi \left(\frac{h(t)}{5}\right)^2 \cdot h(t) = \frac{\pi}{25} h^3(t)$$

Differentiating,

$$V'(t) = \frac{3\pi}{25} h^2(t) \cdot h'(t)$$

Thus, at t_0 (the time at which $h = 8$)

$$V'(t_0) = \frac{3\pi}{25} h^2(t_0)h'(t_0)$$

Since $V'(t) = 3$, we see that

$$3 = \frac{3\pi}{25} \cdot 64 \cdot h'(t_0)$$

and so

$$h'(t_0) = \frac{25}{64\pi} \qquad \text{(feet/minute)}$$

Problems   ✓ **1.**   Consider the equation $(1 + x)y^2 - x = 0$.

a.   Solve this explicitly for y. Observe that you obtain two functions. Write them both explicitly and graph them. Find the derivative of each function.

b.   Use implicit differentiation to find $F'(x)$, assuming that y can be written as $F(x)$. Find $F'(1)$. Explain your answer in terms of (a).

In problems 2–13, use implicit differentiation to find dy/dx, and leave your answer in terms of x and y:

2. $x^2 + xy = 2$ $\checkmark$ 3. $y^2 = x^5$

4. $y^2 = x^2 + x$ 5. $y^5 = (x)^{1/2}$

6. $y^4 + x^2 y^2 = 10$

7. $(x)^{1/2} + (y)^{1/2} = (a)^{1/2}$, where a is any positive real number.

8. $y^2 = (x + 1)/(x - 1)$ 9. $x^3 + y^2 = 1$

10. $x^2 y + xy^2 = 6$ $\checkmark$ 11. $x^2 + y^2 = xy$

12. $x + (xy)^{1/2} = 2y$ 13. $1/y - 2y/x = 1$

In problems 14–18, find the slope of the tangent line to each of the following curves at the point or points indicated:

14. $x^2 - 2xy + y^2 + 2x + y - 6 = 0$ at $(2,2)$

$\checkmark$ 15. $x^2 y^2 - y - 2 = 0$ at $(1,-1)$ and $(1,2)$

16. $x^3 + y^3 = 18xy$ at $(8,4)$

$\checkmark$ 17. $xy^{1/2} - yx^{1/2} = 0$ at $(1,1)$

18. $\dfrac{x - y}{x - 2y} = 2$ at $(3,1)$

$\checkmark$ 19. A ladder 5 feet long rests against a vertical wall. Assume that the ground is horizontal. The ladder slides down the wall in such a way that, at the instant the foot of the ladder is 4 feet from the wall, the foot of the ladder is moving 2 feet/second. How fast is the top of the ladder descending the wall at this instant?

20. A particle moves in the circular path $x^2 + y^2 = 25$ in such a way that when the particle is at the point $(3,4)$, the component of the velocity in the direction of the x-axis is -10 feet/second. Find the component of the velocity in the direction of the y-axis at this point.

$\checkmark$ 21. A man 6 feet tall walks at the rate of 4 feet/second toward a streetlight that is 24 feet tall. How fast is the length of his shadow changing?

22. Water is withdrawn from a conical reservoir 6 feet in diameter and 8 feet deep at the constant rate of 3 cubic feet/minute. The vertex of the cone faces down. How fast is the water level falling at the instant the level is 4 feet deep?

23. Two trucks have a common starting point, but one moves southward at 30 miles per hour and the other westward at 40 miles per hour. At the end of 1 hour, how fast are the trucks moving away from each other?

24. A piece of ice is in the shape of a sphere.
    a. If the ice melts at the rate of 5 cubic inches/minute, how fast is the radius changing at the instant that the radius is 4 inches?
    b. If the surface area is decreasing by 2 square inches/minute, how fast is the radius changing?

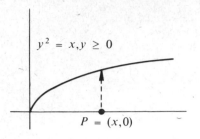

$y^2 = x, y \geq 0$

$P = (x,0)$

Figure 2-13

25. A light source hangs from the parabola $y^2 = x$ in such a way that it always casts a thin beam of light perpendicular to the x-axis, striking the x-axis at a point P (see Figure 2-13). The light source slides down the parabola so that the velocity of the light source in the y-direction is -10 feet/second. Find the velocity of the point of light on the x-axis (that is, the velocity of the point P) at the instant $x = 4$.

26. A point moves along the curve $y^2 = x^3$ in such a way that its distance from the origin increases at the constant rate of 2 units/second. Find dx/dt at $(2, 2\sqrt{2})$.

27. A particle is projected horizontally from a platform 2 feet above the ground. Directly above the platform and 6 feet from the ground is a light source. Assuming that the velocity of the particle at time $t = 1$ is 2 feet/second, find the velocity of the shadow of the particle at time $t = 1$. Assume that the particle moves only horizontally.

28. The marginal revenue corresponding to a demand of $x = 2$ is known to be 10. If the price corresponding to demand $x = 2$ is \$8, what is the rate of change of price (with respect to demand) corresponding to demand $x = 2$?

SECTION 11

THE CHAIN RULE

In section 5 of Chapter 1, we discussed the composition of functions. Recall that, given a function f that depends on u, and then, in turn, u depends on x, we formed the composition $f \circ u$, where $(f \circ u)(x) = f(u(x))$. To find the derivative of $f \circ u$ (that is, to find $Df \circ u$), we could, of course, try to explicitly evaluate $(f \circ u)(x)$, and then differentiate it. Very often the algebra involved makes such an explicit evaluation difficult, sometimes impossible. The question then is whether there is a way to find the derivative of the composition *without* first explicitly finding the composition. The answer is yes, and the result that shows how to do this is called the chain rule. Examples 1 and 2 review the composition of functions.

EXAMPLE 1 Let $F(u) = 2u(u - 1)$ and let $u(x) = 6x + 1$. Find $F \circ u$, and $D(F \circ u)$.

SOLUTION

$$
\begin{aligned}
F[u(x)] &= 2 \cdot u(x)[u(x) - 1] \\
&= 2(6x + 1)[(6x + 1) - 1] \\
&= 12x(6x + 1) \\
&= 72x^2 + 12x
\end{aligned}
$$

Thus, $F \circ u$ is the function given by

$$(F \circ u)(x) = 72x^2 + 12x$$

$$\frac{d}{dx}(F \circ u)(x) = \frac{d}{dx}(72x^2 + 12x) = 144x + 12$$

EXAMPLE 2   Fill in the missing exponents: If $f(u) = u^5$ and $u(x) = x^2$, then

$$(f \circ u)(x) = f[u(x)] = [u(x)]^\square$$

$$= x^\square$$

Notice in Example 1 that in order to find $Df \circ u$, we first had to compose f with u explicitly. The next result, the chain rule shows that *it is possible to find $Df \circ u$ without explicitly knowing $f \circ u$.* Specifically,

---

THE CHAIN RULE   *Let u have a derivative at the point x, and let f have a derivative at the point $u(x)$. Then, at the point x,*

$$\frac{d(f \circ u)(x)}{dx} = \frac{df(u)}{du} \cdot \frac{du(x)}{dx}$$

where $df(u)/du$ is to be evaluated at the point $u(x)$.

Equivalently, if we define $D_x f(x)$ to be $df(x)/dx$ (for example, $D_u f(u) = df(u)/du$), the chain rule can then be written

$$D_x(f \circ u)(x) = D_u f(u) \cdot D_x u(x)$$

where $D_u f(u)$ is to be evaluated at $u(x)$.

We shall not prove the theorem. The next examples illustrate applications of the chain rule to find derivatives.

EXAMPLE 3   Let $f(u) = u^2 + 3$ and let $u(x) = 2x - 1$. Use the chain rule to find $D(f \circ u)(x)$.

SOLUTION

$$\frac{d}{dx}(f \circ u)(x) = \frac{df(u)}{du}\frac{du(x)}{dx}$$

$$\left(\text{where } \frac{df(u)}{du} \text{ is evaluated at } u(x) = 2x - 1\right)$$

$$= 2u \cdot 2$$

$$= 4u$$

$$= 4(2x - 1)$$

EXAMPLE 4   If $F(u) = 2u(u - 1) = 2u^2 - 2u$ and $u(x) = 6x + 1$, use the chain rule to find $(d/dx)(F \circ u)(x)$.

SOLUTION

$$\frac{d}{dx}(F \circ u)(x) = \frac{dF(u)}{du} \cdot \frac{du(x)}{dx}$$

$$= (4u - 2) \cdot 6$$

$$= 24u - 12$$

$$= 24(6x + 1) - 12$$

$$= 144x + 12$$

Compare this with Example 1.

EXAMPLE 5   Let $f(u) = u^3 - 3u^2 + 2u + 1$, and $u = g(x) = x^2 + x - 1$. Then, if we form $f[g(x)]$, we have

$$\frac{d}{dx}(f \circ g)(x) = \frac{df(u)}{du} \cdot \frac{dg(x)}{dx}$$

$$= (3u^2 - 6u + 2)(2x + 1)$$

$$= [3(x^2 + x - 1)^2 - 6(x^2 + x - 1) + 2][2x + 1]$$

The last expression can be simplified, of course.

Another fact worth noting about the composition of functions is that *if u is continuous at c and if f is continuous at u(c), then f ∘ u is continuous at c*; that is,

$$\lim_{x \to c}(f \circ u)(x) = (f \circ u)(c) = f[u(c)]$$

Answer   Example 2: 5; 10

Problems   In problems 1–4, find $f \circ u$ explicitly, and then find $D(f \circ u)$ by explicitly differentiating $f \circ u$. Then use the chain rule to find $Df \circ u$.

1. $f(u) = u^2 + 3u - 1$, $u(x) = 2x + 1$
2. $f(u) = u^2$ and $u(x) = x^2 + 2x + 3$
3. $f(u) = u^2 + 2u - 1$ and $u(x) = x^2 + 1$
4. $f(u) = u + 1$ and $u(x) = x^3 + 3x^2 + x - 5$

In problems 5–10, first find $(f \circ u)(x)$ explicitly, and then use the chain rule to find $D(f \circ u)(x)$.

5. $f(u) = u^{25}$ and $u(x) = 4x^2 + x - 1$
6. $f(u) = u^{100}$ and $u(x) = x + \sqrt{x}$
7. $f(u) = \sqrt{u}$ and $u(x) = x^2 + 1$
8. $f(u) = u^{3/2}$ and $u(x) = x^4 + 1$
9. $f(u) = u^{30}$ and $u(x) = x/(x + 1)$
10. $f(u) = \sqrt{u^2 + 2}$ and $u(x) = (x + 1)/(x - 1)$
11. Let $f(x) = 5x + 2$. Find all x for which $(f \circ f)(x) = 0$.
12. Let $f(x) = x^2 - 1$. Find all x for which $(f \circ f)(x) = 0$.
13. Let f be a function such that $f'(x) = x^5$.
   a. Find $(d/dx) f(x^2)$.   [*Hint:* Let $g(x) = x^2$. Then, $(d/dx) f(x^2) = (d/dx)(f \circ g)(x)$. Now use the chain rule.]
   b. Find $(d/dx) f(x^2 + 1)$
   c. Find $(d/dx) f(x^2 - 5x + 1)$

3

APPLICATIONS OF THE DERIVATIVE

The general setting of this chapter is the use of the derivative to obtain information about the shape of the graph of a function. For example, we show how the signs of the first and second derivatives are used to find, respectively, where a graph is rising (or falling), and where a graph is convex (or concave). The principal application of this is in locating high and low points on graphs, and, in turn, these points are used to determine the maximum and minimum values attained by the function. For example, if we are dealing with a profit function in terms of price, we can determine the maximum (or minimum) profit and the price at which these profits are attained; similarly, if we have a population function in terms of time, we can determine when the population is largest, and the largest value of the population.

SECTION 1

STRICTLY INCREASING AND STRICTLY DECREASING FUNCTIONS

Suppose we are investigating a population of objects undergoing a birth-death process; that is, at any positive time t, a certain number of objects is added to the population because of birth, and a certain number is removed because of death. Let $P(t)$ be the size of the population at time t; hopefully, in any experiment, we can come up with a "formula" for $P(t)$. Our question is: over what time intervals is the population increasing (that is, will there be more births than deaths) and over what intervals is it decreasing? If we graph population P as a function of time t, we can settle such questions by observing over what intervals the graph is rising (in which case population is increasing) or falling (in which case population is decreasing). In this section, we shall show that the sign of the derivative will settle questions about intervals over which the graph of a function is rising or falling, or, equivalently, intervals on which a function is increasing or decreasing.

A function f is said to be **strictly increasing (S.I.)** if $f(x)$ increases as x increases; that is, geometrically, if the graph of f rises (from left to right). Thus,

*Whenever $a < b$, f is **S.I.** if we have $f(a) < f(b)$.*

(See Figure 3-1.)

Similarly, f is said to be *strictly increasing on some set I if we have $f(a) < f(b)$ whenever $a < b$ with a and b in I* (see Figure 3-2).

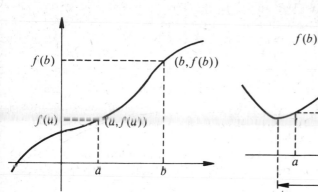

Figure 3-1

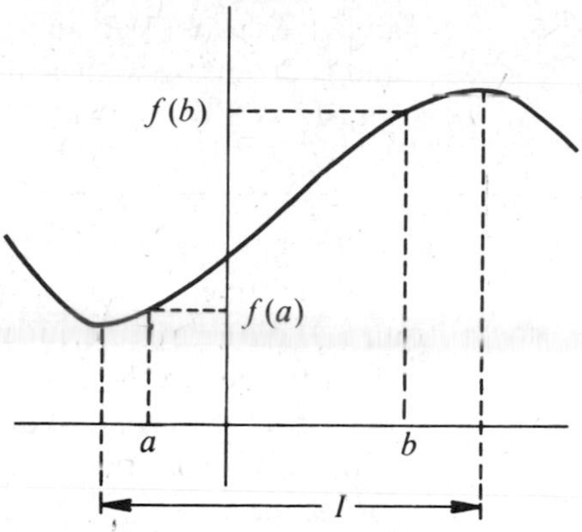

Figure 3-2

EXAMPLE 1   Let $f(x) = 3x - 1$. The graph of f is rising (since it is a straight line with positive slope); hence, f is **S.I.**

Analogously, f is **strictly decreasing (S.D.)** if $f(x)$ decreases as x increases; that is, if the graph of f is falling (from left to right). Thus,

*Whenever $a < b$, f is **S.D.** if we have $f(a) > f(b)$*

with a similar definition for **S.D.** on a set I (see Figures 3-3 and 3-4).

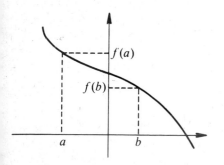

Figure 3-3

Figure 3-4

EXAMPLE 2   Let $f(x) = 1 - 2x$. Since the graph of f is a straight line with negative slope, we see that its graph is falling; hence, f is **S.D.**

EXAMPLE 3   Let $f(x) = x^2$, let N be the set of negative numbers and P be the set of positive numbers (that is, $N = \{x : x \in R \text{ and } x < 0\}$ and $P = \{x : x \in R \text{ and } x > 0\}$. The graph of f is a parabola, and is dropping with respect to N and rising with respect to P. Thus, f is **S.D.** on N and **S.I.** on P.

Figure 3-5

Figure 3-6

We shall see next that the sets on which f is **S.I.** or **S.D.** can be determined from the sign of f'. If we inspect the graph of an **S.I.** differentiable function f (see Figure 3-5), we see that all tangent lines to the graph have positive or zero slope, whereas if we inspect the graph of an **S.D.** differentiable function (see Figure 3-6), we see that the tangent lines to the graph have negative or zero slope.

It can happen that an **S.I.** (or **S.D.**) function will have a finite number of points at which the derivative is zero and at which points the graph will have horizontal tangent lines. For example, the function F given by $F(x) = x^3$ has a graph which is always rising and, thus, is **S.I.**; observe, however, that, although $F'(x) = 3x^2 > 0$ for all $x \neq 0$, $F'(0) = 0$. The horizontal axis is tangent to the graph at $(0,0)$, but this does not interfere with the fact that the curve steadily rises. We summarize these observations below:

1. *Let $f'(x) > 0$ for all x on an interval (or half-line, or the entire line) I, except possibly at a finite number of points at which the derivative is 0, then f is **S.I.** on I. Conversely, if f is **S.I.** and differentiable on I, then $f'(x) \geq 0$ for x in I.*
2. *Let $f'(x) < 0$ for all x on an interval (of half-line, or the entire line) I, except possibly at a finite number of points where the derivative is 0; then f is **S.D.** on I. Conversely, if f is **S.D.** and differentiable on I, then $f'(x) \leq 0$ for x in I.*

EXAMPLE 4   Determine where the function $f(x) = 2x^3 + 9x^2 + 12x + 6$ is **S.I.** and where it is **S.D.**

SOLUTION   First, $f'(x) = 6x^2 + 18x + 12 = 6(x^2 + 3x + 2) = 6(x + 2)(x + 1)$.

Now, $x + 2 > 0$ when $x > -2$, and $x + 2 < 0$ when $x < -2$; $x + 1 > 0$ when $x > -1$, and $x + 1 < 0$ when $x < -1$.

We see from Figure 3-7 that $f'(x) > 0$ when $x > -1$ or

Figure 3-7
```
x + 2 - - - 0 + + + + + + + + +
x + 1 - - - - - - - - 0 + + + + +
          |         |
         -2        -1
```

$x < -2$, and that $f'(x) < 0$ when $-2 < x < 1$.   Hence, $f(x)$ is **S.I.** when $x > -1$ or $x < -2$, whereas $f(x)$ is **S.D.** when $-2 < x < -1$.

The graph of f is sketched in Figure 3-8.

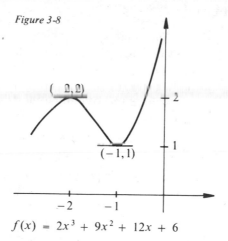

Figure 3-8

$(0, 0)$

$(-1, 1)$

$f(x) = 2x^3 + 9x^2 + 12x + 6$

Problems   Find where each of the following functions is **S.I.** and where each is **S.D.**   Find the location of all horizontal tangents to their respective graphs.

1. $f(x) = x^2 - 1$ 2. $f(x) = x^2 - 4x + 3$
3. $f(x) = x^5 + x^3$ 4. $f(x) = 3x^5 - 5x^3$
5. $f(x) = 2x^3 + 2x^2 - 2x + 1$
6. $f(x) = x^3 - 3x + 4$
7. $f(x) = 2x^3 + 3x^2 - 12x$
8. $f(x) = 3x^4 + 4x^3$
9. $f(x) = 3x^4 + 4x^3 - 36x^2$
10. $f(x) = 3x^4 + 8x^3 + 6x^2$
11. $f(x) = x^4 - 2x^3 - 2x^2$ 12. $f(x) = (2 - x)^3$
13. $f(x) = (x^3 - 3x^2 + 3x)^3$
14. $f(x) = (6x^5 - 15x^4 + 10x^3)^3$
15. $f(x) = (\sqrt{x} - 1)^3$ 16. $f(x) = (x^2 + 3x + 2)^{1/2}$
17. $f(x) = (x^2 - 4)^{1/2}$ 18. $f(x) = x + (2/x)$
19. $f(x) = 1 + 1/x + 1/x^2$
20. $f(x) = x^4 - 4x^2 - 4x$ (*Hint:* $x + 1$ is a factor of f'.)
21. A manufacturer finds that $Q(p)$, the quantity of some item produced if the item is offered at price p, is given by

$$Q(p) = 4p^2 - 15p + 12, \qquad p \geq 0$$

Find the prices for which the total revenue $pQ(p)$ is (a) increasing, (b) decreasing.

$6x^2 + 4x - 2$

$= 2(3x - 1)(x + 1)$

Assume that a manufacturer knows that he will earn $p(x)$ dollars if he produces x items of some commodity. If he produces too few items, he will be unable to meet demand, and if he produces too many, he will be overstocked; thus, he expects $p(x)$ to be small if x is too small or too large. His problem is to determine the number x for which $p(x)$ is maximum. If we call this number x_0, then $p(x_0)$ will be the maximum earnings possible. Such problems which involve maximizing or minimizing functions arise frequently in applications. We shall study them in the more general setting of finding the maximum and minimum values (if they exist) of a function f on its domain S. We say that

1. If $(d, f(d))$ is the highest point on the graph of f (with respect to S)—that is, if $f(d) \geq f(x)$ for all x in S then $f(d)$ is the maximum value of f (on its domain S).
2. If $(c, f(c))$ is the lowest point on the graph of f (with respect to S)—that is, if $f(x) \geq f(c)$ for all x in S then $f(c)$ is the minimum value of f (on S).
3. If $f(w)$ is either a maximum or a minimum value of f, then f has an extremum at w.

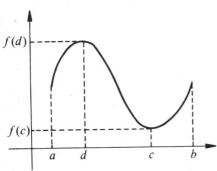

Figure 3-9

For the function graphed in Figure 3-9 (call it f), $f(d)$ is the maximum of f and $f(c)$ is the minimum of f (with respect to the interval $[a,b]$). We also say that f has a maximum at d to indicate that d is the x-coordinate of the point at which the graph of f is highest (with a similar definition for minimum).

There is no guarantee, of course, that a function f will have either a maximum or a minimum. We will illustrate functions which fail to have extrema in subsequent examples. If, however, our function is continuous and its domain is a closed interval, say $[a,b]$, then there are numbers c and d in $[a,b]$ at which f takes on its minimum and maximum values, respectively. If f is not continuous or if its domain is not a closed interval, anything can happen as far as extrema are concerned. The next few examples will illustrate some possibilities.

EXAMPLE 1   Let

$$f(x) = \begin{cases} 1/x & \text{if} \quad 0 < x \leq 1 \\ 0 & \text{if} \quad x = 0 \end{cases}$$

The domain of f is then the closed interval $[0,1]$ (see Figure 3-10). This function is discontinuous at $x = 0$. We see directly that f has no maximum (the closer x is to 0, $x > 0$, the larger is the value of $f(x) = 1/x$, and there is no largest value). The function, however, has a minimum at $x = 0$, and $f(0) = 0$ is the minimum of f on $[0,1]$.

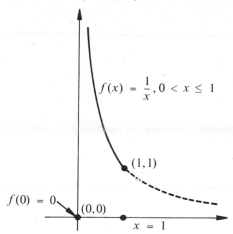

$$f(x) = \frac{1}{x}, 0 < x \le 1$$

$(1,1)$

$f(0) = 0$ $(0,0)$

$x = 1$

Figure 3-10

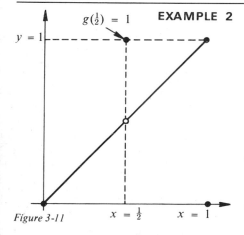

$g(\frac{1}{2}) = 1$

$y = 1$

Figure 3-11 $x = \frac{1}{2}$ $x = 1$

EXAMPLE 2   Let

$$g(x) = \begin{cases} x & \text{if} & 0 \le x \le 1 & \text{but} & x \ne 1/2 \\ 1 & \text{if} & x = 1/2 \end{cases}$$

The graph of the function is illustrated in Figure 3-11. The domain of the function is the closed interval $[0,1]$, and it is discontinuous at $x = 1/2$. (Why?)

Observe that $g(1/2) = g(1) = 1$ is the maximum value of g and $g(0) = 0$ is the minimum value of g. Even though g is not continuous at all points of $[0,1]$, direct investigation shows that both a maximum and minimum for g can be found. Observe that $g(x)$ has a maximum at $x = 1/2$ and $x = 1$, but there is only one maximum value for g; namely, $g(1/2) = g(1) = 1$.

EXAMPLE 3   Let $F(x) = |x|$. The domain of this function is the entire line. This time, f is continuous but its domain is not a closed interval. Observe that F has a minimum at $x = 0$ and that $F(0) = 0$ is the minimum value of F, but that F has no maximum (see graph in Figure 3-12).

Figure 3-12

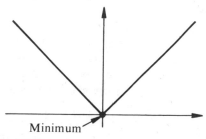

Minimum

EXAMPLE 4   Let $G(x) = x$. The domain of G is the entire line. This function is, of course, continuous everywhere and has no extrema.

EXAMPLE 5   Consider $u(x) = x$, where $0 < x < 1$. This function, of course, is not the function $G(x)$ of Example 4, since, for example, $G(10) = 10$, but $x = 10$ is not in the domain of $u(x)$; that is, $u(10)$ is not defined. The graph of u is pictured in Figure 3-13. The function u is continuous everywhere in its domain, the open interval $(0,1)$. The closer x is taken to 1, the closer $u(x)$ is to 1, but there is no largest value of u ($u(1)$ is not defined).

Hence, there is no maximum value for $u(x)$. Similarly, the closer x is to 0, the smaller is the value of $u(x)$, but there is no smallest value; hence, $u(x)$ has no minimum.

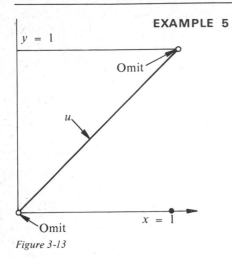

Figure 3-13

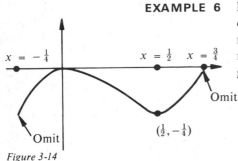

Figure 3-14

EXAMPLE 6   Let $w(x) = 4x^3 - 3x^2$, $-\frac{1}{4} < x < \frac{3}{4}$. This function is continuous on the *open* interval $-\frac{1}{4} < x < \frac{3}{4}$. It can be shown that there is a maximum at 0, and $w(0) = 0$ is the maximum of w; and w has a minimum at $x = \frac{1}{2}$, and $w\left(\frac{1}{2}\right) = -\frac{1}{4}$ is the minimum of w (see the graph of w in Figure 3-14). We shall return to this example later.

We next turn our attention to functions whose domains are closed intervals, and that are continuous on their domains. Furthermore, we shall require that each function has a derivative everywhere on this domain except possibly at a finite number of points. Such functions necessarily have both a maximum and a minimum. Various possibilities are illustrated in Figure 3-15.

In Figure 3-15, we graph five functions. From Figure 3-15(a) we see that f has a maximum at the left endpoint of its domain, a, and a minimum at the right endpoint of its domain b. Thus, $f(a)$ is the maximum value of f and $f(b)$ is the minimum value of f.

From Figure 3-15(b), $g(a)$ is the maximum value of g and $g(b)$ is the minimum value of g. Observe that $g(s)$ is smaller than all

Figure 3-15

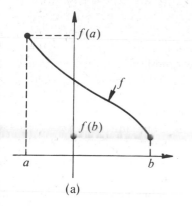

(a)

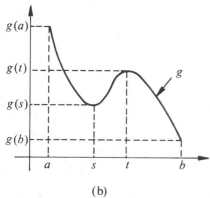

(b)

values $g(x)$ provided x is taken "near" s. Similarly, $g(t)$ is larger than all values $g(x)$ provided x is taken "near" t. The values $g(s)$ and $g(t)$ are certainly not extrema, but they can be considered extrema relative to functional values nearby. We are led to define:

The function f has a local (or, relative) maximum at p if $f(p) \geq f(x)$ for all x near p; similarly, f has a local (or, relative) minimum at q if $f(q) \leq f(x)$ for all x near q.

Combining the two cases, we say that f has a local (or, relative) extremum at r if f has either a local maximum or a local minimum at r. Furthermore, let us agree that a function cannot have a local extremum at an endpoint of its domain.

It is geometrically clear from Figure 3-15 that a continuous function defined on a closed interval attains its extremal values at a local extremum or at an endpoint (or possibly both). For example, the function h graphed in Figure 3-15(c) has both a minimum and a local minimum at c, and for the function F graphed in Figure 3-15(d), the local extrema at c and d are also the extrema of F on $[a,b]$.

Thus, it should be clear that to find the extrema of a continuous function f whose domain is closed interval $[a,b]$, the following steps are necessary:

1. Find the location of all local extrema in (a,b); assume that these local extrema are found at $x = r, x = s, x = t, \ldots$.
2. Evaluate $f(r), f(s), f(t), \ldots$.
3. Evaluate $f(a)$ and $f(b)$.

The largest of the numbers computed in (2) and (3) then is the maximum of f and the smallest is the minimum of f.

The question then is: how do we find the *local* extrema of f, a function? The derivative is the clue; we shall, therefore, suppose that f' exists on $[a,b]$.

Let us assume that f has a local maximum at the point $(c, f(c))$,

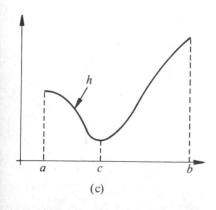

(c)

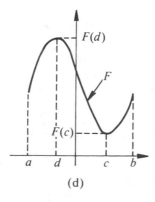

(d)

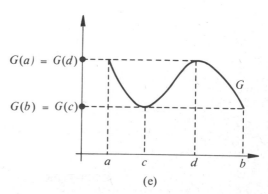

(e)

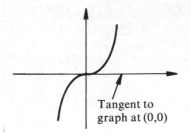

Figure 3-16

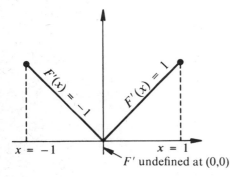

Figure 3-17

and that there is a tangent line to the graph of f at $(c, f(c))$. Then the graph of f must be rising to $(c, f(c))$, must turn smoothly at $(c, f(c))$, and then must fall; thus, we would expect (and it is, in fact, the case) that there is a horizontal tangent line at $(c, f(c))$. There will also be a *horizontal tangent line* if f has a local minimum $(c, f(c))$. See, for example, the functions graphed in Figure 3-15(b) and 3-15(c). Thus, *if f has a local extremum at c, and if $f'(c)$ exists, then $f'(c) = 0$.*

Unfortunately, our result does not assert that if $f'(d) = 0$, a relative extremum will be found at d. The function $f(x) = x^3$ has as its derivative $f'(x) = 3x^2$. Consequently, $f'(0) = 0$; however, since f is **S.I.**, it cannot have any local extrema (see Figure 3-16). Our result also does not tell us if it is possible for f to have a local extremum at a point at which f' does not exist. A simple illustration—the function $F(x) = |x|$ with domain $[-1, 1]$—shows that it is possible, since $F(x)$ has a local minimum at $x = 0$ although $F'(0)$ does not exist (see Figure 3-17). We can conclude that the values of x, at which $f'(x) = 0$ or $f'(x)$ is undefined, include all possible local extrema.

The solutions to $f'(x) = 0$ and the x's at which $f'(x)$ is undefined may give extraneous values (that is, values of x at which a local extremum is not found) as in $f(x) = x^3$ at $x = 0$. Their inclusion, however, will cause no harm, since the functional values obtained by evaluating the function at these points will be discarded when, in the last instruction, we keep only the largest and smallest values obtained in (2) and (3). Thus, we have:

Procedure for finding the extrema of a continuous function f that is differentiable at all but perhaps a finite number of points on its domain, the closed interval $[a, b]$:

1. Find all x in $a < x < b$ that satisfy the equation $f'(x) = 0$ or at which $f'(x)$ is undefined; let $x = r$, $x = s$, $x = t, \ldots$ be such x.
2. Evaluate $f(r), f(s), f(t), \ldots$.
3. Evaluate $f(a)$ and $f(b)$.
4. The largest number of the numbers computed in (2) and (3) is the maximum of $f(x)$ in $[a, b]$ and the smallest number is the minimum.

EXAMPLE 7   Find the extrema of the function

$$f(x) = x^4 - 2x^2 + 5 \qquad (-1.2 \le x \le 1)$$

SOLUTION   Observe that $f'(x) = 4x^3 - 4x = 4(x^3 - x) = 4x(x + 1)(x - 1)$.

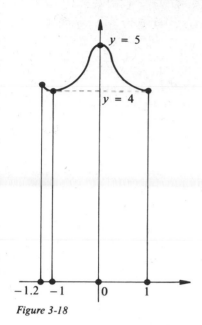

Following (1) of the procedure, we see that $f'(x) = 0$ at $x = 0$, $x = +1$ and $x = -1$; $f'(x)$ is defined for all x in $[-1.2, 1]$. Following (2), we find that $f(0) = 5$, $f(1) = 4$, and $f(-1) = 4$. (Since $x = 1$ is the endpoint of the domain, we must compute $f(1)$ either in this step or in step (3) of the procedure.) Following (3), $f(-1.2) = 4.1976$ and $f(1) = 4$. Clearly, $4 < 4.1976 < 5$. Thus, by (4), $f(-1) = f(1) = 4$ is the minimum of $f(x)$, and $f(0) = 5$ is the maximum of $f(x)$ (see the graph of the functions in Figure 3-18).

Figure 3-18

EXAMPLE 8   Let $v(x) = 4x^3 - 3x^2$, $-\frac{1}{4} \le x \le \frac{3}{4}$. To find the absolute extrema, we follow the procedure above.

1. $v'(x) = 12x^2 - 6x = 6x(2x - 1)$; therefore $v'(x) = 0$ when $x = 0$ or $x = \frac{1}{2}$.
2. $v(0) = 0$, $v(\frac{1}{2}) = -\frac{1}{4}$.
3. $v(-\frac{1}{4}) - -\frac{1}{4}$, $v(\frac{3}{4}) = 0$.
4. $-\frac{1}{4} < 0$; thus, $v(\frac{1}{2}) = v(-\frac{1}{4}) = -\frac{1}{4}$ is the minimum of $v(x)$ and $v(0) = v(\frac{3}{4}) = 0$ is the maximum of $v(x)$.

Compare these results with w of Example 6. For w, the domain is the open interval $(-\frac{1}{4}, \frac{3}{4})$, which contains fewer points than the domain of v which is $[-\frac{1}{4}, \frac{3}{4}]$. Hence, w cannot have any absolute extrema that v does not have. Consequently, $w(0) = 0$ is the absolute maximum of w and $w(\frac{1}{2}) = -\frac{1}{4}$ is the absolute minimum of w; since $-\frac{1}{4}$ and $\frac{3}{4}$ are not in the domain of w, we do not evaluate w at these points (see Figure 3-14).

EXAMPLE 9   Find the extrema of the function

$$g(x) = x^4 - 8x^2 + 12 \qquad (-1 \le x \le 2)$$

SOLUTION
1. $g'(x) = 4x(x + 2)(x - 2)$; therefore, $g'(x) = 0$ when $x = 0$, $x = +2$, or $x = -2$. However, -2 is not in the domain of $g(x)$, and is discarded.
2. $g(0) = 12$, $g(2) = -4$.
3. $g(-1) = 5$ and $g(2) = -4$.
4. $-4 < 5 < 12$.

Hence, $g(2) = -4$ is the minimum of g and $g(0) = 12$ is the maximum of g.

EXAMPLE 10   Let $f(x) = 6x^{2/3}$, $-\frac{1}{2} \le x \le 1$. Find the local extrema.

SOLUTION
1. $f'(x) = 4x^{-1/3}$. There are no solutions to $f'(x) = 0$, but $f'(x)$ is undefined if $x = 0$.
2. $f(0) = 0$.
3. $f(-\frac{1}{2}) = (6/\sqrt[3]{4})$ $f(1) = 6$.
4. $0 < (6/\sqrt[3]{4}) < 6$.

Hence, $f(0) = 0$ is the absolute minimum of f and $f(1) = 6$ is the maximum of f (see Figure 3-19).

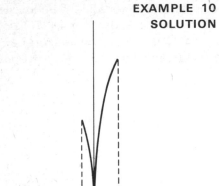

Figure 3-19

Problems   For each of the following functions, find the maximum and minimum in the specified domains.

1. $f(x) = x^3 - 3x^2,$ $-1 \le x \le 3$
2. $f(x) = x^4 - 8x^2,$ $-3 \le x \le 3$
3. $f(x) = 2x^3 - 7x^2 + 8x + 2,$ $0 \le x \le 3$
4. $f(x) = \frac{1}{6}(x^3 - 6x^2 + 9x + 1),$ $0 \le x \le 2$
5. $f(x) = x^{1/3},$ $-1 \le x \le 2$
6. $f(x) = 1 - x^{2/3},$ $-1 \le x \le 1$
7. $f(x) = x^3,$ $\frac{1}{2} \le x \le 1$
8. $f(x) = x^4 - 8x^2,$ $-3 < x < 3$ (See problem 2.)
9. $f(x) = x^3 - 3x^2,$ $-1 < x < 3$ (See problem 1.)
10. $f(x) = 8x^3 - 4x^2 + 72x,$ $0 \le x \le 4$
11. $f(x) = x^3 + 3x + 7$ $0 \le x \le 1$
12. $f(x) = x^4 - 4x^2 - 4x,$ $-1 \le x \le 2$
     [*Hint:* $(x + 1)$ is a factor of $f'(x)$.]
13. $f(x) = x^5 - 5x + 1,$ $-2 \le x \le 2$
14. $f(x) = \sqrt{x}(x - 5)^{1/3},$ $0 \le x \le 6$
15. $f(x) = \sqrt[3]{x}\sqrt[3]{(x - 3)^2},$ $-1 \le x \le 4$
16. $f(x) = x^4 - 8x^3 + 22x^2 - 24x + 100,$ $0 \le x \le 5$
     [*Hint:* $(x - 1)$ is a factor of $f'(x)$.]
17. $f(x) = 3x^4 - 18x^2 - 24x + 160,$ $-3 \le x \le 3$
     [*Hint:* $(x + 1)$ is a factor of $f'(x)$.]
18. $f(x) = 3x^5 - 25x^3 + 60x + 10,$ $-2 \le x \le 2$
19. $f(x) = 3x^5 - 25x^3 + 60x + 10,$ $-3 < x \le 2$

SECTION 3

LOCAL EXTREMA AND CONCAVITY. THE SECOND DERIVATIVE TEST

Using the first derivative, we are able to tell when the graph of a function is rising and when it is falling. A graph can be rising (or, respectively, falling) and be concave up (see Figure 3-20) or it can be rising (or, respectively, falling) and be concave down (see Figure 3-21), but the first derivative cannot distinguish between these possibilities. We shall now see that the second derivative can.

Loosely speaking, we shall say that the graph of a function f is concave up at c if, near c, the graph of the function looks like one of the figures in Figure 3-20; and concave down at c if, near c, the graph of the function looks like one of the figures in Figure 3-21. The second derivative test for the concavity of the graph of f is as follows:

Assume that c is in the domain of f and f' exists near c. Then:

1. *If $f''(c) > 0$, the graph of f is concave up at c, whereas*
2. *If $f''(c) < 0$, the graph of f is concave down at c.*

To see this, we shall rely on the following geometrical considerations. Let us consider, for example, the case that $f''(c) > 0$. Recall that if $f' > 0$, then f is **S.I.** If we apply this same fact to the function f', we see that, if $f'' > 0$, then f' is increasing; that is, if $f''(c) > 0$, then the derivative f' is increasing near c.

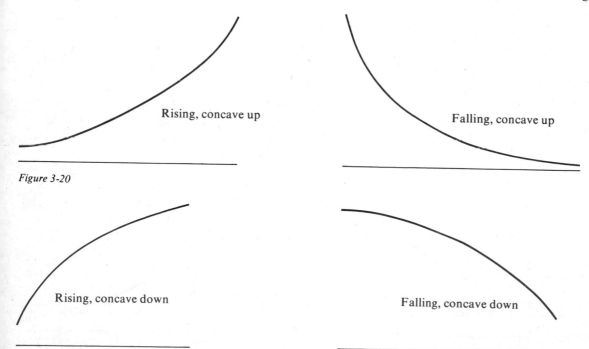

Rising, concave up

Falling, concave up

Figure 3-20

Rising, concave down

Falling, concave down

Figure 3-21

Recall that $f'(x)$ is the slope of the line tangent to the graph of f at $(x, f(x))$. From this, we see that the geometrical significance of the fact that f' is increasing breaks down into two cases:

In the first instance, $f'(x) > 0$ near c, in which case $f'(x)$ increasing implies that the tangent lines are becoming steeper and steeper. For example, in Figure 3-22,

$$0 < f'(x_1) < f'(x_2) < f'(x_3)$$

In this case, f must be increasing and concave up.

In the second instance, $f'(x) < 0$, near c in which case f' decreasing implies that the tangent lines are becoming less and less steep (keeping in mind that negative numbers become larger as they are taken closer to 0). For example, in Figure 3-23,

$$f'(x_4) < f'(x_5) < f'(x_6) < 0$$

In this case, the graph of f must be falling and concave up.

These observations about concavity have an interesting application to the problem of locating local extrema. Observe from Figure 3-24 that, if f is defined and differentiable near c, if $f'(c) = 0$, and if the graph of f is concave up at c, then f necessarily has a local minimum at c; whereas, from Figure 3-25, if the graph of f is concave down at c, then f has a local

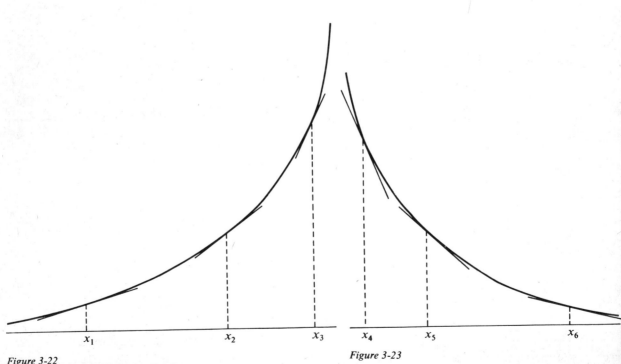

x_1 x_2 x_3 x_4 x_5 x_6

Figure 3-22

Figure 3-23

maximum at c. Equivalently, in terms of derivatives, we have the following result. Assume that f is defined and differentiable near c. Then

1.   If $f'(c) - 0$ and $f''(c) > 0$, f has a local minimum at c.
2.   If $f'(c) = 0$ and $f''(c) < 0$, f has a local maximum at c.

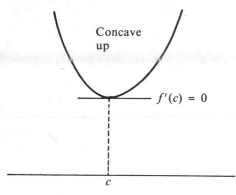

Local minimum at c

Figure 3-24

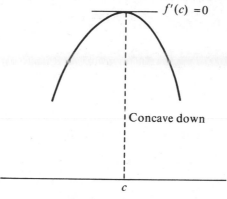

Local maximum at c

Figure 3-25

EXAMPLE 1   Let $f(x) = x^3 - 3x + 1$. Find all the local extrema for f. Sketch the graph of f.

SOLUTION   Note that

$$f'(x) = 3x^2 - 3x = 3(x^2 - 1) = 3(x + 1)(x - 1)$$

and

$$f''(x) = 6x - 3$$

We see that $f'(x) = 0$ for $x = 1$ or -1. Thus, 1 and -1 are possible locations of local extrema. Now, $f''(1) = 3$; hence, f has a local minimum at 1. Next, $f''(-1) = -9$, and f has a local maximum at -1.

Moreover, the graph of f is concave up at all x for which $f''(x) = 3(2x - 1) > 0$, that is, for x in $(\frac{1}{2}, \infty)$; and the graph of f is concave down at all x for which $f''(x) = 3(2x - 1) < 0$, that is, for x in $(-\infty, \frac{1}{2})$.

To get more information in order to graph f, note that

$$f \text{ is } \textbf{S.I. when } f'(x) = 3(x + 1)(x - 1) > 0$$

that is, for x in $(-\infty, -1)$ or $(1, \infty)$ [that is, $(-\infty, -1) \cup (1, \infty)$], and

$$f \text{ is } \textbf{S.D. when } (x + 1)(x - 1) < 0$$

that is, for x in $(-1, 1)$ (see Figure 3-26).

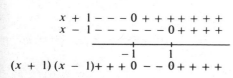

Figure 3-26

The information about f is summarized in the following table:

f	S.I.	S.D.	Location of possible extrema	Concave up	Concave down
Derivatives of f	$f'(x) > 0$	$f'(x) < 0$	$f'(x) = 0$	$f''(x) > 0$	$f''(x) < 0$
Location of x	$(-\infty, -1)$ or $(1, \infty)$	$(-1, 1)$	$1, -1$	$(\frac{1}{2}, \infty)$	$(-\infty, \frac{1}{2})$

The graph of f is sketched in Figure 3-27. Note that the graph of f changes its concavity at $\frac{1}{2}$.

A point at which a graph changes its concavity is called a **point of inflection**.

We have seen that, if f is differentiable near c, if $f'(c) = 0$, and if $f''(c)$ is either positive or negative, f has a local extremum at c. What happens if we keep these assumptions except for one change: now $f''(c) = 0$? As the next example illustrates, there may be a local extremum or a point of inflection, but there is no way to predict just from this information what happens at c; hence, additional information must be gathered.

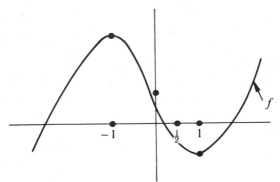

Figure 3-27

EXAMPLE 2   Let $f(x) = x^4$, $g(x) = -x^4$ and $h(x) = x^3$. Verify that

$$f'(0) = g'(0) = h'(0) = 0$$

and

$$f''(0) = g''(0) = h''(0) = 0$$

Hence, these are all functions with the property that, at some point (namely, 0) the first and second derivatives are 0. From

Figures 3-28, 3-29, and 3-30, we see that f has a local minimum at 0, g has a local maximum at 0, and h has a point of inflection at 0.

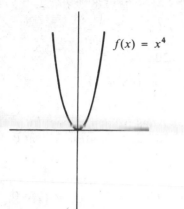

$f(x) = x^4$

Figure 3-28

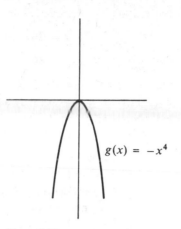

$g(x) = -x^4$

Figure 3-29

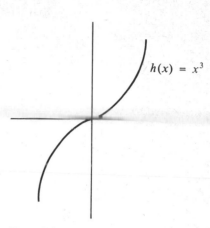

$h(x) = x^3$

Figure 3-30

If the second derivative test for local extrema fails to give information (or, if for some reason we just do not want to use it), we can go back to first principles and observe:

Suppose $f'(c) = 0$, or $f'(c)$ is undefined. Then, near c,

1. If f is **S.D.** to the left of c and **S.I.** to the right of c, f has a local minimum at c (see Figure 3-31) whereas
2. If f is **S.I.** to the left of c and **S.D.** to the right of c, f has a local maximum at c (see Figure 3-32).

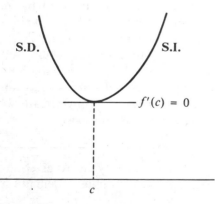

S.D.   S.I.

$f'(c) = 0$

c

Figure 3-31

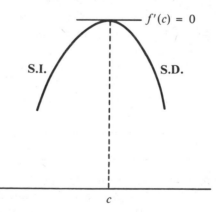

$f'(c) = 0$

S.I.   S.D.

c

Figure 3-32

EXAMPLE 3   Let $f(x) = 4x^5 + 5x^4 + 1$. Find the local extrema for f.

SOLUTION   Note that

$$f'(x) = 20x^4 + 20x^3 = 20x^3(x + 1)$$

and

$$f''(x) = 80x^3 + 60x^2 = 20x^2(4x + 3)$$

Hence, $f'(x) = 0$ if $x = 0$ or -1. Since $f''(-1) = 20(-1)^2(-4 + 3) < 0$, it follows that f has a local maximum at -1. Next, $f''(0) = 0$; hence, the second derivative test fails at 0. To see what happens at 0, note that

1. If x is near 0 but to the left of 0 (that is, negative), then $x^3 < 0$ and $x + 1 > 0$; hence $f'(x) = 20x^3(x + 1) < 0$. Thus, f is **S.D.** near 0, but to the left of 0.
2. If x is near 0 but to the right of 0 (that is, positive), then $x^3 > 0$ and $x + 1 > 0$; hence, $f'(x) > 0$. Thus, f is **S.I.** near 0, but to the right of 0.

From the above, we see that f has a local minimum at 0.

Problems   Use the second derivative test to find all the local maxima and minima for the functions given in problems 1–13, 16–19 of the previous section. Assume now that the domain of each function is the line.

SECTION 4

MORE PROBLEMS IN EXTREMA

Problems involving extrema often arise in applications. If an object is in motion, the physicist often asks for the maximum velocity of the object or the maximum distance traveled by an object. In business, we might inquire about minimum average cost or maximum revenue. We shall illustrate with some examples. It will sometimes be necessary to investigate functions whose domain is a half-line. Strictly speaking, our previous observations for continuous functions on closed intervals do not apply, but for the particular functions we shall investigate we shall see that a minimum or maximum will usually exist.

EXAMPLE 1   Assume that $\mathbf{p}(x)$, the price for the quantity x, is given by $\mathbf{p}(x) = 6 - x^2$. Find the maximum total revenue if demand x varies so that $0 \le x \le \frac{5}{2}$.

SOLUTION   The total revenue $\mathbf{R}(x)$ is given by:

$$\mathbf{R}(x) = x\mathbf{p}(x) = 6x - x^3$$

We see that $R'(x) = 6 - 3x^2 = 3(2 - x^2)$; thus, $R'(x) = 0$ when $x = \pm(2)^{1/2}$. Again, $x = -(2)^{1/2}$ can be ignored since it is not in the domain of $R(x)$. We evaluate $R(0) = 0$, $R[(2)^{1/2}] = 4(2)^{1/2}$, and $R\left(\frac{5}{2}\right) = -\frac{5}{8}$. The maximum total revenue then occurs when demand is $(2)^{1/2}$, and the maximum total revenue is $4(2)^{1/2}$.

EXAMPLE 2   A photographer has a thin piece of wood 16 inches long. How should he cut the wood to make a rectangular picture frame that encloses the maximum area?

SOLUTION   Call the length of the required frame x and the width y. We wish to maximize xy, the area enclosed. The area becomes progressively larger as x and y are made larger, but we are restricted here by the total of 16 inches of wood to make up all the sides. The 16 inches then become the perimeter of the frame; hence, $2x + 2y = 16$ or, $x + y = 8$. The equation $x + y = 8$ is, therefore, a constraint on the possible dimensions of the rectangular frame.

The original problem can be rephrased to read: Maximize the product xy subject to the constraint $x + y = 8$. We see next that our procedures are for maximizing functions of one variable, whereas xy depends on both x and y. We must either eliminate the x or y. Since $x + y = 8$, we have that $y = 8 - x$; and hence xy becomes $x(8 - x)$. The area A subject to the given constraint is

$$A(x) = x(8 - x) = 8x - x^2$$

Since there are two sides of length x, we see that $0 \leq 2x \leq 16$, or $0 \leq x \leq 8$. If $x = 0$, we have 0 area (if the rectangle has 0 length, it has 0 area), and if $x = 8$, we have 0 area (that is, if $x = 8$, then $2x = 16$ so that $y = 0$; hence, the rectangle has 0 width and, consequently, 0 area). We also observe this analytically by seeing that $A(0) = A(8) = 0$. Hence, $[0,8]$ is the domain of A.

Finally, our problem is to find the maximum of the function $A(x) = 8x - x^2$ in the closed interval $[0,8]$. Clearly, $A'(x) = 8 - 2x$, so that $A'(x) = 0$ when $8 - 2x = 0$, $2x = 8$, or $x = 4$. $A'(x)$ is always defined. By the procedure of the last section, we evaluate $A(0) = 0$, $A(8) = 0$, and $A(4) = 16$. Hence, the maximum area is 16, and it occurs when $x = 4$. Since $x + y = 8$, then $y = 4$. Hence, the rectangle with maximum area subject to the fixed perimeter is a square.

EXAMPLE 3   Find the two nonnegative numbers whose sum is 18 if the product of one by twice the square of the other is to be maximum.

SOLUTION   Call one number y; the other, x. Then we wish to maximize $y \cdot 2x^2 = 2yx^2$ subject to the constraint $x + y = 18$.   Since $y = 18 - x$, the problem is to maximize

$$f(x) = 2(18 - x) x^2 = 36x^2 - 2x^3$$

We next seek the domain of f. Since x is assumed nonnegative we have $x \geq 0$. Since $y \geq 0$ by assumption, we see from $x + y = 18$ that the largest value that x can have is 18.   Hence, $0 \leq x \leq 18$.

We must maximize f in the closed interval $[0, 18]$. Note that

$$f'(x) = 72x - 6x^2 = 6x(12 - x)$$

and thus $f'(x) = 0$ has solutions $x = 0$ and $x = 12$. We see that $f(0) = 0$, $f(12) = 1728$, and $f(18) = 0$.   Consequently, f has the maximum value of 1728 attained when $x = 12$; hence, since $x + y = 18$, when $y = 6$.

EXAMPLE 4   Find the two positive numbers whose sum is 18 if the product of one by twice the square of the other is to be maximum. The only difference between this example and Example 3 is that the word "nonnegative" in Example 3 has been replaced by the word "positive" in this example.   The functions involved are very similar except that now the domain is the open interval $(0, 18)$ since x positive does not permit $x = 0$, and if $x = 18$, then $y = 0$, which is not permitted. When we considered the closed interval in Example 3, we found the maximum occurs at $x = 12$, and the minimum occurs at $x = 0$ and $x = 18$, with $f(0) = 0 = f(18)$ and $f(12) = 1728$, the values of the extrema. From Figure 3-33 (completely out of scale!), we see that deleting the points $x = 0$ and $x = 18$ in no way alters the maximum value of f; hence, $36x^2 - 2x^3, 0 < x < 18$ assumes its maximum at the same place as the function $f(x) = 36x^2 - 2x^3$, $0 \leq x \leq 18$.   Thus, the answer to Example 4 is the same as the answer to Example 3.

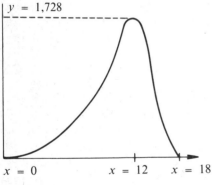

$y = 1{,}728$

$x = 0$ $x = 12$ $x = 18$

Figure 3-33

EXAMPLE 5   Find the largest right circular cylinder of largest volume that can be inscribed in a sphere of radius R.

PARTIAL SOLUTION   Now refer to Figure 3-34. Let the cylinder have radius r and height $2h$ (using $2h$ is a convenience that will become more apparent in a moment).   The volume of the cylinder is then $\pi r^2 (2h) = 2\pi r^2 h$. The problem is to maximize this volume. The problem is to find a relationship between r and h, and thus eliminate one of them. Again from Figure 3-34,

$$h^2 + r^2 = R^2 \qquad \text{and} \qquad r^2 = R^2 - h^2$$

Thus, substituting, we see that $v(h)$ the volume of the cylinder corresponding to height h, is

$$v(h) = 2\pi r^2 h = 2\pi(R^2 - h^2)h = 2\pi R^2 h - 2\pi h^3$$

(Of course, R is the fixed radius of the sphere.) The reader is invited to complete the problem.

Figure 3-34

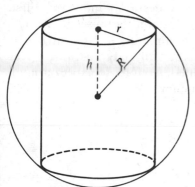

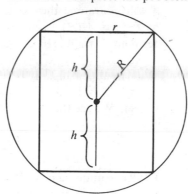

EXAMPLE 6   A manufacturer plans to construct a cylindrical can to hold 1 cubic foot of liquid. If the cost of constructing the top and bottom of the can is twice the cost of constructing the side, what are the dimensions of the most economical can?

SOLUTION   Call c the cost of construction of a square foot of the side of the can. Then $2c$ is the cost of construction of a square foot of the top or bottom of the can. Let R be the radius of the can and H the height of the can, measured in feet. Then, since the volume of the can, a right circular cylinder, is 1 cubic foot, we see that $\pi R^2 H = 1$. The combined area of the base and top of the can is $2\pi R^2$, and the total cost of constructing the top and bottom of the can is $4c\pi R^2$; the total surface area of the cylinder is $2\pi RH$, and the cost of constructing this surface area is $2\pi cRH$. The total cost then is $4c\pi R^2 + 2\pi cRH$. The problem is to minimize $4c\pi R^2 + 2\pi cRH$, subject to the constraint $\pi R^2 H = 1$. Solving the constraint equation for H yields

$$H = \frac{1}{\pi R^2}$$

so that the quantity we wish to minimize, the total cost, becomes

$$G(R) = 4c\pi R^2 + \frac{2c}{R}$$

$$= 2c\left(2\pi R^2 + \frac{1}{R}\right)$$

where c is the constant cost of construction. If we analyze the relation between H and R, it should be clear that the values of R can be chosen near 0 by taking H large although $R \neq 0$, and that the values of R can become quite large by taking H sufficiently small. Hence, the domain of $\mathbf{G}(R)$ is the open half-line, $(0, \infty)$ which we shall denote by J. The results of the last section do not directly apply, but a direct analysis of $\mathbf{G}(R)$ on its domain J will guarantee the existence of the desired minimum. First, as $R \to 0$, $\mathbf{G}(R)$ becomes arbitrarily large because of the $1/R$ term appearing. Hence, there is no minimum to $\mathbf{G}(R)$ at $R = 0$. Next, as $R \to \infty$, $\mathbf{G}(R)$ becomes arbitrarily large because of the R^2 term appearing. It is reasonable, therefore, to expect a minimum which is also a local minimum. We shall next analyze $\mathbf{G}'(R)$. We see that

$$\mathbf{G}'(R) = 2c\left(4\pi R - \frac{1}{R^2}\right)$$

$\mathbf{G}'(R)$ is undefined at $R = 0$. By what we have said, however, this value certainly will not contribute a minimum to $\mathbf{G}(R)$ nor is it even in the domain of $\mathbf{G}(R)$. We set $\mathbf{G}'(R) = 0$. Finding a common denominator, and dividing by $2c$, this becomes

$$\frac{4\pi R^3 - 1}{R^2} = 0$$

or

$$4\pi R^3 - 1 = 0$$

Thus,

$$R = \frac{1}{\sqrt[3]{4\pi}}$$

Observe that $\mathbf{G}'(R) < 0$ when

$$2c\,\frac{4\pi R^3 - 1}{R^2} < 0$$

Since R^2 and c are positive, we divide by $2c$ and multiply by R^2 to obtain $\mathbf{G}'(R) < 0$ when $4\pi R^3 - 1 < 0$, or $R < (1/\sqrt[3]{4\pi})$. Hence $\mathbf{G}(R)$ is **S.D.** for $R < (1/\sqrt[3]{4\pi})$. Similarly, the reader should convince himself that $\mathbf{G}'(R) > 0$ when $R > (1/\sqrt[3]{4\pi})$, so that $\mathbf{G}(R)$ is **S.I.** for $R > (1/\sqrt[3]{4\pi})$. Hence, $R = (1/\sqrt[3]{4\pi})$ is the position of the minimum of $\mathbf{G}(R)$, and

$$\mathbf{G}(1/\sqrt[3]{4\pi}) = 3c\sqrt[3]{4\pi}$$

is the minimum total cost (see Figure 3-35). Finally, since $H = (1/\pi R^2)$, we obtain dividing by R,

$$\frac{H}{R} = \frac{1}{\pi R^3}$$

Since $R = (1/\sqrt[3]{4\pi})$, we have $R^3 = 1/4\pi$; substituting, $H/R = 4$;

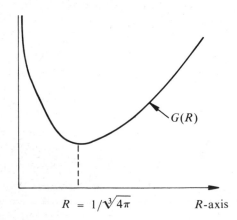

$R = 1/\sqrt[3]{4\pi}$ R-axis

Figure 3-35

that is, the most economical can has a height H four times its radius R.

In many applications, we deal with functions whose domain and/or range is some set of integers. For example, the cost of manufacturing n articles is a function of the integer n. Sometimes the methods outlined in the previous section can be applied to such functions to locate extrema. Call the function whose extrema we seek $F(n)$, where n is a positive integer, $k \leq n \leq m$ (that is, the domain of F is the set of positive integers n which lie between the positive integers k and m, so that the domain of F is the set of integers $k, k + 1, k + 2, \ldots, m$). What we hope to do is to find a continuous function G with domain $[k,m]$, and for which $G(n) = F(n)$ for all integer values n in $[k,m]$; that is, we hope to find a continuous function G which agrees with F at all integer values in the domain of F. If such a G exists, we find its absolute extrema, and observe that the values $F(n)$ which are nearest the values of the extrema of G must be the extrema for $F(n)$ (note that we are not saying the value of n that is closest to where the extrema occurs, but the value of $F(n)$). In Figure 3-36, $G(A)$ is the maximum of $G(x)$, and $F(I)$ is the value of $F(n)$ nearest $G(A)$ so that $F(I)$ is the maximum of $F(n)$. Similarly, $G(B)$ is the absolute minimum of $G(x)$, and therefore, $F(J)$ is the absolute minimum of $F(n)$.

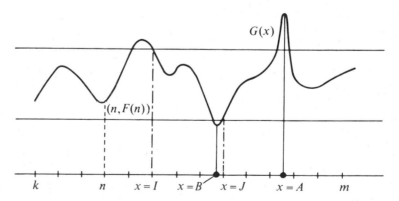

Figure 3-36

EXAMPLE 7   A manufacturer determines that the demand for n lamps if each is priced at $\$p$ is given by the function $n = f(p) = 60 - \left(\frac{1}{2}\right)p$, where p is an even integer in the interval $[0, 120]$. Hence, the range of $f(p)$ is a set of integers; and if $n = f(p)$ for p an even integer in $[0, 120]$, we see that n is an integer in $[0, 60]$. We suppose next that our lamp manufacturer has determined the cost of producing n lamps to be $\$(140 + 25n)$. How many lamps should he manufacture to maximize his profit?

SOLUTION   The total profit $F(n)$ obtained by producing n lamps at $\$p$ is

$$F(n) = np - (140 + 25n)$$
$$= n(120 - 2n) - (140 + 25n)$$
$$= -2n^2 + 95n - 140$$

We consider the function $G(x) = -2x^2 + 95x - 140$, $0 \leq x \leq 60$, and observe that whenever $x = n$ is an integer, $G(n) = F(n)$, and thus all points $(n\ F(n))$ are on the graph of $G(x)$. Observe also that $G(\frac{1}{2}) = -43$, but $F(\frac{1}{2})$ is undefined. Since $G'(x) = -4x + 95 > 0$ if $x < \frac{95}{4}$, and $G'(x) < 0$ if $x > \frac{95}{4}$, we see that $G(x)$ is **S.I.** in $[0, \frac{95}{4}]$ and **S.D.** in $[\frac{95}{4}, 60]$.

Hence, G has its maximum at $x = \frac{95}{4}$. A graph of G appears in Figure 3-37. The dots on the graph indicate the points on the graph of $G(x)$ corresponding to integer values of x; that is, for integers n, they are the points $(n, G(n)) = (n, F(n))$. We calculate to find $G(23) = F(23) = 987$ and $F(24) = G(24) = 992$. Since $G(24)$ is larger, a maximum profit of $\$992$ results from the production of 24 sets. It is only necessary to consider $F(23)$ and $F(24)$, since the function $G(x)$, hence the function $F(n)$ is increasing for $x < 23$ so that no value of $n < 23$ can possibly exceed $F(23)$; a similar remark holds for $n > 24$. Finally, since $n = 60 - (\frac{1}{2})p$, we see that, for $n = 24$, the price that must be charged to yield a maximum profit is $\$72$.

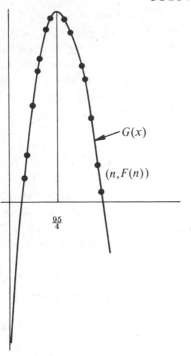

$G(x)$

$(n, F(n))$

$\frac{95}{4}$

Figure 3-37

Problems

1. Assume that the demand law is given by $p = (5 - x)^3$. Determine when the total revenue and marginal revenue are maximum and minimum in $[0, 5]$.

2. Assume that the total cost of producing a quantity x of a commodity is given by $S(x) = 2x^3 - 9x^2 + 12x$. Determine where the total cost and marginal cost are least and greatest on $[0, 3]$.

3. Assume that the total cost $S(x)$ of producing a quantity x is given by $S(x) = x^3 - 5x^2 + 6x$. Determine the minimum average cost for x in $[1, 4]$. What is the minimum average cost if x is an integer and $1 \leq x \leq 4$?

4. Suppose that the total cost $S(x)$ of producing a quantity x is given by $S(x) = 2x^4 - 6x^3 + 12x^2 - 2x + 1$. Determine the minimum and maximum marginal cost if x is in $[0, 3]$.

5. Determine the maximum total revenue if the demand law is given by $p(x) = 8/(4 + x^2)$, $x \geq 0$. Sketch the graph of $p(x)$, and the total revenue function. Justify your answer.

6. A commuters' train carries 600 passengers per day from a suburb to a city. It costs $1 per person to ride the train. It is observed that 50 additional people will ride the train for each 5-cent decrease in fare. What fare should be charged to make the largest profit possible?

7. A toy manufacturer produces an inexpensive doll and an expensive doll in units of x hundreds and y hundreds, respectively. Assume that it is possible to produce the dolls according to the relationship

$$y = \frac{50 - 10x}{10 - x}, \qquad x \neq 10$$

If the expensive dolls cost twice as much as the inexpensive dolls,

a. Find the function R whose value at x is the total revenue obtained by selling x hundred of the inexpensive dolls,

b. Find when R is largest (that is, find the number of expensive and inexpensive dolls that should be manufactured to maximize total revenue).

8. Let x be the number of units produced of some commodity (expressed in thousands), and let $C(x)$ and $R(x)$ be the total costs and total revenue to produce x (in thousands), items, and let $P(x)$ be the corresponding profit. Suppose that for $x \geq 0$,

$$C(x) = \frac{x^3}{3} - 4x^2 + 10x$$

$$R(x) = 10x - 2x^2$$

then

$$P(x) = R(x) - C(x)$$

Find when P is largest (that is, find the optimal production level).

9. A manufacturer with a known demand and cost for his product wishes to maximize his profits. Let x be the demand for his product, p be the price at which the product is offered, A the total cost of production, and R the revenue received from the sale. Hence his profit P is given by

$$P = R - A$$

If we now suppose that he knows $x = 400 - 20p$ and that the average cost is $5 + (x/50)$, then he knows his total cost, A, is $A = 5x + (x^2/50)$.

a. Determine P as a function of demand x.

b. Find the demand at which P is maximum.

c. What is the price he should ask to maximize his profits, and what is his maximum profit?

10. Show that the rectangle with fixed perimeter which encloses the most area is a square.

11. The sum of one number, x, and three times a second number, y, is 80. Find the numbers if the product of x by the cube of y is maximum.

12. A triangular area is enclosed on two sides by a fence and on the third side by the straight edge of a river. The two sides of fence have equal length. Find the maximum area enclosed.

13. An open rectangular box is to be made from a sheet of tin 10 inches by 12 inches by cutting out a square from each corner and bending up the sides. Find how large a square should be cut from each corner if the volume of the box is to be a maximum.

14. Find the dimensions of the right circular cylinder of largest volume that can be inscribed in a sphere of radius R.

15. Find the area of the rectangle of maximum area that can be inscribed in the semicircle given by $f(x) = (R^2 - x^2)^{1/2}$ for fixed $R > 0$.

16. A post 6 feet high stands 4 feet from a wall. We want to place a ladder against the wall touching the ground and have the ladder rest on the post. What is the shortest ladder that can be used?

17. A north-south street and an east-west street meet at a point P. We wish to construct a diagonal road from a point R north of P to a point Q east of P, but with the restriction that this road must pass through a point 2 miles east of P and 1 mile north of P. Where should R and Q be located so that the area of RPQ is a minimum (see Figure 3-38).

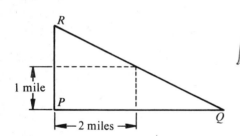

Figure 3-38

18. A thin square piece of tin 100 square inches in size is to be cut so as to construct an open box with a square base and sides perpendicular to the base. What should be the dimensions of the box if the box is to enclose the largest volume (ignore the waste in cutting)?

19. Bush and Mosteller proposed a learning model based on probability theory (see "A Stochastic Model With Applications to Learning," *Annals of Math. Stat.* 24 (1953): 559–685). Roughly, the probability of a particular event occurring when all events are equally likely is the frequency of that event relative to all possible outcomes; for example, the "probability that a six appears on the roll of a die is $\frac{1}{6}$" means that we expect the six to appear about one-sixth of the time relative to all possible outcomes of the roll of an "honest" die (more will be said of this in Chapter 4). For the learning model, we wish to estimate the probability p that a subject will make a certain response A. Assume that we have M observations, and of these the response A occurred n times (and failed to occur $M - n$ times). Then, if p denotes the probability of A (we wish to estimate somehow what this p should be), the likelihood that the response A occurs n times, $F(p)$ is given by $F(p) = p^n(1 - p)^{M-n}$, $0 \le p \le 1$. The maximum likelihood estimate of p is the value of p at which $F(p)$ assumes its absolute maximum on its domain, $0 \le p \le 1$. Show that the maximum likelihood estimate of p is $p = (n/M)$. Interpret this result.

20. A manufacturer finds that his costs to produce x articles per day, $0 \le x \le 100,000$, break down into (a) a fixed cost to employees of $\$1,200$ per day, (b) a fixed unit production cost of $\$1.20$ per day per article, and (c) a maintenance cost of $x^2/10^5$ per day. Hence, the total cost $F(x)$ (in dollars) per article produced is

$$F(x) = \frac{1200}{x} + 1.2 + \frac{x}{100,000}$$

(Why?) How many articles should be produced each day to minimize cost?

21. In an experiment conducted by Neifeld and Poffenberger ("A Math Analysis of Work Curves," *J. of Psych.* 1 (1928): 448–456), it was found that the number of units of work per second y performed in n seconds by a subject making a series of arm contractions is given by the equation

$$y = an^3 + bn^2 + cn + d, \qquad n \ge 0$$

where a, b, c, and d are empirical constants that vary with the subject. We suppose for this problem and for a particular subject that $a = 4$, $b = -30$, $c = 27$, and $d = 130$. (Decide at the conclusion of the problem whether these are realistic constants.) Find the maximum work per second done in the first 5 seconds.

Supplementary Topics:

APPLICATIONS OF THE DERIVATIVE TO ECONOMICS

The three supplementary sections will provide additional insights and information on the application of the derivative to some problems in economics. The material included is self-contained and independent of the remaining portions of the book.

SECTION 1

SOME ASPECTS OF THE THEORY OF DEMAND*

As before, p will denote the price of a commodity; and $\mathbf{x}(p)$ the corresponding demand function.

What is desired is a measure of the responsiveness of the consumer to price changes. Since there is no uniform set of units used in business (land is measured in acres, milk in quarts or gallons, and steel by the ton), such a measure should be independent of particular units if it is to be applicable to all business areas. Next, percentage changes might be preferred to absolute changes. To illustrate, if the price of penny bubble gum drops one penny, the demand for the gum by bubble gum devotees would no doubt swell to great proportions, whereas a one penny drop in the retail price of a Rolls-Royce would probably draw only smiles. It is more reasonable, however, to suppose that bubble gum devotees and Rolls-Royce fans would be stirred more equally if the price of each product were to be reduced by 50 percent.

We are led to use as our measure of responsiveness what we shall call average (price) elasticity of demand for an item w, call it Q, defined by

$$(3.1) \qquad Q = -\frac{\text{Percentage change in the demand for } w}{\text{Percentage change in the price of } w}$$

Remark   Since a rise in price is accompanied by a drop in demand for demand functions $\mathbf{x}(p)$ that are **S.D.**, the numerator and denominator in (3.1) are of opposite sign, so that $Q \geq 0$.

*For a more complete discussion, see W. J. Baumol, *Economic Theory and Economics* (Englewood Cliffs, New Jersey: Prentice-Hall Inc., 1961) especially chap. 8.

EXAMPLE 1 Suppose that a demand function is given by

$$\mathbf{x}(p) = 10 - p^2$$

with the price $p = 2$; hence, demand is $\mathbf{x}(2) = 6$. Suppose that the price is decreased by 5 percent. We wish to find the average percentage change in demand relative to the percentage change in price as the price is decreased; that is, we wish to find Q.

SOLUTION The new price is $2 - (5\ \text{percent}) \cdot 2 = 2 - 0.1 = 1.9$, and the new demand is $\mathbf{x}(1.9) = 6.39$. Hence, the change in demand is

$$\mathbf{x}(1.9) - \mathbf{x}(2) = 6.39 - 6 = 0.39$$

and the percentage increase in demand is

$$\frac{\mathbf{x}(1.9) - \mathbf{x}(2)}{\mathbf{x}(2)} = \frac{0.39}{6} = 6.5\ \text{percent}$$

The change in price is $1.9 - 2 = -0.1$; therefore, the average change in price is $(-0.1)/2 = -5$ percent. Hence, $Q = -6.5$ percent$/-5$ percent $= 1.3$. Observe that Q has no units; that is, it is a pure number.

We seek next a general formula for Q as price changes from p to $p + h$. This results in a change of price equal to h, with a corresponding change of demand $\mathbf{x}(p + h) - \mathbf{x}(p)$, and thus the percentage change in demand is $[x(p + h) - x(p)]/x(p)$. Since the "new" price is $p + h$ and the original price is p, the difference in price is $(p + h) - p$; that is, it is h, and hence the percentage change of price is h/p. Thus, from (3.1)

$$Q = -\frac{[x(p + h) - x(p)]/x(p)}{h/p}$$

that is,

$$(3.2) \qquad Q = -\frac{p}{\mathbf{x}(p)} \frac{\mathbf{x}(p + h) - \mathbf{x}(p)}{h}$$

We define elasticity of demand with respect to price when the price is p and the corresponding demand is $\mathbf{x}(p)$ as the limit of Q as $h \to 0$. If we designate the elasticity of demand by N, we have at any arbitrary price p (assuming $\mathbf{x}(p)$ differentiable),

$$(3.3) \quad \mathbf{N}(p) = -\lim_{h \to 0} \frac{p}{\mathbf{x}(p)} \frac{\mathbf{x}(p + h) - \mathbf{x}(p)}{h} = -\frac{p}{\mathbf{x}(p)} \frac{d\mathbf{x}(p)}{dp}$$

Price is sometimes expressed as a function of demand, and is then written $\mathbf{p}(x)$. Our object is to reformulate (3.1) in terms of the function $\mathbf{p}(x)$. As demand changes from x to $x + h$, the corresponding prices change from $\mathbf{p}(x)$ to $\mathbf{p}(x + h)$. Hence, (3.1) becomes

$$(3.4) \qquad A = -\frac{\mathbf{p}(x)}{x} \frac{h}{\mathbf{p}(x + h) - \mathbf{p}(x)}$$

thus, in the limit, we get the elasticity of demand corresponding to demand x is

$$\mathbf{N} = -\lim_{h \to 0} \frac{\mathbf{p}(x)}{x} \left[\frac{1}{\dfrac{\mathbf{p}(x + h) - \mathbf{p}(x)}{h}} \right]$$

that is

$$(3.5) \qquad \mathbf{N}(x) = -\frac{\mathbf{p}(x)}{x} \frac{1}{\mathbf{p}'(x)}$$

where $'$ represents differentiation with respect to x.

EXAMPLE 2   Let a demand curve be given by $p = [8/(2 + x)]$, $0 \leq x \leq 6$. Find the elasticity of demand (with respect to demand) at the endpoints of the given interval. Here,

$$\frac{dp}{dx} = \frac{-8}{(2 + x)^2}$$

and hence by (3.5),

$$\mathbf{N}(x) = \frac{-8}{x(2 + x)} \frac{(2 + x)^2}{-8} = \frac{2 + x}{x}$$

We see that $\mathbf{N} = 4/3$ at $x = 6$; $\mathbf{N}$ is undefined at $x = 0$, and $\mathbf{N}(x) \to +\infty$ as $x \to 0$, with $x \geq 0$. Next, we compute $\mathbf{N}$ in terms of p, using (3.3). Solving $p = 8/(2 + x)$ for x, we obtain $2p + xp = 8$; thus, $\mathbf{x}(p) = (8 - 2p)$ and $\mathbf{x}'(p) = -8/p^2$. Hence, by (3.3),

$$\mathbf{N}(p) = -\frac{p}{(8 - 2p)/p} \left(-\frac{8}{p^2} \right) = \frac{4}{(4 - p)}$$

Observe that $0 \leq x \leq 6$ implies $2 \leq 2 + x \leq 8$, which implies

$$\frac{1}{2} \geq \frac{1}{(2 + x)} \geq \frac{1}{8}$$

which, in turn, implies $4 \geq 8/(2 + x) \geq 1$; that is, the domain of $\mathbf{x}(p)$ is $4 \geq p \geq 1$. At $p = 1$, $\mathbf{N} = 4/3$ (and $x = 6$); and as $p \to 4$, $\mathbf{N} \to \infty$ (and $x \to 0$). We have measured the *same* responsiveness in terms of different variables. In fact, setting $p = 8/(2 + x)$ in the expression for $\mathbf{N}$ yields

$$N = \frac{4}{4 - p} = \frac{4}{4 - (8/[2 + x])}$$

$$= \frac{4}{(8 + 4x - 8)/(2 + x)} = \frac{2 + x}{x}$$

which was our initial value for N.

Suppose the demand function $x(p)$ has, as its graph, a straight line parallel to the price axis (see Figure 3-39(a)), with equation $x(p) = a$, $0 \leq p \leq b$. For this demand function, $dx/dp = 0$; hence, $N = 0$ for all p in $[0,b]$. No matter what the price may be in this situation the demand does not change; thus, the public is completely unresponsive to price changes; that is, elasticity of demand is 0.

In Figure 3-39(b), we see that the price is not responsive to demand; that is, demand changes, but price does not; and from (3.5) with $p(x) = b$, we see that, since $p'(x) = 0$, elasticity of demand is undefined (some economists call this "infinite elasticity").

It is interesting to observe that for the general nonhorizontal, nonvertical straight-line demand curve, given by the function

$$x(p) = -(a/b)p + a$$

$0 \leq p \leq b$ (see Figure 3-39(c)), N is not constant. In fact, since $x'(p) = -a/b$, we may simplify (3.3) to

$$N = p/(b - p)$$

At the left endpoint $p = 0$ of the domain of $x(p)$, we see that $N = 0$, and as $p \rightarrow b$, the right endpoint of the domain of $x(p)$, we see that $N \rightarrow \infty$ (observe that although $p = b$ is in the domain of $x(p)$, it is not in the domain of N).

What must be the structure of the demand function if the corresponding elasticity of demand is constant? We have the following result:

1. Let $x(p)$ be the demand function for some commodity, with

Figure 3-39

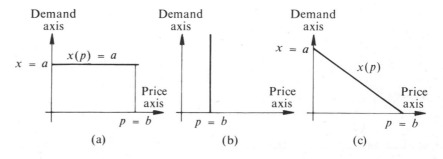

(a)   (b)   (c)

domain $[a,b]$. *We conclude that the product* $p \cdot \mathbf{x}(p)$ *is constant throughout the interval* $[a,b]$ *if and only if* $\mathbf{N} = 1$ *throughout the interval* $[a,b]$.

Before verifying this result, let us investigate its implications. The product $p \cdot \mathbf{x}(p)$ represents the price of a commodity times its demand at that price (the total amount the consumers would spend and, therefore, that which the producer will receive if the commodity is offered at price p). The quantity $a \cdot \mathbf{p}(a)$ (the value of $p \cdot \mathbf{x}(p)$ when $p = a$, the left endpoint of the domain of $\mathbf{x}(p)$, that is, the lowest price at which the commodity might be offered) is the consumers' initial outlay for the commodity. The theorem now asserts that, *if* $\mathbf{N}(p) = 1$ *for all* p *in* $[a,b]$, *then* $p \cdot \mathbf{x}(p)$ *is constant, and equal to* $a \cdot \mathbf{p}(a)$; *that is, a fall in prices must result in an increase in purchase by exactly that amount needed to keep the consumers' total outlay the same as their initial outlay.*

Verification   We shall not prove the entire result. We prove here that if $p \cdot \mathbf{x}(p)$ is a constant c, then $\mathbf{N} = 1$ for all p in $[a,b]$. Suppose now that $p \cdot \mathbf{x}(p) = c$; then $\mathbf{x}(p) = c/p$, and hence $\mathbf{x}'(p) = -c/p^2$. By (3.3)

$$\mathbf{N}(p) = -\frac{p}{\mathbf{x}(p)}\,\mathbf{x}'(p) = \frac{c}{p \cdot \mathbf{x}(p)} = \frac{c}{c} = 1$$

Another interesting result is the following:

2. *Let* $\mathbf{x}(p)$ *be a demand curve with domain* $[a,b]$ *and let* $\mathbf{N}(p)$ *be the corresponding elasticity of demand at price* p. *Then: (a)* $\mathbf{N} < 1$ *on* $[a,b]$ *(in which case* $\mathbf{x}(p)$ *is called* **inelastic**) *implies that a rise in prices will increase consumer expenditure* $p \cdot \mathbf{x}(p)$; *and, if a rise in prices will increase consumer expenditure, then* $\mathbf{N} \leq 1$; *but, in view of (1),* $\mathbf{N}$ *cannot be identically one for all* p *on* $[a,b]$. *(b)* $\mathbf{N} \geq 1$ *on* $[a,b]$ *(in which case* $\mathbf{x}(p)$ *is called* **elastic**) *implies that a rise in prices will decrease consumer expenditure; and if a rise in prices decreases consumer expenditure, then* $\mathbf{N} > 1$, *but, in view of (1),* $\mathbf{N}$ *cannot be identically one for all* p *on* $[a,b]$.

Remark   Now, we can rephrase this second result as follows: On the interval $[a,b]$, (a) if $\mathbf{N} < 1$, then $p \cdot \mathbf{x}(p)$ is an **S.I.** function of p; and if $p \cdot \mathbf{x}(p)$ is an **S.I.** function of p, then $\mathbf{N} \leq 1$, but not identically one; (b) If $\mathbf{N} > 1$, then $p \cdot \mathbf{x}(p)$ is an **S.D.** function of p; and if $p \cdot \mathbf{x}(p)$ is an **S.D.** function of p, then $\mathbf{N} \geq 1$, but not identically one.

Verification

We shall investigate only part of the implication in (b) above. In the spirit of the remark, we prove in (b) that $p \cdot \mathbf{x}(p)$ is an **S.D.** function of p implies $\mathbf{N} \geq 1$. Since $p \cdot \mathbf{x}(p)$ is **S.D.**, it follows that $D[p \cdot \mathbf{x}(p)] \leq 0$; and hence, by the product rule,

$$p \cdot \mathbf{x}'(p) + \mathbf{x}(p) \leq 0$$

or, equivalently, $p \cdot \mathbf{x}'(p) \leq -\mathbf{x}(p)$. Dividing by $-\mathbf{x}(p)$ yields

$$-[p/\mathbf{x}(p)]\mathbf{x}'(p) \geq 1$$

that is, $\mathbf{N} \geq 1$. (Recall that multiplying an inequality by a negative number changes the sense of the inequality.)

The elasticity result (2) has an interesting application for a country X suffering from a "dollar shortage." The prime minister of country X might suggest as a solution that his nation's currency be devalued, thus making products cheaper to the American consumer to create an increased flow of dollars to country X. An astute economist in country X might caution his prime minister that, among other problems, if the elasticity of demand for country X's exports is smaller than one, a decrease in prices will decrease consumer expenditure, and country X may end up obtaining fewer dollars than before.

Problems

1. Assume that the demand law is given by $\mathbf{x}(p) = 3/p^4$. Show that $\mathbf{N} = 4$. In general, if the demand law is given by $\mathbf{x}(p) = ap^{-m}$, $m > 0$, show that $\mathbf{N} = m$.

2. In Example 1, the demand curve is given by $\mathbf{x}(p) = 10 - p^2$. Find $\mathbf{N}$ at $p = 2$.

3. Find the elasticity of demand in terms of x if the demand law is given by $p = 1/(1 + x^2)$.

4. Assume that price and demand are connected by the relation $p^3 + x^3 = 9$. Find $\mathbf{N}(p)$ when $p = 2$.

5. Assume that the demand law is given by $x = 100(30 - 2p)^3$. Find the average percentage change in demand relative to the percentage change in price as price changes from 1 to $\frac{3}{2}$. Find $\mathbf{N}$ at $p = 1$.

6. In an agricultural industry such as the citrus industry, a frost can do great harm to the citrus crop, affect the size of the crop, and so affect the selling price. If $\mathbf{x}(p)$ denotes the demand as a function of price and $p = f(s)$ indicates that price as a function of the crop size s, express the elasticity of demand $\mathbf{N}$ in terms of s. Explain why you would expect f to be **S.D.**

7. The demand law $\mathbf{x}$ in terms of price p is given by $\mathbf{x}(p) = 5/p^6$. Determine when the elasticity of demand is greatest and when it is least on $p \geq 1$.

8. Assume that elasticity of demand N and price p are related by the equation $N^2 + p^2 - 2p = 3$. Determine when the elasticity of demand is greatest and when it is least on $[0, 3]$ (keep in mind that $N \geq 0$).

9. A manufacturer finds that the demand law for his product is given by $x(p) = 100/p^7$. What should he do to the price of his product if he wishes to increase consumer expenditure? Repeat the problem for $x(p) = 73/\sqrt{p}$. In general, if $x(p) = ap^{-m}$, $m > 0$, for what m should he increase prices and for what m should he decrease prices to increase consumer expenditure?

10. Assume that we know that the supply of a quantity is functionally related to the interest rate i by the function $S(i)$. Define average interest elasticity of supply, and interest elasticity of supply in terms of i and $S(i)$. Assume that interest changes from $i = A$ to $i = B = A + h$.

SECTION 2

AN INVENTORY PROBLEM

We will next apply our methods of finding extrema to investigate the following inventory problem. A retailer is faced with the problem of deciding how much inventory he should stock. If he carries a small inventory, he saves money on so-called carrying costs, such as storage costs, maintenance costs, and, if his goods are perishable, the cost of ruined articles, etc. If his inventory is too small he will have to reorder frequently, and pay a penalty for many small orders; such reorder costs can become prohibitive if our retailer consistently understocks. Our retailer has the following problem: What is the optimal inventory? We shall use this notation:

x = the number of units of the commodity delivered per shipment.

y = the total carrying cost (say, in dollars) spent in holding one unit of the commodity one year.

s = the number of units the retailer plans to sell over the entire year.

To use methods of calculus, we assume x is defined in an interval or half-line, so that our units for x must be pounds or fluid ounces or the like. We first analyze the carrying costs of the inventory. Suppose that our retailer expects to sell 10,000 units annually and accepts one initial shipment of 10,000 units. Assuming a uniform demand, his inventory should drop to 0 by the year's end (this, of course, is a convenient fiction since no businessman would allow all his stock to be sold before reordering), and thus his average inventory stock is $(10,000 + 0)/2 = 5,000$ units. If he accepts two shipments of 5,000 units, one shipment per 6-month period, his inventory will drop uniformly from 5,000 to 0 in 6 months (over each 6-month period) so that his average inventory is $(5,000 + 0)/2 = 2,500$ units. In general then, if our retailer orders x units per shipment, his average inventory level will be $(x + 0)/2 = x/2$ units. Since it costs u dollars to hold each unit, his total carrying cost will be $xu/2$ dollars for the year.

Next, if the retailer expects to sell s units, and receives x units per shipment, he will require s/x shipments. For example, if 10,000 units are to be sold annually, and each shipment involves 1,000 units, then $10,000/1,000 = 10$ shipments will be required.

Let a represent all fixed shipping costs (that is, those costs that are independent of x), and let b be the cost of shipment per unit so that the cost of shipping x units is bx. The cost of delivering x units therefore, is $a + bx$ dollars. The total annual reordering cost will equal the number of shipments, s/x multiplied by the total cost per shipment, $a + bx$, and hence equals

$$\frac{s(a + bx)}{x} = \left[\frac{sa}{x}\right] + sb$$

The total cost, $G(x)$, to the retailer is the sum of the total carrying cost and reordering cost, and hence, for $x > 0$

$$G(x) = \tfrac{1}{2}xu + \frac{sa}{x} + sb$$

To minimize the total cost, $G(x)$ (that is, to find the number of units per shipment to order which will minimize total cost), we solve $G'(x) = (\tfrac{1}{2})u - (sa)/x^2 = 0$ to obtain $x = \sqrt{2sa/u}$. Since $G'(x) < 0$ for $0 \le x < \sqrt{2sa/u}$ (why is this so?), we see that $G(x)$ is **S.D.** for $0 \le x < \sqrt{2sa/u}$; and since $G'(x) > 0$ for $x > \sqrt{2sa/u}$, we see that $G(x)$ is **S.I.** for $x > \sqrt{2sa/u}$, so that $G\sqrt{2sa/u}$ is the minimum of $G(x)$. Observe that the inventory should increase in proportion to the square root of projected total sales.

✓ **Problem**   1. Verify that $G'(x) < 0$ for $0 \le x < \sqrt{2sa/u}$ and that $G'(x) > 0$ for $x > \sqrt{2sa/u}$ and that $G'(x) = 0$ for $x = \sqrt{2sa/u}$.

A FIXED-POINT THEOREM AND AN APPLICATION TO GENERAL EQUILIBRIUM

Let f be a continuous function on $[0,1]$ whose range is a subset of $[0,1]$. Such a function is interesting in that its graph must touch or cross the line given by $y = x$ for some x in $[0,1]$; that is, its graph must touch or cross the diagonal from $(0,0)$ to $(1,1)$ of the square $\{(x,y): 0 \leq x \leq 1, 0 \leq y \leq 1\}$. (See Figure 3-40)

Your belief can be increased in this result by constructing for yourself various continuous functions on $[0,1]$ (with range a subset of $[0,1]$). Suppose the graph of f crosses the line given by $y = x$ at (u,u); then $f(u) = u$ (see Figure 3-40), and the point u is called a fixed point of f (that is, it is a point that is left unchanged, or fixed by f).

Summarizing these geometrical observations is the **fixed-point theorem**:

Let f be defined and continuous on the interval $[0,1]$, and let the range of f be some subinterval of the interval $[0,1]$ (hence, $0 \leq f(x) \leq 1$ for all x in $[0,1]$. There is then at least one point u in the interval $[0,1]$ at which $f(u) = u$.

A perhaps unexpected application of a fixed-point theorem is to be found in economics, although the particular fixed theorem used is a streamlined version of the one presented here. The ideas involved can, nevertheless, be outlined.

Assume we have an economy with 5,276 commodities; our list of commodities would include money, of course, as well as stocks and bonds, houses, consumer goods, etc. A typical commodity on this list, say houses, might be item 35. We will assume (the mathematical model we are creating to describe the economy has built into it this assumption) that the demand for houses depends upon the prices of each of the 5,276 items on the market as well as a few noncommodity-type influences. Call $D(35)$ the demand for commodity 35; that is, $D(35)$ is the demand for houses. We are asserting that $D(35)$ depends upon $P(1)$, the price of commodity 1, $P(2)$, the price of commodity $2,\ldots,$ to $P(5,276)$, the price of commodity 5,276, $P(5,277)$, the price of the stock of cash in existence, and perhaps a few more. Hence, $D(35)$ depends on at least 5,276 variables! Similar statements apply to $D(1)$, the demand for commodity 1, $D(2)$, the demand for commodity $2,\ldots,$ to $D(5,276)$, the demand for commodity 5,276.

Next, we have a similar set of supply equations. If $S(35)$ denotes the supply of houses, we shall assume that $S(35)$ depends on $P(1)$, $P(2)$, $P(3),\ldots,P(5,277)$, where the P's have the same meaning as above. This gives us a list of supply equations in which each of $S(1)$, $S(2),\ldots,S(5,276)$ depends on $P(1)$, $P(2),\ldots,$ $P(5,277)$, and perhaps a few more.

The economy is said to be in general equilibrium if the supply of each item and its demand are equal; that is, if we look at each

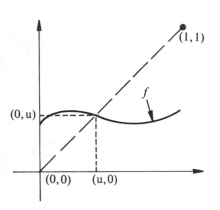

Figure 3-40

commodity, k, we have general equilibrium if $D(k) = S(k)$.

If we look at the set of equations $D(k) = S(k)$, and observe that each of the values $D(k)$ and $S(k)$ depends on $P(1)$, $P(2)$, ..., $P(5,277)$, ..., we are facing a massive collection of 5,276 equations in at least 5,277 unknowns.

Using some assumptions from economics, one of them being Walras' law (for a more complete discussion, see Baumol, 1961, chap. 12, sec. 3), we can reduce the system to as many unknowns as there are equations; in our particular example, we would have 5,276 equations in 5,276 unknowns. General equilibrium will be realized if there are numbers $P(1)$, $P(2)$, ..., $P(5,276)$ which will satisfy every one of the equations $D(1) = S(1)$, $D(2) = S(2)$, ..., $D(5,276) = S(5,276)$, simultaneously!

The question of whether such numbers $P(1)$, $P(2)$, ..., $P(5,276)$ exist which simultaneously satisfy the 5,276 equations $D(k) = S(k)$ is far from trivial.

Consider a simple one-commodity situation, and label this commodity 1; assume that at price $P(1) = p$ the demand $D(1)$ is given by $D(1) = 900 - p$ and the supply $S(1)$ is given by $S(1) = 1,000 - p$. Since our equilibrium equation is simply $D(1) = S(1)$, equilibrium occurs if $900 - p = 1,000 - p$. This equation, however, states that $100 = 0$, which is absurd, and the assumption that we can have equilibrium is untenable.

We shall next consider equations under which general equilibrium is realized. We shall assume in our purely fictional model that we have a two-commodity economy, with $P(1) = p$, and $P(2) = q$, the prices of commodities 1 and 2, respectively. We assume that our supply functions $S(1)$ and $S(2)$ each depend on the prices p and q according to the equations

$$S(1) = 12 - 5p + q \qquad \text{and} \qquad S(2) = 5 + 2p + q$$

Similarly, our demand functions $D(1)$ and $D(2)$ each depend on the same prices p and q according to the equations

$$D(1) = 2 - 3p + 5q \qquad \text{and} \qquad D(2) = 2 + p + 3q$$

Equilibrium is realized if there exist prices p and q such that $S(1) = D(1)$ and $S(2) = D(2)$ simultaneously. This leads to the equations

$$12 - 5p + q = 2 - 3p + 5q$$
$$5 + 2p + q = 2 + p + 3q$$

or

$$p + 2q = 5$$
$$-p + 2q = 3$$

If we add these equations, we obtain $4q = 8$ or $q = 2$. Substituting into either equation gives $p = 1$. Hence, at prices $p = 1$ and $q = 2$, general equilibrium is realized.

In general, then, we expect that more assumptions will be necessary to insure a set $P(1), \ldots, P(5,276)$ which solves the equilibrium equations. We shall outline a proof of the existence of a solution to the equilibrium equations based on a modification of a model displayed by Lionel McKenzie (see "On the Existence of a General Equilibrium for a Competitive Market," *Econometrica* 27 [1959]: 54–71). To simplify the model, we shall restrict our attention to the one variable case, although the McKenzie model assumes a system of demand equations in many variables. Let

p_d = demand price
p_s = supply price (that is, the price at which the seller is willing to supply)
Y_d = demand quantity
Y_s = supply quantity

The economic assumptions are sets of inequalities which assert that (1) production is proportional to demand using no more than available resources, and (2) we are assuming a "perfect competition" economy; and thus, profits are zero. Essentially, we assume that the demand quantity is a continuous function of the demand price and write $Y_d = D(P_d)$; and we assume that the supply quantity is a continuous function of the supply price and write $Y_s = S(P_s)$. Our object is to show that the economy is in general equilibrium; that is, there is a value $P = P_s = P_d$ such that $S(P) = D(P)$. We assume that it is possible to solve $Y_s = S(P_s)$ for P_s to obtain the continuous function $P_s = G(Y_s)$. This leads us to the equations

$$Y_d = D(P_d) \qquad P_s = G(Y_s)$$

Our model guarantees that there are values of Y_d and Y_s that coincide. Then

$$G[D(P_d)] = P_s$$

Since $G[D(P_d)]$ is continuous, an appeal to the appropriate fixed-point theorem guarantees a value P such that $G[D(P)] = P$. At this price, the demand conditions, the production conditions, and the no-profit conditions are satisfied.

4

THE INTEGRAL AS AN AREA

In this chapter we introduce the **integral** via the geometrical problem of finding the area under the graph of a "nice" function. The integral joins the derivative as one of the major concepts in calculus. The introduction of the integral is geometric, and does not fully display the usefulness of the concept. In fact, we can only see its value in applications when the integral is connected to a certain limit of a sum (in Chapter 6).

The chapter begins with a discussion of **antiderivatives**, a mathematical object closely related to the derivative (in fact, almost the "inverse" of the derivative). The integral is then introduced via the area problem. The Fundamental Theorem of Calculus which follows is an amazing theorem—very important for the computation of integrals—that connects the integral (that is, the problem of finding area) to the derivative, thus bringing together what conceptually would seem to be two totally distinct concepts. A brief discussion of the use of symmetry of the graph of a function in finding the area under the graph concludes the chapter.

SECTION 1
ANTI-DERIVATIVES

So far, our problem has been of the following type. Given a (differentiable) function G, fill in the blank:

$$D G(x) = \underline{\hspace{2cm}}$$

What fills the blank is, of course, some function, say f, that is the derivative of G. For example, if $G(x) = x^5$, we would fill in the blank with "$5x^4$"; that is, if $G(x) = x^5$, then *the only* function f for which $DG = f$ is given by $f(x) = 5x^4$. Another way to think of our problem is as follows: Given a function G which depends on x, we have been asked to find the function f where $f(x)$ is the rate of change of G at x.

We now consider a closely related problem. Given some function f, fill in the blank:

$$D \underline{\hspace{2cm}} = f(x)$$

that is, *find a function, say G, whose derivative is the given function* *f*. For example, if $f(x) = 5x^4$, our problem of filling in the blank looks like

$$D \underline{\hspace{1cm}} = 5x^4$$

and as we have seen, we can fill in the blank with "x^5"; that is, if $f(x) = 5x^4$, then a function *G* for which $DG = 5x^4$ is the function $G(x) = x^5$. Notice that other functions also correctly fill in the blank; for example, any one of the following will work:

$$D\underline{(x^5 + 1)} = 5x^4$$

$$D\underline{(x^5 + 10)} = 5x^4$$

$$D\underline{(x^5 - \sqrt{2})} = 5x^4$$

In fact, for *any number c*,

$$D\underline{(x^5 + c)} = 5x^4$$

Notice that this general problem can be viewed as follows: Find the functions *G* which have a *given* rate of change, *f*. From this point of view, it is not very surprising that more than one function will be subject to the same rate of change. For example, suppose two cars start traveling at the same time on the same straight road at the same constant speed of say 10 miles per hour (mph), but the cars start traveling 2 miles apart from each other. For each car, the rate of change of distance (that is, velocity) is the *same*, namely, 10 mph, but the distance they have traveled will at any time *t* differ by 2 miles. Thus, different distance functions give rise to the *same* velocity function, where one distance function is simply a constant added to the second distance function.

The problem we are now considering is called **antidifferentiation**. Specifically,

Given a function f, we shall call a function G for which DG = f an antiderivative of f, and denote any such function *G* by

$$\int f(x)\, dx$$

Summarizing, a *function G(x) is said to be an* **antiderivative** *of f(x), written*

$$G(x) = \int f(x)\, dx$$

if the derivative of G(x) is f(x); that is,

$$\int f(x)\, dx = G(x) \qquad \text{means} \qquad DG(x) = f(x)$$

The symbol $\int \cdots dx$ is read "an antiderivative with respect to *x*," *dx* indicating only that antidifferentiation is to take place with respect to x just as the symbol d/dx means that differentiation is to take place with respect to x. Some authors prefer to replace the symbol $\int f(x)\, dx$ by the symbol $D^{-1}f(x)$.

EXAMPLE 1 $\int 5x^4\, dx = x^5 + c$ (for any number c) since $D(x^5 + c) = 5x^4$

EXAMPLE 2 $$\int x\, dx = x^2/2$$

that is, $x^2/2$ is an antiderivative of $f(x) = x$. This is because $D\left(\frac{1}{2}\right)x^2 = x$. Observe that $D\left(\frac{1}{2}x^2 + 1\right) = x$ and $D\left(\frac{1}{2}x^2 - 7\right) = x$, so that $\frac{1}{2}x^2 + 1$ and $\frac{1}{2}x^2 - 7$ are also antiderivatives of $f(x) = x$. In fact, if c is any number, $D\left(\frac{1}{2}x^2 + c\right) = x$, so that $G(x) = \frac{1}{2}x^2 + c$ is an antiderivative of $f(x) = x$.

EXAMPLE 3 $$\int (x^2 + x - 3)\, dx = \tfrac{1}{3}x^3 + \tfrac{1}{2}x^2 - 3x + c$$

because $D\left(\tfrac{1}{3}x^3 + \tfrac{1}{2}x^2 - 3x + c\right) = x^2 + x - 3$, where c is any number.

From the previous examples it is apparent that, if

$$G(x) = \int f(x)\, dx$$

[and, hence, $D\,G(x) = f(x)$], then for any number c

$$G(x) + c = \int f(x)\, dx$$

because $D[G(x) + c] = D\,G(x) = f(x)$. In fact, let $G(x)$ be any antiderivative of $f(x)$; then the "most general" antiderivative is given by $G(x) + c$ where c is any number. Equivalently, if $G(x)$ and $F(x)$ are any two antiderivatives of $f(x)$ (that is, if $G'(x) = F'(x)$), then there is a number c for which

$$F(x) = G(x) + c$$

Remark   What we are asserting is that, if G is any particular antiderivative of f, any antiderivative of f is of the form $G + c$, where c is any real number. Also observe that the most general antiderivative is not one function but a set of functions $G + c$, depending on c. This number c is often referred to as an "arbitrary constant" arising from the antidifferentiation.

EXAMPLE 4

$$\int x(x^2 + 1)^{1/2} dx = \tfrac{1}{3}(x^2 + 1)^{3/2} + c$$

because

$$D\{\tfrac{1}{3}(x^2 + 1)^{3/2} + c\} = x(x^2 + 1)^{1/2}$$

Note that

$$\int x^n dx = \frac{x^{n+1}}{n+1} + c, \qquad n \neq -1$$

since, for any number c,

$$D\left[\frac{x^{n+1}}{n+1} + c\right] = \frac{1}{n+1} D x^{n+1} = \frac{n+1}{n+1} x^{(n+1)-1} = x^n$$

EXAMPLE 5   (4.1)

$$\int x \, dx = \frac{x^2}{2} + c, \qquad c \text{ in } R$$

(4.2)

$$\int x^2 \, dx = \frac{x^3}{3} + c, \qquad c \text{ in } R$$

(4.3)   Complete:

$$\int x^3 \, dx = \frac{\Box}{4} + c, \qquad c \text{ in } R$$

$$\int x^{10} \, dx = \frac{x^{\Box}}{\Box} + c, \qquad c \text{ in } R$$

$$\int \sqrt{x} \, dx = \int x^{\Box} \, dx = \Box \; x^{\Box} + c, \qquad c \text{ in } R$$

Since $D x = 1$, we have that

$$\int 1 \, dx = x + c, \qquad c \text{ in } R$$

$\int 1 \, dx$ is usually denoted by $\int dx$; that is,

$$\int dx = x + c, \qquad c \text{ in } R$$

You are probably aware that all the antiderivatives we have evaluated so far were found by knowing derivatives first, and that the calculations often seem to be good guesses verified by differentiation. In fact the techniques for antidifferentiation are not as easy or as systematic or as comprehensive as the techniques for differentiation. Nevertheless, we have: *Given $\int f(x) \, dx$ and $\int g(x) \, dx$, and any number A,*

(4.4) $$\int [f(x) + g(x)] \, dx = \int f(x) \, dx + \int g(x) \, dx$$

and

(4.5) $$\int Af(x)\,dx = A\int f(x)\,dx$$

In words, this result states that the antiderivative of a sum is the sum of antiderivatives, and that the antiderivative of a number times a function is that number times the antiderivative of the function.

The verification of this depends on the fact that the antiderivative and derivative are closely related, and that the derivative has similar properties. Specifically, to verify (4.4) let

$$F(x) = \int f(x)\,dx \qquad \text{and} \qquad G(x) = \int g(x)\,dx$$

[We will suppose the arbitrary constants are already in the functions F and G.] Then $D\,F(x) = f(x)$ and $D\,G(x) = g(x)$. Hence,

$$D[F(x) + G(x)] = f(x) + g(x)$$

We see that $D[F(x) + G(x)] = f(x) + g(x)$ implies that

$$F(x) + G(x) = \int (f(x) + g(x))\,dx$$

and substituting the values of $F(x)$ and $G(x)$, we finally obtain

$$\int f(x)\,dx + \int g(x)\,dx = \int [f(x) + g(x)]\,dx$$

We shall omit the verification of (4.5).

EXAMPLE 6 $\int (3x^7 - 4x^5 + 2x^2 - 7)\,dx$

$$= 3\int x^7\,dx - 4\int x^5\,dx + 2\int x^2\,dx - 7\int dx$$

$$= 3\left(\frac{x^8}{8}\right) - 4\left(\frac{x^6}{6}\right) - 2\left(\frac{x^3}{3}\right) - 7x + c$$

$$= \tfrac{3}{8}x^8 - \tfrac{2}{3}x^6 + \tfrac{2}{3}x^3 - 7x + c$$

Observe that, although each antiderivative gave rise to an "arbitrary constant," all the constants have been combined in the single number c.

In section 6, chapter 2, we established the relationship between distance at time t, $s(t)$, instantaneous velocity at time t, $v(t)$, and instantaneous acceleration at time t, $a(t)$, given by

$$v(t) = D_t s(t) = D_t v(t)$$

Thus, we get the corresponding antiderivatives

(4.6) $$s(t) = \int v(t)\,dt$$

(4.7) $$v(t) = \int a(t)\,dt$$

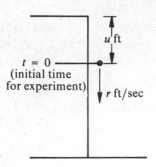

$t = 0$
(initial time
for experiment)

Figure 4-1

EXAMPLE 7   An early experiment on the effect of gravity on objects falling under free-fall indicated that (close to the surface of the earth) all objects, from lead bricks to small pebbles (ignoring air resistance, etc.), experience the same acceleration, designated by g. It was measured to be about 32 feet/second^2; that is, $g \sim 32$ feet/second^2.

Let us now suppose that an object is dropped off a platform so that at time $t = 0$, the object moves r feet/second and is u units from the top of the building; that is, $v(0) = r$ and $s(0) = u$ (see Figure 4-1). We shall determine the laws of motion of the object. From (4.7) since $a(t) = g$, a constant, we have

$$v(t) = \int a(t)\, dt = \int g\, dt$$

But since g is constant, $\int g\, dt = g \int dt$.

From Example 5, $\int dt = \int 1\, dt = t + c$, since $D_t(t + c) = 1$. (Observe that t may systematically replace x used throughout our previous work.) Hence, from (4.7),

$$v(t) = \int g\, dt = g \int dt = gt + k$$

where $k = gc$. Since $v(0) = r$, we have

$$r = v(0) = g \cdot 0 + k = k$$

and $k = r$, the velocity at time $t = 0$. Hence,

$$v(t) = gt + r$$

Next from (4.6), we have

$$s(t) = \int v(t)\, dt = \int (gt + r)\, dt$$

Using (4.4) and (4.5) (keep in mind that g and r are known constants),

$$s(t) = \int (gt + r)\, dt = g \int t\, dt + r \int dt$$
$$= g(t^2/2) + rt + A$$

Since $s(0) = u$, we have $u = s(0) = g \cdot (0^2/2) + r \cdot 0 + A$, so that $u = A$. Hence,

$$s(t) = \left(\tfrac{1}{2}\right) gt^2 + rt + u$$

where $u = s(0)$ and $r = v(0)$

Answer   Example 5, equation (4.3):

$$\frac{x^4}{4} + c; \qquad \frac{x^{11}}{11} + c; \qquad \int x^{1/2}\, dx = \frac{2}{3} x^{3/2} + c$$

Problems   In problems 1–12, compute the given antiderivative.

1. $\int 2x\,dx$

2. $\int 7\,dx$

3. $\int (t^3 - 1)\,dt$

4. $\int (u^{10} - 5u^7 + 4u)\,du$

5. $\int (x + 1)^2\,dx$

6. $\int [u^{3/2} - (u)^{1/2} + (1/u^{10})]\,du$

7. $\int x^2(x^2 - 1)\,dx$

8. $\int x(x)^{1/2}\,dx$

9. $\int (\sqrt[5]{x} + \sqrt[2]{x})\,dx$

10. $\int (1/x^2 + 1/x^3)\,dx$

11. $\int [(x^2 + 1)/x^2]\,dx$ (*Hint:* $(x^2 + 1)/x^2 = 1 + (1/x^2)$, $x \neq 0$)

12. $\int [(x^{1/2} + 1)/x^{1/2}]\,dx$

For problems 13–16, find the velocity $v(t)$ and the distance traveled $s(t)$ at the end of t seconds if an object moves in a straight-line path.

13. $a(t) = t$ with $v(1) = 2$ and $s(2) = 4$

14. $a(t) = 5t$ with $v(1) = 1$ and $s(1) = 5$

15. $a(t) = 3t^2 - 2t$ with $v(1) = 5$ and $s(2) = 10$

16. $a(t) = t^2 - t(t)^{1/2}$ with $v(0) = 1$ and $s(1) = 2$

17. Find $f(x)$ if $f'(x) = x^5 - x^{\sqrt{2}} + x^{-(3/2)} - 5/x^{25}$ and $x > 0$.

18. Suppose the marginal cost of an item is given by $(x^2 - 1)/x^2$ and the total cost is known to be 1 when one unit has been produced. Find the total cost of production.

19. Find $f(x)$ if $f''(x) = x^3 + x$, given that $f'(1) = 1$ and $f(0) = 2$.

20. For $x > 0$, find $f(x)$ if $f''(x) = (x)^{1/2} + [1/(x)^{1/2}]$, given that $f'(1) = 2$ and $f(1) = 0$.

21. Starting from rest, with what constant acceleration must a car proceed to go 360 feet in 6 seconds?

22. a. If an object is projected vertically upward, it is clear that, for the first moments of flight, the velocity is strictly decreasing, since the object will be slowing down. What does this imply about the sign of g (the acceleration due to gravity)?

   b. Suppose an object is projected vertically upward from rest with an initial velocity of 16 feet/second. What is the maximum height attained? When does it reach the ground again?

23. What is the total revenue accrued from the manufacture of a commodity if the marginal revenue at the demand level d is $d^2 + 2d + 1$. Assume that there is no total revenue when there is no demand.

24. A sense organ receives a stimulus in such a way that at time t, the total number of action potentials produced is $N(t)$. An experiment finds that the *rate* at which action potentials are produced at time t is given by $t^4 + t^2 + 2$. We also know that $N(0) = 0$ (that is, at time 0, there are no action potentials produced). Find $N(t)$.

SECTION 2

THE INTEGRAL AS AN AREA: AN INTRODUCTION

A curious fact is that it is possible to find the area under certain curves by using the antiderivative (hence, indirectly, the derivative). We shall see in our subsequent work that the ideas we will present have more content than simply the geometrical notion of "area under a curve." The fact that we are looking toward more applications than simply the one of area will help explain the somewhat unusual notation we shall use.

The curves we shall consider are graphs of function, and the area is the one "between" the graph and some closed interval, say $[a,b]$, on the horizontal axis (strictly speaking, it is the area bounded by the graph, the horizontal axis, and the lines given by $x = a$ and $x = b$) (see Figure 4-2). To make sure there is such an area, we want the graph of f to be "nice" in the following sense:

1. The graph should be a continuous curve except possibly at a finite number of points, but not blow up (that is, have a vertical asymptote) anywhere in $[a,b]$. One guarantee is to consider only functions that are continuous on $[a,b]$, or graphs that can be constructed by pasting together, side by side on abutting closed intervals, a finite number of graphs of continuous functions, each of which can be made continuous on its closed interval (see Figure 4-3). We shall call such functions **piecewise continuous.**

2. The graph should be above or touch the horizontal axis; that is, we want $f(x) \geq 0$ for $x \in [a,b]$. Our reasoning is

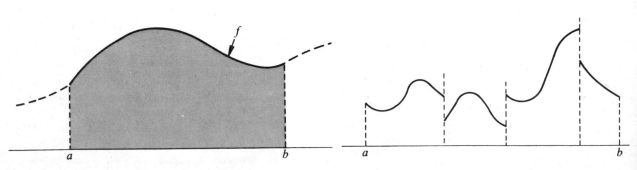

Figure 4-2   *Figure 4-3*

that if we allowed the graph to dip below the horizontal axis, then the part below the axis would subtract off area. We will pursue this later.

We are about to introduce what may seem to be rather curious notation (the objects we are now defining will ultimately have more application than area, and we need a sufficiently general notation).

Let f be a function satisfying the preceding conditions 1 and 2. For any x in $[a,b]$, we shall denote the *area under the graph of f from a to x* (see Figure 4-4) by

$$\int_a^x f$$

or

$$\int_a^x f(t)\,dt$$

and call this new symbol the **integral** *of f from a to x*. Thus, for functions f satisfying 1 and 2, the integral of f from a to x is the area under the graph of f from a to x (an analogue to this is that the derivative of f is the slope of a tangent line). Although the symbol for integral and antiderivative are similar, they should not be confused; the underlying concepts they represent are quite distinct. And we shall see later an amazing relationship between these concepts. Note that the integral depends on x (as well as a and f, of course); thus, to keep the bookkeeping straight, the function f will be in terms of something other than x (in fact, t in the above definition), and it will be convenient to think of the horizontal axis as, say, a t-axis, as in Figure 4-4. In terms of our new notation,

$$\int_a^b f$$

is the area under the graph of f from a to b.

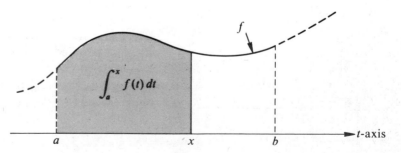

Figure 4-4

EXAMPLE 1 Interpret and compute:

$$\int_0^1 t\, dt$$

SOLUTION The integral is the area under the graph of $f(t) = t$ from $t = 0$ to $t = 1$. (See Figure 4-5.) Since this is the area of a triangle whose base and height are each 1, we see that (since area = $\frac{1}{2} \cdot$ base $\cdot$ height)

$$\int_0^1 t\, dt = \tfrac{1}{2} \cdot 1 \cdot 1 = \tfrac{1}{2}$$

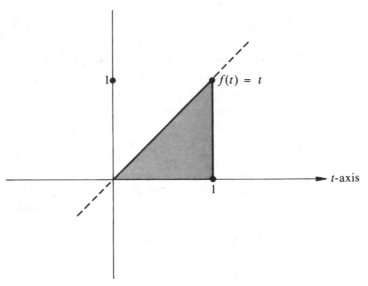

Figure 4-5

For functions satisfying conditions 1 and 2, the integral has an area interpretation. For an arbitrary continuous, or piecewise continuous, function (arbitrary in the sense that the function may be positive for some domain values, and negative or zero for others, or perhaps negative or zero for all domain values), a useful interpretation of the integral is as a "signed" area, treating as negative those areas that arise when the graph is below the horizontal axis. For example, if f is the function graphed in Figure 4-6, then

$$\int_a^b f = 0$$

since the area under the graph of f from a to c is equal to the

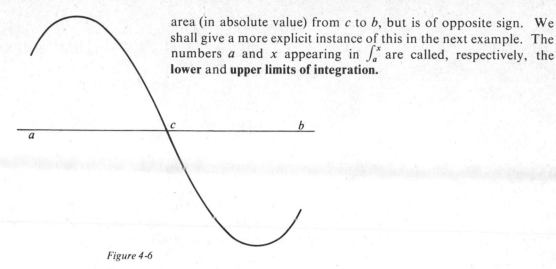

area (in absolute value) from c to b, but is of opposite sign. We shall give a more explicit instance of this in the next example. The numbers a and x appearing in $\int_a^x$ are called, respectively, the **lower** and **upper limits of integration.**

Figure 4-6

EXAMPLE 2   Interpret and compute:

$$\int_0^x t\,dt, \quad x > 0$$

SOLUTION   The integral is the area under the graph of $f(t) = t$ from $t = 0$ to $t = x$ (with $x > 0$) (see Figure 4-7). Since the region is a triangle whose base is x and whose height is $f(x)$, where $f(x) = x$, we see that the area is $\frac{1}{2}x^2$; thus,

$$\int_0^x t\,dt = \frac{1}{2}x^2$$

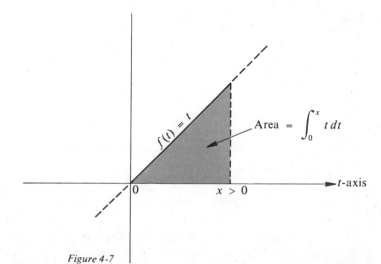

Figure 4-7

Thus, for example,

$$\int_0^1 t \, dt = \tfrac{1}{2} \cdot 1^2 = \tfrac{1}{2} \qquad \text{(as in Example 1)}$$

$$\int_0^3 t \, dt = \tfrac{1}{2} \cdot 3^2 = \tfrac{9}{2}, \qquad \text{and so on.}$$

Let us compute

$$\int_{-1}^1 t \, dt$$

Notice in Figure 4-8 that the area from -1 to 0 is equal to the area from 0 to 1, both equal to $\tfrac{1}{2}$. However, since the graph from -1 to 0 is below the horizontal axis, the integral from -1 to 0 is treated as a signed area, and thus

$$\int_{-1}^0 t \, dt = -\tfrac{1}{2}$$

Hence,
$$\int_{-1}^1 t \, dt = -\tfrac{1}{2} + \tfrac{1}{2} = 0$$

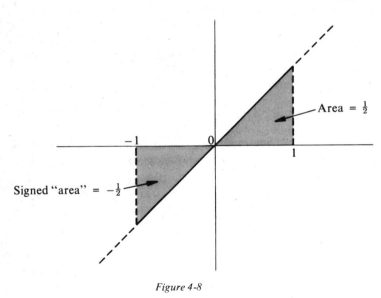

Figure 4-8

Remark   In the symbol

$$\int_a^x f$$

the function f appearing in the integral is called the **integrand**, and the process of evaluating the integral of f is called **integrating** f.

For instance, in Example 2, with $f(t) = t$, the function f is the integrand; and when we integrate f from 0 to x, we get $x^2/2$.

Note that all we have done so far is to create a symbol, the integral, whose *interpretation* for functions (satisfying 1 and 2 at the beginning of this section) is that of area. Except for certain elementary integrals, we still face the problem of computing integrals. The next section begins the search for ways to evaluate integrals.

Problems   In problems 1–8, interpret and use plane geometry to compute:

1. $\displaystyle\int_0^1 3t\,dt$ 2. $\displaystyle\int_1^2 3t\,dt$

3. $\displaystyle\int_1^2 (3t + 1)\,dt$ 4. $\displaystyle\int_{-1}^0 (t + 1)\,dt$

5. $\displaystyle\int_{-2}^0 (-3t + 4)\,dt$ 6. $\displaystyle\int_{-2}^{-1} (-2t + 4)\,dt$

7. $\displaystyle\int_0^x 5t\,dt,\quad x > 0$ 8. $\displaystyle\int_0^x (5t + 2)\,dt,\quad x > 0$

In problems 9–12, sketch the region whose area is given by each of the following integrals.

9. $\displaystyle\int_{-1}^1 t^2\,dt$ 10. $\displaystyle\int_{-1}^1 |\,t\,|\,dt$

11. $\displaystyle\int_1^4 (t^2 + 2t + 1)\,dt$ 12. $\displaystyle\int_1^2 \frac{1}{t}\,dt$

13. $\displaystyle\int_{-1}^1 t^3\,dt$ 14. $\displaystyle\int_{-3}^3 \sqrt[3]{t}\,dt$

SECTION 3

THE FUNDAMENTAL THEOREM OF CALCULUS

We turn now to the problem of evaluating integrals (and, consequently, of finding areas under graphs). The main result will be an amazing relationship between the integral and the antiderivative. As powerful as this result is, however, do not expect to evaluate all integrals with it.

Let f be a continuous and positive function defined on an interval $[a,b]$. Let us denote by $A(x)$ the area under the graph of f from a to x; that is,

$$A(x) = \int_a^x f$$

Thus, the area under the graph of f from a to b is $A(b)$, since

$$A(b) = \int_a^b f$$

We want to find explicitly the function (that is, the integral) A. A preliminary result exhibits a surprising relationship between the integral and the derivative; specifically, *if we differentiate the integral of f from a to x, we get f itself evaluated at x*. That is, we have the result that for any x in $[a, b]$

$$A'(x) = f(x)$$

or, in terms of the integral directly (remembering that the integral is a function of the upper limit of integration, x)

$$D \int_a^x f = f(x)$$

Furthermore, this result remains valid regardless of whether or not f is a positive function; if f is not positive, of course, the function A must be interpreted as a signed area. In any case, we see that *A is one of the antiderivatives of f* (stated another way, $A'(x) = f(x)$). The question is as follows: For a given f, which antiderivative of f is A? If we knew which antiderivative, we would know the integral, since $A(x)$ is $\int_a^x f$. In a moment we shall show that *any* antiderivative of f can be used to evaluate $\int_a^x f$; that is, *any* antiderivative of f can be used to find A. First, however, we shall show that for any x in $[a, b]$, $A'(x) = f(x)$.

To see this, note that for a positive number h, $A(x + h) - A(x)$ is the area under the graph of f from x to $x + h$ (see Figure 4-9), with $[x, x + h] \subset [a, b]$.

Since f is continuous on $[a, b]$, it is certainly continuous in the subinterval $[x, x + h]$. Hence, there are numbers u and v in $[x, x + h]$ for which $f(u)$ and $f(v)$ are, respectively, the maximum and minimum values of f in $[x, x + h]$. Thus, looking at the areas

Figure 4-9

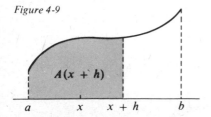

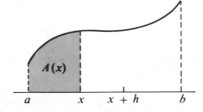

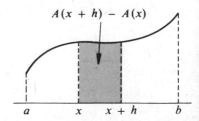

of the rectangles with base h and heights, respectively, $f(v)$ and $f(u)$, we see that

$$f(v) \cdot h \leq A(x + h) - A(x) \leq f(u) \cdot h$$

That is, $$f(v) \leq \frac{A(x + h) - A(x)}{h} \leq f(u)$$

Since $x \leq u \leq x + h$ and $x \leq v \leq x + h$, we see that as h tends to 0, u and v each tend to x. Thus (since f is continuous), $f(u)$ and $f(v)$ tend to $f(x)$; hence,*

$$f(x) \leq \lim_{h \to 0} \frac{A(x + h) - A(x)}{h} \leq f(x)$$

that is, $A'(x) = f(x)$, which is the desired result.

EXAMPLE 1   (4.8)   If $F(x) = \displaystyle\int_{-5}^{x} (t^8 + 3t^2 + 5)\, dt$, then for $x \geq -5$,

$$F'(x) = x^8 + 3x^2 + 5$$

(4.9)   If $G(x) = \displaystyle\int_{8}^{x} \{(t + 1)^3 + \sqrt{t - 1}\}\, dt$, then for $x \geq 8$,

$$G'(x) = (x + 1)^3 + \sqrt{x - 1}$$

(4.10) $\dfrac{d}{dx} \displaystyle\int_{1}^{x} \left(t^{100} + \dfrac{1}{t^{10}}\right) dt = \underline{\hspace{2cm}}$ (for $x \geq 1$)

Although the next result is a consequence of the previous one, it is the main result of the section, and is called the **fundamental theorem of calculus**—the theorem which shows that *any* anti-derivative of f can be used to find the integral of f.

Let f be continuous on [a,b], and let G be any antiderivative of f on [a,b] (that is, $G'(x) = f(x)$ for x $\in$ [a,b]). Then

$$\int_{a}^{b} f = G(b) - G(a)$$

Remark   Notice that the theorem does not require f to be positive.

---

*We are using the fact that, if three functions f, g, h are defined in some open interval I containing a point c, if $f(x) \leq g(x) \leq h(x)$ for all x in I, and if $\lim_{x \to c} f(x) = \lim_{x \to c} h(x) = A$, then $\lim_{x \to c} g(x)$ exists and has value A.

For convenience, we introduce the following notation:

$$G(x) \Big|_a^b = G(b) - G(a)$$

With this notation, we see that if $G'(x) = f(x)$ for x in $[a,b]$, the fundamental theorem can be rephrased as

$$\int_a^b f = G(x) \Big|_a^b$$

or, equivalently (and note here how nice it is to use a symbol for the integral that resembles the one for the antiderivative)

$$\int_a^b f(x)\,dx = \int f(x)\,dx \Big|_a^b$$

since by definition $\int f(x)\,dx$ is an antiderivative of f.

We shall now illustrate the power of the fundamental theorem of calculus.

EXAMPLE 2   (4.11) $\displaystyle \int_0^x t\,dt = \int t\,dt \Big|_0^x = \frac{t^2}{2}\Big|_0^x = \frac{x^2}{2} - \frac{0^2}{2} = \frac{x^2}{2}$

(4.12) $\displaystyle \int_1^2 x^5\,dx = \int x^5\,dx \Big|_1^2 = \frac{x^6}{6}\Big|_1^2$

$$= \frac{2^6}{6} - \frac{1^6}{6} = \frac{64}{6} - \frac{1}{6} = \frac{63}{6}$$

(4.13) $\displaystyle \int_{-1}^1 x^{100}\,dx = \int x^{100}\,dx \Big|_{-1}^1 = \frac{x^{101}}{101}\Big|_{-1}^1$

$$= \frac{1}{101} - \frac{-1}{101} = \frac{2}{101}$$

(4.14) $\displaystyle \int_{-2}^1 x^3\,dx = \int x^3\,dx \Big|_{-2}^1 = \frac{x^4}{4}\Big|_{-2}^1$

$$= \frac{1}{4} - \frac{(-2)^4}{4} = -\frac{15}{4}$$

(4.15) $\displaystyle \int_{-1}^1 x^3\,dx = \frac{x^4}{4}\Big|_{-1}^1 = 0$

(4.16) $\displaystyle \int_a^b 1\,dx = x\Big|_a^b = b - a$

Note: $\int_a^b 1 \, dx$ is usually denoted by $\int_a^b dx$. Thus, from (4.16),

$$\int_a^b dx = b - a$$

Generalizing, we see that for $n \neq -1$,

$$\int_a^b x^n \, dx = \frac{b^{n+1} - a^{n+1}}{n + 1}$$

since

$$\int_a^b x^n \, dx = \int x^n \, dx \, \Big|_a^b$$

$$= \frac{x^{n+1}}{n + 1} \Big|_a^b$$

$$= \frac{b^{n+1} - a^{n+1}}{n + 1}$$

EXAMPLE 3   (4.17) $\displaystyle\int_1^4 \sqrt{x} \, dx = \int x^{1/2} \, dx \, \Big|_1^4 = \frac{2}{3} x^{3/2} \Big|_1^4 = \frac{2}{3}(4^{3/2} - 1) = \frac{14}{3}$

(4.18) $\displaystyle\int_2^3 (1/x^2) \, dx = \int x^{-2} \, dx \, \Big|_2^3 = -(1/x) \Big|_2^3$

$$= \left(-\tfrac{1}{3}\right) - \left(-\tfrac{1}{2}\right) = \tfrac{1}{6}$$

EXAMPLE 4   Fill in the missing information.

$$\int_0^1 (3x^2 - 2x + 1) \, dx = \int (3x^2 - 2x + 1) \, dx \, \Big|_0^1$$

$$= 3 \int x^2 \, dx \, \Big|_0^1 - 2 \int x \, dx \, \Big|_0^1 + \int dx \, \Big|_0^1$$

$$= 3 \underline{\hspace{1cm}} \Big|_0^1 - 2 \underline{\hspace{1cm}} \Big|_0^1 + \underline{\hspace{1cm}} \Big|_0^1$$

$$= \underline{\hspace{1cm}} - \underline{\hspace{1cm}} + \underline{\hspace{1cm}}$$

$$= \underline{\hspace{1cm}}$$

Example 4 generalizes to the following: *If f and g are continuous functions on [a,b] and if A and B are numbers, then*

$$\int_a^b [Af(x) + Bg(x)] \, dx = A \int_a^b f(x) \, dx + B \int_a^b g(x) \, dx$$

since

$$\int_a^b [Af(x) + Bg(x)] \, dx = \int [Af(x) + Bg(x)] \, dx \Big|_a^b$$

$$= A \int f(x) \, dx \Big|_a^b + B \int g(x) \, dx \Big|_a^b$$

$$= a \int_a^b f(x) \, dx + B \int_a^b g(x) \, dx$$

To verify the fundamental theorem, again let $A(x) = \int_a^x f$; by our previous result, A is an antiderivative of f. Let G be any other (presumably known) antiderivative of f. Then, from our results on antiderivatives in the previous section, there is a real number k for which

$$A(x) = G(x) + k$$

Since $A(a) = \int_a^a f = 0$, it follows that

$$0 = A(a) = G(a) + k$$

or

$$k = -G(a)$$

Hence,

$$G(x) - G(a) = A(x) = \int_a^x f$$

and finally,

$$G(b) - G(a) = \int_a^b f$$

Answers   Example 1, equation (4.10): $x^{100} + \dfrac{1}{x^{10}}$

Example 4: $\dfrac{x^3}{3}; \dfrac{x^2}{2}; x; 1; 1; 1; 1$

Problems $\checkmark$**1.**   Find $f'(x)$ if

a. $f(x) = \displaystyle\int_0^x (t^{15} - t^3 + t - 1) \, dt$ (with $x \geq 0$)

b. $f(x) = \displaystyle\int_1^x \sqrt[3]{t^2 + t + 1} \, dt$ (with $x \geq 1$)

c. $f(x) = \displaystyle\int_1^x (t^2 + 1)^3 \cdot (t^4 + 1)^{1/2} \, dt$ (with $x \geq 1$)

In problems 2–25, use the fundamental theorem to evaluate the given definite integrals.

$\checkmark$**2.** $\displaystyle\int_{-3}^5 2x \, dx$ **3.** $\displaystyle\int_{-10}^{10} 7 \, dx$

4. $\displaystyle\int_0^2 (t^3 - 1) \, dx$ $\checkmark$**5.** $\displaystyle\int_0^1 (u^{10} - 5u^7 + 4u) \, du$

6. $\displaystyle\int_0^1 (x^2 - 2x + 5)\, dx$ **7.** $\displaystyle\int_1^3 (x - 1)^2\, dx$

8. $\displaystyle\int_1^2 \frac{1}{x^3}\, dx$ **9.** $\displaystyle\int_1^3 \frac{3}{x^4}\, dx$

10. $\displaystyle\int_{-1}^1 x^5(1 + x + x^2)\, dx$ **11.** $\displaystyle\int_0^1 x^2(1 + x)^2\, dx$

12. $\displaystyle\int_1^2 \frac{(x + 1)}{(x)^{1/2}}\, dx$ $\left(Hint: \dfrac{a + b}{c} = \dfrac{a}{c} + \dfrac{b}{c}\right)$

13. $\displaystyle\int_1^4 \frac{x^2 + 1}{(x)^{1/2}}\, dx$ **14.** $\displaystyle\int_1^2 \frac{(x^3 + 1)}{x^2}\, dx$

15. $\displaystyle\int_0^1 (x)^{1/2}[1 - (x)^{1/2}]\, dx$ **16.** $\displaystyle\int_0^1 x(x^2 - 3)\, dx$

17. $\displaystyle\int_{-1}^1 (\sqrt[3]{x} + \sqrt[5]{x})\, dx$ **18.** $\displaystyle\int_0^1 \sqrt[3]{x^2}\, dx$

19. $\displaystyle\int_0^1 \sqrt[5]{x^4}\, dx$ **20.** $\displaystyle\int_{-1}^0 \sqrt[5]{x^2}\, dx$

21. $\displaystyle\int_a^b (Ax^2 + Bx + C)\, dx$ (A, B, C are any numbers)

22. $\displaystyle\int_{-2}^0 (-x)^{1/2}\, dx$ **23.** $\displaystyle\int_1^2 \left(\frac{1}{(x)^{1/2}} + \frac{1}{x^2} + \frac{1}{x^4}\right)\, dx$

24. $\displaystyle\int_1^8 \frac{1}{\sqrt[3]{x^2}}\, dx$ **25.** $\displaystyle\int_1^{32} \frac{1}{\sqrt[5]{x^4}}\, dx$

In problems 26–30, find the area of the region bounded by the graphs of the following:

26. $f(x) = x^2 + 1,$ $x = -1,$ $x = 1,$ and   the x-axis.
27. $f(x) = \sqrt{x},$ $x = 0,$ $x = 1,$ and   the x-axis.
28. $f(x) = x^3,$ $x = -1,$ $x = 1,$ and   the x-axis (caution: area is positive).
29. $f(x) = \sqrt[3]{x},$ $x = -2,$ $x = 2,$ and   the x-axis.
30. $f(x) = 1/\sqrt[3]{x},$ $x = -2,$ $x = 2,$ and   the x-axis.

SECTION 4
SYMMETRY

In this section, we shall see that the graph of f having certain symmetry properties will often aid in the evaluation of the integral of f. First, note that, if $a < c < b$, the area under the graph of f from a to b is just the sum of the areas from a to c and from c to b (see Figure 4-10); that is,

$$\int_a^b f(x)\,dx = \int_a^c f(x)\,dx + \int_c^b f(x)\,dx$$

Assume that the graph of f on the interval $[-a,a]$ (with $a > 0$) can be obtained by reflecting in the y-axis the graph of f on the interval $[0,a]$ (see Figure 4-11). Such functions have the property that for each x in $[-a,a]$, $f(x) = f(-x)$. It is possible to prove (as Figure 4-11 leads us to believe) that

$$\int_{-c}^0 f = \int_0^c f = \tfrac{1}{2}\int_{-c}^c f$$

Another type of symmetry occurs if the graph of f on $[-a,a]$ can be obtained by reflecting the graph of f on $[0,a]$ first in the y-axis and then in the x-axis (see Figure 4-12). Such functions are characterized by the relationship $f(x) = -f(-x)$. It is possible to prove (as Figure 4-12 leads us to believe) that

$$\int_{-a}^0 f = -\int_0^a f \qquad \text{and} \qquad \int_{-a}^a f = 0$$

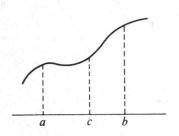

Figure 4-10

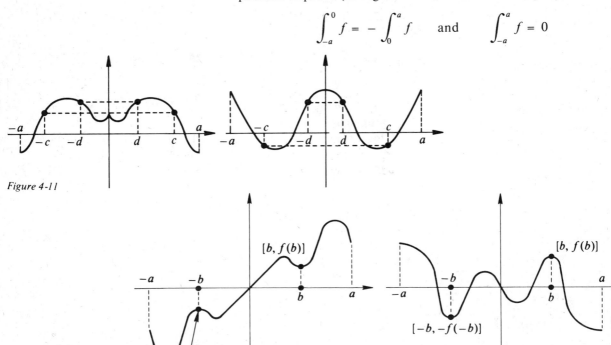

Figure 4-11

Figure 4-12

We formalize these observations as follows: *Let f be continuous in* $[-a,a]$ $(a > 0)$.

1. If $f(x) = f(-x)$ *for every x in* $[-a,a]$ *(such functions are called* **even** *on the interval), then*

$$\int_{-a}^{a} f = 2 \cdot \int_{0}^{a} f = 2 \cdot \int_{-a}^{0} f$$

2. If $f(x) = -f(-x)$ *for every x in* $[-a,a]$ *(such functions are called* **odd** *on the interval), then*

$$\int_{-a}^{a} f = 0$$

Note that each of the functions (for $c \in R$). $f(x) = c$, $f(x) = x^2$, $f(x) = x^4, \ldots, f(x) = x^{2n}$ (where n is a positive integer) is an even function. To see that this is true in general, let $f(x) = x^{2n}$; we must find $f(-x)$ and see if $f(x) = f(-x)$. Note that $f(-x) = (-x)^{2n} = [(-x)^2]^n = [x^2]^n = x^{2n}$; thus, $f(x) = f(-x)$ for every number x.

Similarly, each of the functions

$$f(x) = x, f(x) = x^3, f(x) = x^5, \ldots, f(x) = x^{2n+1}$$

(where n is a positive integer or zero) is an odd function. To see this is true in general, note that, if $f(x) = x^{2n+1}$, then

$$f(-x) = (-x)^{2n+1} = (-x)^{2n}(-x) = x^{2n} \cdot (-1) \cdot x = -x^{2n+1}$$

thus, $$f(x) = -f(-x)$$

for every number x.

From these observations, we see that (for $a > 0$)

$$\int_{-a}^{a} x \, dx = \int_{-a}^{a} x^3 \, dx = \int_{-a}^{a} x^5 \, dx = \int_{-a}^{a} x^{2n+1} \, dx = 0$$

where n is any positive integer or zero.

Similarly, for $a > 0$,

$$\int_{-a}^{a} x^2 \, dx = 2 \int_{0}^{a} x^2 \, dx = \tfrac{2}{3} x^3 \Big|_{0}^{a} = \tfrac{2}{3} a^3$$

$$\int_{-a}^{a} x^4 \, dx = 2 \int_{0}^{a} x^4 \, dx = \tfrac{2}{5} x^5 \Big|_{0}^{a} = \tfrac{2}{5} a^5$$

and so on.

EXAMPLE 1   Find $\int_{-3}^{3} | x | \, dx$.

SOLUTION   Let $f(x) = |x|$. We shall compute $f(-x)$. Note that $f(-x) = |-x| = |x|$. Hence, $f(x) = f(-x)$, and thus f is even. Therefore,

$$\int_{-3}^{3} |x| \, dx = 2 \int_{0}^{3} |x| \, dx$$

$$= 2 \int_{0}^{3} x \, dx \qquad \text{(since } |x| = x \quad \text{if } x \geq 0)$$

$$= x^2 \big|_{0}^{3} = 9$$

EXAMPLE 2   Evaluate $\int_{-2}^{2} (3x^{101} - 2x^{35} + 6x^2) \, dx$.

SOLUTION   Because the appropriate functions are odd, we see that

$$\int_{-2}^{2} x^{101} \, dx = 0 \qquad \text{and} \qquad \int_{-2}^{2} x^{35} \, dx = 0$$

Thus,

$$\int_{-2}^{2} (3x^{101} - 2x^{35} + 6x^2) \, dx$$

$$= 3 \int_{-2}^{2} x^{101} \, dx - 2 \int_{-2}^{3} x^{35} \, dx + 6 \int_{-2}^{2} x^2 \, dx$$

$$= 12 \int_{0}^{2} x^2 \, dx \qquad \text{(Why?)}$$

$$= 4x^3 \big|_{0}^{2}$$

$$= 32$$

Problems   Compute the following integrals.

√1. $\displaystyle\int_{-2}^{2} (x^6 + x^2) \, dx$ 2. $\displaystyle\int_{-1}^{1} (x^{10} - 5x^2) \, dx$

3. $\displaystyle\int_{-25}^{25} (x^{15} - 4x^{11}) \, dx$ 4. $\displaystyle\int_{-1}^{1} (x^{15} + x^3) \, dx$

5. $\displaystyle\int_{-1}^{1} (x^{17} + x^2) \, dx$ 6. $\displaystyle\int_{-100}^{100} (x^{121} + 2) \, dx$

7. $\displaystyle\int_{-7}^{7} x^{1/19} \, dx$ √8. $\displaystyle\int_{-2}^{2} x^{2/3} \, dx$

9. $\displaystyle\int_{-3}^{3} (x^5 + x^3)^{21} \, dx$ 10. $\displaystyle\int_{-5}^{5} [(x^{21} - x^{17})^{31} + x^2] \, dx$

11. $\displaystyle\int_{-4}^{4} \sqrt{|x|} \, dx$ 12. $\displaystyle\int_{-2}^{2} |x|^{3/4} \, dx$

5

SEQUENCES. EXPONENTIAL AND LOGARITHMIC FUNCTIONS

The chapter begins with a brief investigation of sequences, and after introducing sigma notation, the notion of sequence is applied to sums and series. The usefulness of sequences will become more and more apparent as the next chapters develop (for example, in the construction of series and in a new interpretation of the integral). An immediate application of sequences is in the construction of the exponential function. From the exponential function, we obtain the logarithm function. These two nonpolynomial functions turn out to be very important in applications. The chapter ends with an application of the exponential and logarithm functions to the problem of exponential growth and decay.

SECTION 1
SEQUENCES

So far we have considered functions whose domains are intervals, half-lines, or the line. We shall now briefly investigate functions whose domain is the set of positive integers (occasionally with 0 as well), and whose range is some set of real numbers. Such a function is called a **sequence**. We shall denote the set of positive integers by I^+.

For example, the function A given by

$$A(n) = \frac{1}{n}, \qquad n \in I^+$$

is a sequence. It is customary to write A_n in place of $A(n)$, with $n \in I^+$ left implicit. Thus, the sequence is written: $A_n = 1/n$. Exploiting the fact that the positive integers are ordered (that is, 1 comes before 2, 2 before 3, etc.), we often denote this sequence by the (ordered) array

$$1, \tfrac{1}{2}, \tfrac{1}{3}, \tfrac{1}{4}, \ldots$$

or, more briefly,

$$\left\langle \frac{1}{n} \right\rangle$$

In general, we can denote any sequence A by the array

$$A_1, A_2, A_3, A_4, \ldots, A_n, A_{n+1}, \ldots$$

or, more briefly,

$$\langle A_n \rangle$$

and A_n is called the **nth term of the sequence**.

EXAMPLE 1 A question of considerable interest to the psychologist is the reliability of a test; that is, whether a test measures what it is intended to measure. A test is said to be *lengthened by a factor of n* if $(n - 1)$ new tests, each one somehow equivalent to the original test, have been added to the original test, and the entire aggregate is given as a single new test. Hence, a test is lengthened by a factor of n if it is replaced by a new but equivalent test n times as long as the original test. Let r be the measure of the reliability of the original test; the Spearman-Brown formula asserts that reliability of the test after it has been lengthened by a factor of n is given by a sequence f_n, and that under suitable conditions,

$$f_n = \frac{nr}{1 + (n - 1)r}, \qquad 0 \leq r \leq 1$$

EXAMPLE 2 Write out the first five terms of the sequence

$$\left\langle \frac{1}{2^{n-1}} \right\rangle$$

SOLUTION If $n = 1$, then $\dfrac{1}{2^{n-1}} = \dfrac{1}{2^0} = 1$; thus, 1 is the first term.

If $n = 2$, then $\dfrac{1}{2^{n-1}} = \dfrac{1}{2^{2-1}} = \dfrac{1}{2}$; thus, $\dfrac{1}{2}$ is the second term.

If $n = 3$, then $\dfrac{1}{2^{n-1}} = \dfrac{1}{4}$; thus, $\dfrac{1}{4}$ is the third term.

If $n = 4$, then _____ ; thus, _____ is the fourth term.

If $n = 5$, then _____ ; thus, _____ is the fifth term.

EXAMPLE 3 Let $A_n = (1 + n)/(2 + n^2)$. Find the second, fifth, and seventh terms of this sequence.

SOLUTION The second term is A_2, and $A_2 = (1 + 2)/(2 + 4) = 1/2$. The fifth term, A_5, is $(1 + 5)/(2 + 25) = 6/27$. The seventh term is _____.

EXAMPLE 4 The sequence, $1, 1, 1, 1, 1, 1, \ldots 1, \ldots$, is the infinite sequence $<B_n>$, where $B_n = 1$ for all n; this is the analogue of the constant function $f(x) = 1$, and is an infinite sequence all of whose terms are identical.

EXAMPLE 5 The sequence $2, 0, 2, 0, \ldots$ is the infinite sequence $<D_n>$ with $D_n = 1 + (-1)^{n+1}$.

If a and r are nonzero real numbers, then the sequence

$$a, ar, ar^2, ar^3, \ldots, ar^n, \ldots$$

(that is, the sequence with the property that, except for the first term, each term is r times the previous term) is called a **geometric sequence** (or, **geometric progression**) with **first term** a and **ratio** r. Note that

$$a_n = ar^{n-1}$$

EXAMPLE 6 Each of the following is a geometric sequence:

(5.1) $1, \dfrac{1}{3}, \dfrac{1}{3^2}, \dfrac{1}{3^3}, \ldots$ $\left(\text{the first term is 1; the ratio is } \dfrac{1}{3}\right)$

(5.2) $5, 5\cdot7, 5\cdot7^2, 5\cdot7^3, \ldots$ (the first term is 5; the ratio is 7)

(5.3) $\dfrac{1}{5}, \dfrac{1}{5^2}, \dfrac{1}{5^3}, \ldots$ $\left(\text{the first term and the ratio are } \dfrac{1}{5}\right)$

(5.4) $1, -\dfrac{1}{3}, \dfrac{1}{3^2}, -\dfrac{1}{3^3}, \ldots$ $\left(\text{the first term is 1; the ratio is } -\dfrac{1}{3}\right)$

In our earlier discussion of function, we discussed the limit

$$\lim_{x \to \infty} f(x)$$

This type of limit has an analog in sequences. Specifically, let $< A_n >$ be an infinite sequence. We define

$$\lim_{n \to \infty} A_n = L$$

to mean that the numbers A_n can be made to stay arbitrarily close to L by taking n sufficiently large.

Just as in the case of functions with domain **R**, we have:

Let $\{A_n\}$ and $\{B_n\}$ be sequences such that $\lim_{n\to\infty} A_n$ and $\lim_{n\to\infty} B_n$ exist, and let c be any number. Then

(5.5)
$$\lim_{n\to\infty} cA_n = c \lim_{n\to\infty} A_n$$

(5.6)
$$\lim_{n\to\infty} (A_n \pm B_n) = \lim_{n\to\infty} A_n \pm \lim_{n\to\infty} B_n$$

(The limit of a sum (or difference) is the sum (or difference) of the individual limits.)

(5.7)
$$\lim_{n\to\infty} A_n B_n = \lim_{n\to\infty} A_n \cdot \lim_{n\to\infty} B_n$$

(The limit of a product is the product of the limits.)

(5.8) $$\lim_{n\to\infty} A_n/B_n = \lim_{n\to\infty} A_n \Big/ \lim_{n\to\infty} B_n \quad \text{if } \lim_{n\to\infty} B_n \neq 0$$

(The limit of a quotient is the quotient of the limits, provided the limit in the denominator is not zero.)

EXAMPLE 7   With $A_n = 1/n$, we see that

$$\lim_{n\to\infty} A_n = \lim_{n\to\infty} 1/n = 0$$

since $1/n$ can be made arbitrarily close to 0 by taking n sufficiently large. Hence, $\lim_{n\to\infty} 1/n^2 = \lim_{n\to\infty} (1/n \cdot 1/n) = \lim_{n\to\infty} 1/n \cdot \lim_{n\to\infty} 1/n = 0 \cdot 0 = 0$. Similarly, $\lim_{n\to\infty} c/n = 0$ where c is any number.

EXAMPLE 8   With $B_n = 1$ for all n, we see that $\lim_{n\to\infty} B_n = 1$, since B_n can be made to stay arbitrarily close to 1 by taking n sufficiently large; in fact, $B_n = 1$ for *every n*.

EXAMPLE 9   Find $\lim_{n\to\infty} (n^2 - 1)/(n^2 + 2)$.

SOLUTION   The largest exponent appearing in either the numerator or the denominator is 2 (it happens to appear in both numerator and denominator). We divide the numerator and denominator by n raised to the largest exponent determined, 2, obtaining

$$\frac{(n^2 - 1) \cdot 1/n^2}{(n^2 + 2) \cdot 1/n^2} = \frac{1 - (1/n^2)}{1 + (2/n^2)} \to 1 \qquad \text{as } n \to \infty$$

thus,
$$\lim_{n\to\infty} \frac{n^2 - 1}{n^2 + 2} = 1$$

EXAMPLE 10   Find

$$\lim_{n \to \infty} \frac{n^3 + n - 1}{n^4 + 2n + 1}$$

SOLUTION   This time the largest exponent is 4; we divide numerator and denominator by n^4 to obtain

$$\frac{(n^3 + n - 1) \cdot 1/n^4}{(n^4 + 2n + 1) \cdot 1/n^4} = \frac{(1/n) + (1/n^3) - (1/n^4)}{1 + (2/n^3) + (1/n^4)} \to \frac{0}{1}$$

$$= 0 \qquad \text{as } n \to \infty$$

thus, $$\lim_{n \to \infty} \frac{n^3 + n - 1}{n^4 + 2n + 1} = 0$$

EXAMPLE 11   Fill in the missing information.

$$\lim_{n \to \infty} \frac{3n^4 + n^2 + 3}{-n^4 + n^3 + 7} = \lim_{n \to \infty} \frac{(3n^4 + n^2 + 3) \cdot \dfrac{1}{\boxed{}}}{(-n^4 + n^3 + 7) \cdot \dfrac{1}{\boxed{}}}$$

$$= \lim_{n \to \infty} \frac{\boxed{} + \dfrac{1}{n^{\boxed{}}} + \dfrac{\boxed{}}{n^4}}{\boxed{} + \dfrac{1}{n^{\boxed{}}} + \dfrac{7}{n^{\boxed{}}}}$$

$$= \boxed{}$$

EXAMPLE 12   Find $\lim_{n \to \infty} (\sqrt{n^2 + n} - n)$.

SOLUTION   A technique to compute this limit involves rationalizing as follows:

$$\sqrt{n^2 + n} - n = \frac{(\sqrt{n^2 + n} - n)(\sqrt{n^2 + n} + n)}{\sqrt{n^2 + n} + n}$$

$$= \frac{(n^2 + n) - n^2}{\sqrt{n^2 + n} + n}$$

$$= \frac{n}{\sqrt{n^2 + n} + n}$$

$$= \frac{n}{n\left(\dfrac{\sqrt{n^2 + n}}{n} + 1\right)}$$

$$= \frac{1}{\frac{\sqrt{n^2 + n}}{\sqrt{n^2}} + 1}$$

$$= \frac{1}{\sqrt{1 + \frac{1}{n}} + 1}$$

Hence,

$$\lim_{n \to \infty} (\sqrt{n^2 + n} - n) = \lim_{n \to \infty} \frac{1}{\sqrt{1 + \frac{1}{n}} + 1} = \frac{1}{2}$$

Answers Example 2: $n = 4$: $\frac{1}{2^{n-1}} = 8$; 8

$n = 5$: $\frac{1}{2^{n-1}} = 16$; 16

Example 3: $\frac{8}{51}$

Example 11: $= \lim_{n \to \infty} \dfrac{(3n^4 + n^2 + 3) \cdot \dfrac{1}{n^4}}{(-n^4 + n^3 + 7) \cdot \dfrac{1}{n^4}}$

$= \lim_{n \to \infty} \dfrac{3 + \dfrac{1}{n^2} + \dfrac{3}{n^4}}{-1 + \dfrac{1}{n^1} + \dfrac{7}{n^4}}$

$= -3$

Problems In problems 1–11, write the first five terms (starting with $n = 1$) of the given infinite sequence.

1. $\langle 1 - (1/2^n) \rangle$ 2. $\langle (-1)^n \rangle$
3. $\langle n[1 + (-1)^n] + (1/n) \rangle$
4. $\langle [1 - (-1)^n]/2 + (1/n) \rangle$
5. $\langle (-1)^n/n \rangle$ 6. $\langle (-1/2)^n \rangle$
7. $\langle 1 + (-1)^n \rangle$ 8. $\langle 3^{n-3} \rangle$
9. $\langle 2^{-n+2} \rangle$ 10. $\langle (3n + 1)/(n + 2) \rangle$
11. $\langle (n^2 - n)/(n^2 + n) \rangle$

Evaluate each of the following limits (when they exist).

12. $\displaystyle\lim_{n \to \infty} \frac{n^5 - 4n^6 + 5n^7}{n^7}$ 13. $\displaystyle\lim_{n \to \infty} \frac{-3n^4 + n^2 - 100,000}{-6n^4 - n^2 + 10,000}$

14. $\lim\limits_{n \to \infty} \dfrac{n^{16} - 1}{1 - n^{16}}$ 　　　　**15.** $\lim\limits_{n \to \infty} \dfrac{n^4 - 2n^3 + n^2 + n - 1}{-2n^4 + n^3 - n^2 - n + 5}$

16. $\lim\limits_{n \to \infty} \dfrac{n^5 - 3}{n^4 + 100}$ 　　　　**17.** $\lim\limits_{n \to \infty} \dfrac{n^{10} - n^5 - 1}{n^{11} - n^6 + 3}$

18. $\lim\limits_{n \to \infty} \dfrac{2n^3 + 3n^2 + n - 10}{-n^3 - 4n^2 + 3n + 10}$

19. $\lim\limits_{n \to \infty} \dfrac{5n^4 - 3n + 17}{4n^4 + 10n^2 - 4}$

In problems 20–23, write the first five terms of the geometric progression whose first term a and ratio r are given.

20. $a = 2,$ 　　$r = 2$ 　　　　**21.** $a = 1,$ 　　$r = \frac{1}{2}$

22. $a = 1,$ 　　$r = -\frac{1}{2}$ 　　　　**23.** $a = -2,$ 　　$r = -\frac{1}{2}$

In problems 24–27, find the limit.

24. $\lim\limits_{n \to \infty} \dfrac{2n}{n + 7\sqrt{n}}$ 　$\Bigg($*Hint:* Note that

$$\frac{2n}{n + 7\sqrt{n}} = \frac{2n}{n[1 + (7/\sqrt{n})]} = \frac{2}{1 + (7/\sqrt{n})}.\Bigg)$$

25. $\lim\limits_{n \to \infty} \dfrac{8n - 500\sqrt{n}}{2n + 1{,}000\sqrt{n}}$ 　　**26.** $\lim\limits_{n \to \infty} \dfrac{3\sqrt{n}}{5\sqrt{n} + \sqrt[3]{n}}$

27. $\lim\limits_{n \to \infty} \dfrac{3\sqrt{n} + \sqrt[3]{n}}{2\sqrt{n} - 10\sqrt[3]{n}}$

In problems 28–30, use the rationalization technique outlined in Example 10.

28. $\lim\limits_{n \to \infty} (\sqrt{n + 5n^{1/2}} - \sqrt{n})$

29. $\lim\limits_{n \to \infty} n(\sqrt{n^2 + 1} - n)$

30. $\lim\limits_{n \to \infty} (\sqrt{n + \sqrt{n}} - \sqrt{n - \sqrt{n}})$

31. Recall that, in the learning theory model proposed by Thurstone, the number of successful acts, $G(x)$, accomplished after practicing the act x times is given by

$$G(x) = [a(x + b)]/(x + c)$$

Since x denotes the number of practice acts, x is actually a positive integer; therefore, the learning function is a sequence $<G(n)> = <A_n>$ where

$$A_n = [a(n + b)]/(n + c)$$

Show that $\lim\limits_{n \to \infty} (1/a) \cdot A_n = 1$.

32. Suppose the reliability of a test lengthened by a factor of n, call it (as in Example 1) f_n, is given for fixed reliability r by the Spearman-Brown formula

$$f_n = \frac{nr}{1 + (n - 1)r}, \qquad 0 \le r \le 1$$

Compute $\lim_{n \to \infty} f_n$. What does this limit imply to the experimenter?

SECTION 2

SUMS. SIGMA NOTATION

Assume that we wish to add the first 35 numbers in a sequence $<a_n>$. We could, of course, write the desired sum as

$$a_1 + a_2 + a_3 + \cdots + a_{35}$$

A more compact notation for this sum is:

$$\sum_{n=1}^{35} a_n$$

which is read: the sum of the numbers a_n from $n = 1$ to $n = 35$. Thus, for example,

$$\sum_{n=15}^{23} a_n = a_{15} + a_{16} + \cdots + a_{22} + a_{23}$$

and, in general, for p and q integers with $p < q$,

$$\sum_{n=p}^{q} a_n = a_p + a_{p+1} + \cdots + a_{q-1} + a_q$$

EXAMPLE 1   Write out $\sum_{k=2}^{6} k^2$.

SOLUTION

When $k = 2$, we have $k^2 = 2^2 = 4$. When $k = 3$, we have $k^2 = 3^2 = 9$. Continuing in this way, we have

$$k = 4, \qquad k^2 = 16$$
$$k = 5, \qquad k^2 = 25$$
$$k = 6, \qquad k^2 = 36$$

Hence, $\sum_{k=2}^{6} k^2 = 4 + 9 + 16 + 25 + 36$

EXAMPLE 2

$$\sum_{k=3}^{8} \frac{k+1}{2^k} = \frac{4}{2^3} + \frac{5}{2^4} + \frac{6}{2^5} + \frac{7}{2^6} + \frac{\square}{\square} + \frac{\square}{\square}$$

The following are some useful algebraic facts about sums. *Let $<A_n>$ and $<B_n>$ be sequences. Then,*

$$\sum_{k=1}^{N} (A_k + B_k) = \sum_{k=1}^{N} A_k + \sum_{k=1}^{N} B_k$$

To verify this, observe that

$$\sum_{k=1}^{N} (A_k + B_k) = (A_1 + B_1) + (A_2 + B_2) + (A_3 + B_3)$$
$$+ \cdots + (A_N + B_N)$$
$$= (A_1 + A_2 + A_3 + \cdots + A_N)$$
$$+ (B_1 + B_2 + B_3 + \cdots + B_N)$$
$$= \sum_{k=1}^{N} A_k + \sum_{k=1}^{N} B_k$$

Let c be any number and let $<A_n>$ be a sequence. Then

$$\sum_{k=1}^{N} (cA_k) = c \sum_{k=1}^{N} A_k$$

To verify this, note that

$$\sum_{k=1}^{N} (cA_k) = (cA_1) + (cA_2) + (cA_3) + \cdots + (cA_N)$$
$$= c(A_1 + A_2 + A_3 + \cdots + A_N)$$
$$= c \sum_{k=1}^{N} A_k$$

EXAMPLE 3   Let $<A_n>$ and $<B_n>$ be sequences for which

$$\sum_{n=1}^{10} A_n = 14 \quad \text{and} \quad \sum_{n=1}^{10} B_n = 23$$

Find $\sum_{n=1}^{10} (3A_n - 2B_n)$.

SOLUTION   Using the previous two results, we see that

$$\sum_{n=1}^{10} (3A_n - 2B_n) = \sum_{n=1}^{10} 3A_n + \sum_{n=1}^{10} (-2)B_n$$
$$= 3 \sum_{n=1}^{10} A_n + (-2) \sum_{n=1}^{10} B_n$$
$$= 3(14) - 2(23)$$
$$= -4$$

Recall (see the previous section) that with $a \neq 0$ and $r \neq 1$, the sequence

$$a, ar, ar^2, ar^3, \ldots$$

is a geometric sequence. We shall now show that the sum of the first n consecutive terms of this geometric sequence is

$$a \cdot \frac{1 - r^n}{1 - r}$$

that is,

$$a + ar + ar^2 + \cdots + ar^{n-1} = a \cdot \frac{1 - r^n}{1 - r} \qquad (r \neq 1)$$

To see this, let $s = a + ar + ar^2 + \cdots + ar^{n-1}$. Then, multiplying by r we see that

$$rs = ar + ar^2 + ar^3 + \cdots + ar^n$$

Thus, subtracting,

$$s - rs = (a + ar + ar^2 + \cdots + ar^{n-1}) - (ar + ar^2 + \cdots + ar^n)$$

$$= a - ar^n$$

that is, $\qquad\qquad s(1 - r) = a(1 - r^n)$

or, equivalently, if $r \neq 1$,

$$s = a \frac{1 - r^n}{1 - r}$$

which is what we wanted to show.

Note that with sigma notation, if $r \neq 1$, then

$$\sum_{k=1}^{n} ar^{k-1} = a \frac{1 - r^n}{1 - r}$$

EXAMPLE 4

$$(5.9) \quad 1 + \frac{1}{3} + \frac{1}{3^2} + \cdots + \frac{1}{3^{10}} = 1 \cdot \frac{1 - (1/3)^{11}}{1 - 1/3} = \frac{3}{2} \left(1 - \frac{1}{3^{11}} \right)$$

$$(5.10) \quad 1 - \frac{1}{3} + \frac{1}{3^2} - \frac{1}{3^3} + \cdots + \frac{1}{3^{100}} = 1 \cdot \frac{1 - [-(1/3)]^{101}}{1 + 1/3}$$

$$= \frac{3}{4} \left(1 + \frac{1}{3^{101}} \right)$$

$$(5.11) \quad 5 + 5 \cdot 7 + 5 \cdot 7^2 + \cdots + 5 \cdot 7^{19} = 5 \frac{1 - 7^{20}}{1 - 7} = \frac{5}{6} (7^{20} - 1)$$

$$(5.12) \quad \frac{1}{5} + \frac{1}{5^2} + \frac{1}{5^3} + \cdots + \frac{1}{5^{29}} = \frac{1}{5} \frac{1 - (1/5)^{29}}{1 - 1/5} = \frac{1}{4}\left(1 - \frac{1}{5^{29}}\right)$$

Equivalently, (5.12) can be done as follows:

$$\frac{1}{5} + \frac{1}{5^2} + \frac{1}{5^3} + \cdots + \frac{1}{5^{29}} = \frac{1}{5}\left[1 + \frac{1}{5} + \frac{1}{5^2} + \cdots + \frac{1}{5^{28}}\right]$$

$$= \frac{1}{5}\left[1 \cdot \frac{1 - (1/5)^{29}}{1 - 1/5}\right] = \frac{1}{4}\left(1 - \frac{1}{5^{29}}\right)$$

$$(5.13) \quad 3 + \frac{3}{2^2} + \frac{3}{2^4} + \frac{3}{2^6} + \frac{3}{2^8} = \boxed{} \frac{1 - \boxed{}^5}{1 - \boxed{}}$$

Answers

Example 2: $\left(\dfrac{8}{2^7}\right) + \left(\dfrac{9}{2^8}\right)$

Example 4, equation (5.13): $3 \cdot \dfrac{1 - (1/4)^5}{1 - 1/4}$

Problems   Write out each of the following. Compute the sum when feasible.

1. $\displaystyle\sum_{k=1}^{4} k$ 2. $\displaystyle\sum_{k=3}^{7} \sqrt{k}$

3. $\displaystyle\sum_{k=2}^{8} \frac{1}{2k}$ 4. $\displaystyle\sum_{k=3}^{5} k^k$

5. $\displaystyle\sum_{k=2}^{5} \frac{k}{k+1}$ 6. $\displaystyle\sum_{k=3}^{7} 8^{2k}$

7. $\displaystyle\sum_{k=1}^{10} (2^k - 2^{k-1})$ 8. $\displaystyle\sum_{k=1}^{10} [(k+1)^3 - k^3]$

9. $\displaystyle\sum_{k=2}^{100} (-1)^k$ 10. $\displaystyle\sum_{k=1}^{10} [(-1)^k + 1]$

11. $\displaystyle\sum_{k=3}^{1,000} (-1)^{k+1}$ 12. $\displaystyle\sum_{k=0}^{5} x^k \quad (x \neq 0)$

13. Let $<a_n>$ and $<b_n>$ be sequences for which

$$\sum_{k=1}^{18} a_k = 37 \quad \text{and} \quad \sum_{k=1}^{18} b_k = -83$$

Find:

$$\sum_{k=1}^{18} (-3a_k + b_k)$$

$$\sum_{k=1}^{18} (4a_k - 2b_k)$$

14. Given that

$$\sum_{k=1}^{n} k = \frac{n(n + 1)}{2}$$

and

$$\sum_{k=1}^{n} k^2 = \frac{n(n + 1)(2n + 1)}{6}$$

find

(a) $\displaystyle\sum_{k=1}^{10} k$ (b) $\displaystyle\sum_{k=1}^{10} k^2$ (c) $\displaystyle\sum_{k=1}^{10} (10k + 2k^2)$

Find:

15. $\frac{1}{5} + \left(\frac{1}{5}\right)^2 + \left(\frac{1}{5}\right)^3 + \cdots + \left(\frac{1}{5}\right)^{12}$

16. $\frac{1}{6} + \left(\frac{1}{6}\right)^2 + \left(\frac{1}{6}\right)^3 + \cdots + \left(\frac{1}{6}\right)^{14}$

17. $\frac{1}{3} - \left(\frac{1}{3}\right)^2 + \left(\frac{1}{3}\right)^3 - \left(\frac{1}{3}\right)^4 + \cdots + \left(\frac{1}{3}\right)^{15}$

18. $\dfrac{1}{10} + \dfrac{1}{10^5} + \dfrac{1}{10^9} + \dfrac{1}{10^{13}} + \dfrac{1}{10^{17}}$

19. $8 + 8^6 + 8^{11} + 8^{16} + 8^{21} + 8^{26}$

20. $5 + 5 \cdot 7 + 5 \cdot 7^2 + \cdots + 5 \cdot 7^{10}$

21. $3 + \dfrac{3}{4} + \dfrac{3}{4^2} + \cdots + \dfrac{3}{4^8}$

22. $3 - \dfrac{3}{4} + \dfrac{3}{4^2} - \dfrac{3}{4^3} + \cdots - \dfrac{3}{4^{15}}$

SECTION 3

SERIES

Our concern in this section is to give meaning to "adding up all the terms of a sequence"; that is, given a sequence $<a_n>$, to give meaning to the "infinite sum"

$$a_1 + a_2 + a_3 + a_4 + \cdots.$$

This can be done as follows:

Beginning with a sequence $<a_n>$, we can form a new sequence $\{s_n\}$ by considering the following sums:

$$s_1 = \sum_{n=1}^{1} a_n = a_1$$

$$s_2 = \sum_{n=1}^{2} a_n = a_1 + a_2$$

$$s_3 = \sum_{n=1}^{3} a_n = a_1 + a_2 + a_3$$

$$\vdots$$

(For $m > 3$) $s_m = \displaystyle\sum_{n=1}^{m} a_n = a_1 + a_2 + a_3 + \cdots + a_m$

Our intuition tells us that the larger the value of m, the more closely s_m should be to a description of $a_1 + a_2 + a_3 + \cdots$. With $s_m = \sum_{n=1}^{m} a_n$, we call the sequence $<s_m>$ the *infinite series*, or simply, the *series*, formed from the sequence $<a_n>$.

We simply define the *sum* of the infinite series as follows:

$$a_1 + a_2 + a_3 + a_4 + \cdots = \lim_{m \to \infty} (a_1 + a_2 + a_3 + a_4 + \cdots + a_m)$$

whenever this limit exists. More precisely, if

$$\lim_{m \to \infty} s_m = \lim_{m \to \infty} \sum_{n=1}^{m} a_n = L$$

then we say the series $<s_m>$ *converges to L*, and denote this by

$$\sum_{n=1}^{\infty} a_n = L$$

or, as above, $a_1 + a_2 + a_3 + \cdots = L$

A useful example of this is the *geometric series*:

$$s_1 = a$$
$$s_2 = a + ar$$
$$s_3 = a + ar + ar^2$$
$$\vdots$$
$$s_m = a + ar + ar^2 + \cdots + ar^{m-1}$$

It is possible to show that *if* $|r| < 1$, *then*

$$\lim_{m \to \infty} (a + ar + ar^2 + \cdots + ar^{m-1}) = \frac{a}{1-r}$$

or, equivalently,

$$a + ar + ar^2 + ar^3 + \cdots = \frac{a}{1-r}$$

The idea behind verifying this result is to recall that

$$a + ar + ar^2 + \cdots + ar^{m-1} = a\frac{1-r^m}{1-r} = \frac{a}{1-r} - \frac{ar^m}{1-r}$$

and then noticing that for $|r| < 1$, $\lim_{m \to \infty} r^m = 0$; hence,

$$\lim_{m \to \infty} (a + ar + ar^2 + \cdots + ar^{m-1}) = \frac{a}{1-r}$$

EXAMPLE 1 (5.14) $1 + \dfrac{1}{3} + \dfrac{1}{3^2} + \dfrac{1}{3^3} + \cdots = \dfrac{1}{1 - 1/3} = \dfrac{3}{2}$ $\left(a = 1, r = \tfrac{1}{3}\right)$

(5.15) $1 - \dfrac{1}{3} + \dfrac{1}{3^2} \quad \dfrac{1}{3^3} + \cdots = \dfrac{1}{1 - (-1/3)} = \dfrac{3}{4}$

$\left(a = 1, r = -\tfrac{1}{3}\right)$

(5.16) $\dfrac{1}{5} + \dfrac{1}{5^2} + \dfrac{1}{5^3} + \cdots = \dfrac{1/5}{1 - 1/5} = \dfrac{1}{4}$ $\left(a = \tfrac{1}{5} \text{ and } r = \tfrac{1}{5}\right)$

EXAMPLE 2 Find integers p and q for which $0.3333\cdots = p/q$.

SOLUTION First, $0.3333\cdots$ means the infinite repeating decimal in which 3 is repeated indefinitely; that is,

$$0.3333\cdots = 0.3 + 0.03 + 0.003 + 0.0003 + \cdots$$

$$= \frac{3}{10} + \frac{3}{10^2} + \frac{3}{10^3} + \frac{3}{10^4} + \cdots$$

This is a geometric series with $a = 3/10$ and $r = 1/10$. Thus,

$$0.3333\cdots = \frac{3/10}{1 - 1/10} = \frac{3/10}{9/10} = \frac{1}{3}$$

Problems In problems 1–14, find the "sum."

1. $1 + \tfrac{1}{3} + \left(\tfrac{1}{3}\right)^2 + \left(\tfrac{1}{3}\right)^3 + \cdots$

2. $1 + \tfrac{1}{4} + \left(\tfrac{1}{4}\right)^2 + \left(\tfrac{1}{4}\right)^3 + \cdots$

3. $3 + \dfrac{1}{2} - \dfrac{1}{2^2} + \dfrac{1}{2^3} - \dfrac{1}{2^4} + \cdots$

4. $1 + \dfrac{3}{8} + \dfrac{3^2}{8^2} + \dfrac{3^3}{8^3} + \dfrac{3^4}{8^4} + \cdots$

5. $\left(\tfrac{1}{3}\right)^2 + \left(\tfrac{1}{3}\right)^4 + \left(\tfrac{1}{3}\right)^6 + \left(\tfrac{1}{3}\right)^8 + \cdots$

6. $\tfrac{1}{4} + \left(\tfrac{1}{4}\right)^4 + \left(\tfrac{1}{4}\right)^7 + \left(\tfrac{1}{4}\right)^{10} + \cdots$

7. $\tfrac{2}{3} - \left(\tfrac{2}{3}\right)^3 + \left(\tfrac{2}{3}\right)^5 - \left(\tfrac{2}{3}\right)^7 + \cdots$

8. $\tfrac{1}{2} - \left(\tfrac{1}{2}\right)^2 + \left(\tfrac{1}{2}\right)^4 - \left(\tfrac{1}{2}\right)^6 + \cdots$

9. $\displaystyle\sum_{n=1}^{\infty} \left(\frac{2}{3}\right)^n$ 10. $\displaystyle\sum_{n=1}^{\infty} \left(\frac{2}{5}\right)^n$

11. $\displaystyle\sum_{n=1}^{\infty} \left(\frac{85}{95}\right)^n$ 12. $\displaystyle\sum_{n=1}^{\infty} \left(\frac{73}{81}\right)^n$

13. $\displaystyle\sum_{n=1}^{\infty} \left(-\frac{6}{11}\right)^n$ 14. $\displaystyle\sum_{n=1}^{\infty} \left(-\frac{7}{17}\right)^n$

15. Discuss the following series for the various possibilities for *a*:

$$\frac{a}{5} + \frac{a^2}{5^2} + \frac{a^3}{5^3} + \frac{a^4}{5^4} + \cdots$$

In problems 16–21, find the corresponding rational number for the given decimal:

16. 0.12 12 12 12...   17. 0.123 123 123...
18. 0.01 01 01 01...   19. 0.001 001 001...
20. 0.010 010 010...   21. 0.111 111 111...

22. A ball is dropped from a height of 10 feet. Each time the ball bounces, it rises three-fifths the distance it had previously fallen. What is the total distance traveled by the ball?

SECTION 4

THE EXPONENTIAL FUNCTION

Assume that we have a single bacterium in a culture. After a period of time, say *T*, the bacterium will divide into two "daughter" cells, and after another time period *T*, each of these cells divides again, thus producing four cells. That is, after a time period *T*, the population of bacteria doubles. The following table shows the history of the splittings:

$T = 0$	Population = 1 (the original bacterium)
$T = 1$	Population = 2 (the first splitting)
$T = 2$	Population = $4 = 2^2$
$T = 3$	Population = $8 = 2^3$
$\vdots$	$\vdots$
$T = n$, a positive integer	Population = 2^n

For the sake of illustration, assume that our period of time, *T*, is 1 hour. Thus, at the end of *n* hours, there are 2^n bacteria. We want to make our model a continuous one; that is, it is useful to have a model in which we assume that at *any* $t \geq 0$, there are 2^t bacteria. We see that at time $5\frac{1}{2}$ hours (that is, at time $\frac{11}{2}$ hours), there are

$$2^{11/2} = (\sqrt{2})^{11}$$

bacteria in our cultu.   Question: What is the theoretical size of the population at the end of say π, or $\sqrt{2}$ hours; that is, what is 2^π, or $2^{\sqrt{2}}$? If we can define numbers like 2^π and $2^{\sqrt{2}}$, we shall have succeeded in constructing a "population function" f given by $f(x) = 2^t$ for all real numbers *t* (for our specific population we would take $t \geq 0$, but it will turn out that it is possible to define 2^t for $t < 0$ as well). We proceed by cases.

Case 1 x is a positive integer, say $x = n$. Then $f(n) = 2^n = \underbrace{2 \cdot 2 \cdot 2 \cdots 2}_{n \text{ times}}$

Therefore, $f(1) = 2, f(3) = 8, f(37) = 2^{37}$, etc.

Case 2 x is a nonpositive integer. For $x = -n$, $n > 0$, $f(-n) = 2^{-n} = (1/2)^n$; also $f(0) = 2^0 = 1$.   Hence, $f(-1) = 2^{-1} = (1/2)$, $f(-2) = (1/4), f(-37) = 2^{-37} = (1/2)^{37}$, etc.

Case 3 x is a rational number, say $x = p/q$, where p and q are integers, with $q > 0$.   Recall that $a^{p/q} = \sqrt[q]{a^p}$ when this is meaningful.*
Hence, $f(p/q) = 2^{p/q} = \sqrt[q]{2^p}$. For instance,
$$f(1/2) = 2^{1/2} = \sqrt{2},$$
$$f(-1/2) = 2^{-1/2} = 1/2^{1/2} = 1/\sqrt{2}$$
$$f(5/7) = \sqrt[7]{2^5}, \qquad f(-4/3) = 1/\sqrt[3]{16}$$

Remark   Since every integer is rational, all three cases can be subsumed under the heading "x rational."

Let us graph $f(x) = 2^x$ for Cases 1–3 by plotting a large number of rational values of x. We see from Figure 5-1 that the graph of $f(x) = 2^x$ for x rational is monotone increasing and is above the x-axis.

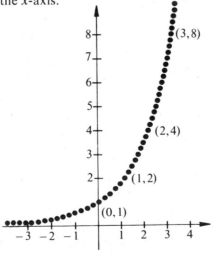

Figure 5-1 $f(x) = 2^x$, x Rational

*A discussion of powers and roots appears in part 2 of the appendix.

Case 4 x is irrational. We define $f(\delta) = 2^\delta$ at an irrational number δ *so as to make the function continuous there.* Roughly, to find 2^δ for an irrational number δ, we can consider a sequence $r_1 < r_2 < \cdots < r_n < \cdots$ of rational numbers such that $r_n \to \delta$ as $n \to \infty$. Since f is a strictly increasing function of x for x rational, we have that $2^{r_1} < 2^{r_2} < \cdots < 2^{r_n} < \cdots$. We define $2^\delta = \lim_{n \to \infty} 2^{r_n}$. That this limit really exists and is independent of the particular choice of sequence, as well as the fact that this makes 2^x continuous, is beyond the scope of this book.

The graph of $f(x) = 2^x$ is now defined for all real x, and is strictly increasing and continuous everywhere (see Figure 5-2).

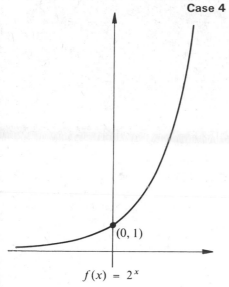

$f(x) = 2^x$

(0, 1)

Figure 5-2

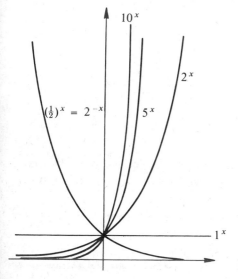

Figure 5-3

For any positive number a, we can investigate the function g given by $g(x) = a^x$ in the same way as $f(x) = 2^x$. The curve for $y = 3^x$ will become steeper more rapidly than the curve for $y - 2^x$ as x increases through positive values; for instance,

$$3^2 = 9 > 4 = 2^2, \qquad 3^3 = 27 > 8 = 2^3$$

and in general,

$$3^x > 2^x \text{ if } x > 0$$

If $x < 0$, we have,

$$3^{-1} = (1/3) < (1/2) = 2^{-1}$$
$$3^{-2} = (1/9) < (1/4) = 2^{-2}$$

in general, $\qquad 3^x < 2^x \text{ if } x < 0$

In all cases, $3^0 = 2^0 = a^0 = 1$, when $a > 0$. Some graphs are drawn in Figure 5-3.

We now turn to an investigation of the derivative of $f(x) = a^x$. We must resort to the definition of derivative as a limit. We have:

$$D\,a^x = \lim_{h \to 0} \frac{a^{x+h} - a^x}{h} = \lim_{h \to 0} a^x \left[\frac{a^h - 1}{h} \right] = a^x \lim_{h \to 0} \left[\frac{a^h - 1}{h} \right]$$

hence,

$$D\,a^x \big|_{x=0} = \lim_{h \to 0} \frac{a^h - 1}{h}$$

There is precisely one number a for which

$$\lim_{h \to 0} \frac{a^h - 1}{h} = 1$$

and that number is denoted by e; thus,

$$\lim_{h \to 0} \frac{e^h - 1}{h} = 1$$

or, equivalently,

$$D\, e^x \big|_{x=0} = 1$$

See Figure 5-4 for the graph of the equation $y = e^x$; note that the tangent line at $(0,1)$ has slope equal to 1. We are deliberately

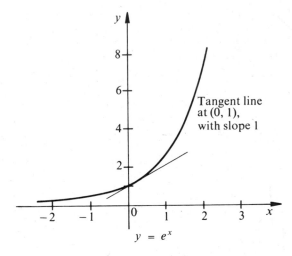

$y = e^x$

Figure 5-4

evading the issue of whether such a number exists at all; we claim that it does; furthermore, there is only one (unique) such number. With this value $a = e$, we have

$$D\, e^x = e^x \qquad \text{or} \qquad \frac{d}{dx}\, e^x = e^x$$

We shall find another expression for e when we discuss the logarithm. We shall anticipate this and state that e is *approximately* 2.7183; observe that e is a real number. The fact that $y = e^x$ is **S.I.** implies that

$$e^a = e^b \text{ if and only if } a = b$$

The function $f(x) = e^x$ satisfies the usual laws for exponents. Specifically, if a and b are any real numbers, then

$$e^a e^b = e^{a+b}$$
$$e^{-a} = 1/e^a$$
$$(e^a)^b = e^{ab}$$

That this holds for rational numbers a and b is well-known from algebra. That the rules are still valid for any real numbers a and b depends on the definition of e^x for arbitrary real x; we shall not give a proof here.

We next consider $e^{f(x)}$ where $f(x)$ is some function of x. Because the notation $e^{f(x)}$ is rather clumsy, we shall often replace it by the symbol **exp** $f(x)$, called **exponential** of $f(x)$. We define the *exponential function* **exp** by

$$\exp(x) = e^x$$

thus,*
$$\exp f(x) = e^{f(x)}$$

Since $D e^x = e^x$, it follows (from the chain rule) that

$$\mathbf{D \exp f(x) = f'(x) \cdot \exp f(x)}$$

EXAMPLE 1 (5.17) $\qquad D e^{ax} = e^{ax} \cdot D(ax) = ae^{ax}, \qquad a \text{ in } \mathbf{R}$

(5.18) $\qquad D e^{2x+1} = e^{2x+1} \cdot D(2x+1) = 2e^{2x+1}$

that is, $\qquad D \exp(2x + 1) = 2 \cdot \exp(2x + 1)$

(5.19) $\qquad D e^{x^2} = e^{x^2} \cdot D x^2 = 2xe^{x^2}$

that is, $\qquad D \exp(x^2) = 2x \exp(x^2)$

(5.20) $\quad D \exp \sqrt{x^2 + 1} = (\exp \sqrt{x^2 + 1}) \cdot D \sqrt{x^2 + 1}$

$$= \underline{\hspace{4cm}}$$

(5.21) $\quad D \exp(ax^2 + bx + c) = [\exp(\underline{\hspace{1.5cm}})] \cdot \underline{\hspace{1.5cm}}$

Since $D e^x = e^x$, we see that

$$\int e^x \, dx = e^x + c \qquad (c \text{ in } \mathbf{R})$$

EXAMPLE 2

$$\int_0^1 e^x \, dx = \int e^x \, dx \Big|_0^1 = e^x \Big|_0^1 = e - 1$$

Summarizing, we see that exp has the following properties:

1. exp is everywhere continuous and differentiable
2. $e^x > 0$, all x; $\quad e^0 = 1$

---

*More precisely, using the composition of functions, $\exp f(x) = (\exp \circ f)(x)$.

3. exp is **S.I.**
4. $\lim_{x \to -\infty} e^x = 0$, $\lim_{x \to \infty} e^x = \infty$
5. $e^{x+y} = e^x e^y$
6. $(e^x)^y = e^{xy}$
7. $e^{-x} = 1/e^x$
8. $D e^x = e^x$
9. $D e^{f(x)} = f'(x) e^{f(x)}$
10. $\int e^x \, dx = e^x + c$, c in **R**

Answers   Example 1, equation (5.20): $(\exp \sqrt{x^2 + 1}) \cdot x (x^2 + 1)^{-(1/2)}$
Example 1, equation (5.21): $ax^2 + bx + c$, $2ax + b$

Problems   In problems 1–14, find f'.

1. $f(x) = e^{-3x}$ 2. $f(x) = 2e^{5x}$
3. $f(x) = \exp(2x + 3)$ 4. $f(x) = \exp(-2x + 3)$
5. $f(x) = \exp(x^2 + x)$ 6. $f(x) = xe^x$
7. $f(x) = \exp \sqrt{x - 1}$
8. $f(x) = [\exp(x + 1)] \cdot [\exp(x - 1)]$
9. $f(x) = \exp[x\sqrt{x + 1}]$ 10. $f(x) = x^2 e^x$
11. $f(x) = \exp(-1/x^2)$ 12. $f(x) = \exp \sqrt{x/(x - 1)}$
13. $f(x) = \exp[(x^2 + 1)^{10}]$ 14. $f(x) = (\exp[x^2 + 1])^{10}$
15. Let a be any real number.
    a. Find $D e^{ax}$
    b. Use (a) to find $\int e^{ax} \, dx$
    c. Find $\int_0^1 (e^{-x} + 2e^{3x}) \, dx$
16. Show that $D \, xe^x = xe^x + e^x$.
    a. Use this to find $\int (xe^x + e^x) \, dx$.
    b. Then find $\int_0^1 xe^x \, dx$.
    c. And $\int_0^1 (x + 5) e^x \, dx$.
17. Find $D \, y$ if $xe^y + ye^x = 2x$.
18. Find $D \, y$ if $(x + 1) e^{y-1} + x^2 = 1$.

In problems 19–23, find the limit,

19. $\lim_{n \to \infty} e^{-n}$

20. $\lim_{n \to \infty} \dfrac{3 + 2e^n}{10 + 4e^n}$

21. $\lim_{n \to \infty} \dfrac{e^n + e^{2n}}{e^n - e^{2n}}$

    (*Hint:* Factor e^{2n} in the numerator and in the denominator.)

22. $\lim_{n \to \infty} \exp\left[\dfrac{1 + n}{1 - n}\right]$

    (*Hint:* Use the fact that if $a_n \to a$ as $n \to \infty$, then $e^{a_n} \to e^a$ as $n \to \infty$)

23. $\lim\limits_{n \to \infty} \exp\left[\dfrac{2n^2 + 8}{n^2 - 5}\right]$

24.   Find the extrema for the function

$$f(x) = (1/s\sqrt{2\pi})\exp\left(-(x - u)^2/2s^2\right)$$

where $s > 0$ and u are fixed constants (this function is important in statistics).

25.   If the demand law for a certain commodity is $p = 9e^{-x/3}$, find the maximum total revenue.

26.   Let $f(x) = e^{-ax}$ and $g(x) = e^{-bx}$. Show that each of these functions satisfies the "differential" equation

$$y''(x) + (a + b)y'(x) + aby(x) = 0$$

where a and b are numbers.

SECTION 5

THE NATURAL LOGARITHM FUNCTION

We define the *natural logarithm function*, which we shall denote by **ln** (or, occasionally, by $\log_e$ or simply **log**), as follows:

$$x = e^y \text{ if and only if } y = \ln x$$

Note that x must be a positive number. For example,

$$1 = e^0 \quad \text{and} \quad \ln 1 = 0$$
$$e = e^1 \quad \text{and} \quad \ln e = 1$$

thus,

ln x is the exponent to which we must raise e to obtain x.

Hence, for example,

$$\ln e^2 = 2 \qquad \ln e^3 = 3 \qquad \ln e^4 = 4$$

and in general, for any number x,

(5.22) $\ln e^x = x$

Furthermore, since $x = e^y$ if and only if $y = \ln x$, we have, by substituting y for $\ln x$,

(5.23) $x = e^{\ln x}$,   with $x > 0$ (since e^y is > 0)

   Notice that if we think of $x = e^y$ as an equation that we want to solve for y, then the solution would be $y = \ln x$. That is, $x = e^y$ and $y = \ln x$ represent the same relation, with the roles of x and y interchanged. Thus, the graph of $x = e^y$ and the graph of $y = \ln x$ must be the same graph but with the axes interchanged. Hence, the graph of ln can be obtained from the graph of exp by revolving 90° counterclockwise the graph of exp, and then, so that the horizontal axis will have the usual orientation, reflecting this graph in the vertical axis.

The graph of ln is given in Figure 5-5. The domain of ln is the positive real line (to see that ln is not defined for $x < 0$, note, for example, that if ln $(-1) = y$, then $e^y = -1$, which is impossible since $e^y > 0$ for all y). It turns out that ln is **S.I.**, continuous and differentiable on $\{x : x > 0\}$. The next result concerns the fundamental algebraic properties of ln.

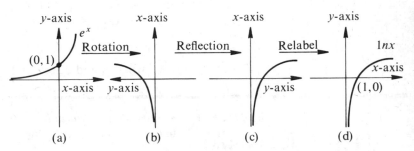

Figure 5-5

Let a and b be positive numbers. Then,

(5.24)
$$\ln ab = \ln a + \ln b$$

(5.25)
$$\ln \frac{1}{a} = -\ln a$$

(5.26)
$$\ln \left(\frac{a}{b} \right) = \ln a - \ln b$$

(5.27)
$$\ln (a^c) = c \ln a \text{ (for any number c)}$$

The verification relies heavily on the properties of exponentials.

(5.24′) Let $u = \ln a$ and $v = \ln b$. Then, $e^u = a$ and $e^v = b$. Thus, $a \cdot b = e^u \cdot e^v = e^{u+v}$. Hence, ln $a \cdot b = \ln e^{u+v} = u + v$, again by (5.22). However, $u + v = \ln a + \ln b$; thus, ln $a \cdot b = \ln a + \ln b$.

(5.25′) Let $w = \ln (1/a)$. Then $e^w = e^{\ln (1/a)} = 1/a$ by (5.23). Hence, $a = 1/e^w = e^{-w}$, and ln $a = \ln e^{-w} = -w = -\ln (1/a)$, and (5.25) is proved.

(5.26′) ln $(a/b) = \ln (a \cdot b^{-1})$, and by (5.24), ln $(ab^{-1}) = \ln a + \ln b^{-1}$. By (5.25), ln $b^{-1} = -\ln b$. Hence, ln $(a/b) = \ln a - \ln b$.

(5.27′) Let $u = \ln a$; then $e^u = e^{\ln a} = a$. Recall that $(e^u)^b = e^{ub}$. Consequently, $e^u = a$ implies $(e^u)^b = e^{ub} = a^b$; therefore, ln $e^{ub} = \ln a^b$. But ln $e^{ub} = ub = b \ln a$; thus, $b \ln a = \ln a^b$, and (5.27) is proved.

EXAMPLE 1 $\ln 16 = \ln \{2^4\} = 4 \ln 2$; $\ln 8 = \ln 4 \cdot 2 = \ln 4 + \ln 2$; with $x > 0$ and $y > 0$, $\ln x^5 y^2 = \ln x^5 + \ln y^2 = 5 \ln x + 2 \ln y$; for $z > 0$, $\ln \sqrt{z} = \ln z^{1/2} = \frac{1}{2} \ln z$.

EXAMPLE 2   An early learning and memory theory model designed by Hermann Ebbinghaus (*Memory*, translated by H. A. Ruger. New York: Teachers College, Columbia University, 1913) states that $y(t)$, the percent of the nonsense syllables out of a list that are retained t minutes after the end of the learning period (with $t \geq 1$), is given by

$$y(t) = \frac{a}{(\ln t)^c + b}$$

where a, b, and c are constants.

We next evaluate $D \ln x$, assuming from our geometrical considerations that it exists for all x in the domain of $\ln x$. From (5.23)

$$x = \exp (\ln x), \qquad x > 0$$

Differentiating yields

(5.28) $D x = D \exp (\ln x)$

Clearly $D x = 1$, and since

$$D e^{f(x)} = e^{f(x)} \cdot f'(x)$$

we see that $D e^{\ln x} = e^{\ln x} \cdot f'(x) = x f'(x)$

(since $e^{\ln x} = x$). Thus,

$$1 = x \cdot f'(x)$$

or $f'(x) = \frac{1}{x}$

that is, $D \ln x = \frac{1}{x}, \qquad x > 0$

It can be shown that

$$D \ln |x| = \frac{1}{x}, \qquad x \neq 0$$

In fact, if $f'(x)$ exists, then

$$D \ln f(x) = \frac{1}{f(x)} \cdot f'(x) \qquad \text{for} \qquad f(x) > 0$$

and also

$$D \ln |f(x)| = \frac{f'(x)}{f(x)} \quad \text{for} \quad f(x) \neq 0$$

Summarizing:

$$D \ln f(x) = \frac{f'(x)}{f(x)}, \quad f(x) > 0$$

$$D \ln |f(x)| = \frac{f'(x)}{f(x)}, \quad f(x) \neq 0$$

$$D \ln x = \frac{1}{x}, \quad x > 0$$

$$D \ln |x| = \frac{1}{x}, \quad x \neq 0$$

Applying our results on derivatives to antiderivatives, we obtain

$$\int \frac{1}{x} \, dx = \ln |x| + c, \quad \int \frac{f'(x)}{f(x)} \, dx = \ln |f(x)| + c$$

EXAMPLE 3   For $x > 0$,

(5.29) $$D \ln (x^{10}) = \frac{D x^{10}}{x^{10}} = \frac{10x^9}{x^{10}} = \frac{10}{x}$$

or, equivalently,

$$D \ln (x^{10}) = D \, 10 \ln x = \frac{10}{x}$$

(5.30) $$D \ln |3x^2 + 2x + 5| = \frac{D(3x^2 + 2x + 5)}{3x^2 + 2x + 5}$$

$$= \frac{6x + 2}{3x^2 + 2x + 5}$$

(5.31) $$D \ln |x^{20} + x^{10} + 2| = \frac{D(\underline{\hspace{1cm}})}{\underline{\hspace{2cm}}} = \underline{\underline{\hspace{2cm}}}$$

(5.32) $$D \ln \sqrt[5]{x} = D \ln (x^{1/5}) = \frac{1}{5} D \ln x = \frac{1}{5x} \quad (x > 0)$$

(5.33) $$D \ln (|x^2 + 1|^{10}) = 10 \, D \ln |x^2 + 1|$$

$$= 10 \frac{D(x^2 + 1)}{x^2 + 1}$$

$$= \frac{20x}{x^2 + 1}$$

$$(5.34) \qquad D \ln\left(\frac{x+2}{x+3}\right) = D\left[\ln(x+2) - \ln(x+3)\right]$$

$$= D\ln(x+2) - D\ln(x+3)$$

$$= \frac{D(\underline{\quad})}{\underline{\qquad}} - \frac{D(\underline{\quad})}{\underline{\qquad}}$$

$$= \frac{\underline{\qquad}}{\underline{\qquad}} - \frac{\underline{\qquad}}{\underline{\qquad}}$$

EXAMPLE 4

$$(5.35) \qquad \int \frac{2x}{x^2+1}\,dx = \int \frac{D(x^2+1)}{x^2+1}\,dx$$

$$= \ln(x^2+1) + c \qquad (c \text{ in } \mathbf{R})$$

$$(5.36) \qquad \int \frac{x^2}{x^3-1}\,dx = \int \frac{\frac{1}{3}D(x^3-1)}{x^3-1}\,dx$$

$$= \frac{1}{3}\int \frac{D(x^3-1)}{x^3-1}\,dx$$

$$= \tfrac{1}{3}\ln|x^3-1| + c \qquad (c \text{ in } \mathbf{R})$$

$$(5.37) \qquad \int \frac{10x^9+2x}{x^{10}+x^2+3}\,dx = \int \frac{D(\underline{\quad})}{x^{10}+x^2+3}\,dx$$

$$= \ln|\underline{\qquad}| + c, \qquad c \text{ in } \mathbf{R}$$

The exponential and logarithmic functions give a technique for differentiating certain functions of a somewhat complex nature. We shall call this technique **logarithmic differentiation**. The next example illustrates the technique.

EXAMPLE 5 Let $G(x) = x^x, x > 0$. Find $G'(x)$.

SOLUTION Since $G(x) = x^x$, we have

$$\ln G(x) = \ln(x^x) = x \ln x$$

Hence,
$$D \ln G(x) = D(x \ln x)$$

$$\frac{G'(x)}{G(x)} = 1 + \ln x$$

$$G'(x) = G(x)[1 + \ln x]$$

that is, for $x > 0$

$$G'(x) = D x^x = x^x(1 + \ln x)$$

Answers   Example 3, equation (5.31): $\dfrac{D(x^{20} + x^{10} + 2)}{x^{20} + x^{10} + 2} = \dfrac{20x^{19} + 10x^9}{x^{20} + x^{10} + 2}$

Example 3, equation (5.34): $\dfrac{D(x + 2)}{x + 2} - \dfrac{D(x + 3)}{x + 3}$

$$= \frac{1}{x + 2} - \frac{1}{x + 3}$$

Example 4, equation (5.36): $\displaystyle\int \frac{D(x^{10} + x^2 + 3)}{x^{10} + x^2 + 3}\, dx$

$$= \ln|x^{10} + x^2 + 3| + c$$

Problems   In problems 1–18, find f'.

1. $f(x) = \ln[(x + 1)^{1/2}]$ 2. $f(x) = \ln[(x + 1)^2]$
3. $f(x) = (\ln x)^2$ 4. $f(x) = x \ln x$
5. $f(x) = x^3 \ln 4x$ 6. $f(x) = \ln(xe^x)$

7. $f(x) = \ln\left(\dfrac{x}{1 + 3x}\right)$ 8. $f(x) = \ln\sqrt{\dfrac{x - 1}{x + 1}}$

9. $f(x) = \ln|x^3 - 3x^2 + 2x + 5|$
10. $f(x) = \ln|x^{10} + x^2 - 4|$
11. $f(x) = \ln[x(x + 1)^5]$
12. $f(x) = \ln[(x)^{1/2}(x^2 + 1)^3]$
13. $f(x) = \ln[x(x + 1)(x + 2)]^{1/2}$
14. $f(x) = \ln(e^x) + e^x \ln x$
15. $f(x) = \ln(\ln x)$
16. $f(x) = \ln(\ln x^2)$

17. $f(x) = \ln\sqrt[7]{\dfrac{(x - 3)(x^2 + 1)}{x^4 + 2}}$

18. $f(x) = \ln\sqrt[10]{\dfrac{(x^{100} + 1)(x + 1)^{100}}{2x + 1}}$

In problems 19–26, use logarithmic differentiation to find dy/dx:

19. $y = x^{x^2 + x - 2}$, $x > 0$ 20. $y = x^{\ln x}$, $x > 0$
21. $y = (\ln x)^{\ln x}$, $x > 0$
22. $y = (x^2 + x + 1)^{2x - 7}$, $x > 0$
23. $y = (\sqrt{x})^{\sqrt{x}}$, $x > 0$
24. $y = (x + 2)^{x + 3}$, $x + 2 > 0$

25. $y = \sqrt[10]{\dfrac{x^8 + 3}{x^{10} - 5}}$

26. $y = \sqrt[18]{(x^{10} + 1)^3(x^7 - 3)^8}$
27. Find when $f(x) = (\ln x)/x$, $x > 0$, is **S.I.** and when it is **S.D.** Discuss the extrema of $f(x)$.

28. Psychologists believe that a person's childhood ability to memorize, measured by $G(t)$ where t is measured in years, is given by

$$G(t) = \begin{cases} t \ln t + 1, & 0 < t \le 4 \\ 1, & t = 0 \end{cases}$$

Find the extrema of G.

Compute 29 and 30:

29. $\displaystyle\lim_{n \to \infty} \ln\left(\frac{n^2 + 2}{n^2 + n + 10}\right)$

(*Hint:* If $a_n \ge 0$ and $a_n \to a$ as $n \to \infty$, then $\ln a_n \to \ln a$ as $n \to \infty$.)

30. $\displaystyle\lim_{n \to \infty} \ln\left(\frac{n^3 + 1}{en^3 - 1}\right)$

31. Find $\displaystyle\sum_{k=2}^{99} \ln\left(\frac{k+1}{k}\right)$. Then find $\displaystyle\sum_{k=2}^{\infty} \ln\left(\frac{k+1}{k}\right)$.

In problems 32–34, find f'.

32. $f(x) = \ln(\ln(\ln x))$

33. $f(x) = (\ln x)^x / x^{\ln x}$

34. $f(x) = x^{x^x}$

SECTION 6

EXPONENTIAL GROWTH AND DECAY

In a vast selection of problems drawn from the physical sciences and the life sciences, experimental evidence based on observations of a given population, such as the amount of radium in a given sample or the size of the population of bacteria in a culture, indicates that the rate of change of the size of the population at any given instant of time is directly proportional to the size of the population at that instant. Mathematically, if we set $y(t)$ equal to the size of the population at time t, then, for a constant of proportionality A,

$$\frac{dy(t)}{dt} = Ay(t)$$

If $A > 0$, then the population is growing, whereas if $A < 0$, the population is decreasing in size with time (this latter is the case in radioactive decay, a situation in which radioactive substances emit particles). Thus, we say that we have **exponential growth** if $A > 0$ and **exponential decay** if $A < 0$.

We shall show that the function given by

$$y(t) = ce^{At} \qquad \text{(with } c \text{ and } A \text{ in } \mathbf{R}\text{)}$$

satisfies the equation $y'(t) = Ay(t)$. To see this, note that

$$y'(t) = D[ce^{At}]$$
$$= ce^{At} \cdot D(At)$$
$$= A(ce^{At})$$
$$= Ay(t)$$

It can be shown that *every function y which satisfies the condition y'(t) = Ay(t) must be of the form y(t) = ce^{At} for some number c*; that is, $y'(t) = Ay(t)$ if and only if $y(t) = ce^{At}$ for some number c.

EXAMPLE 1   Find $y(t)$ if $y'(t) = -2y(t)$ and $y(0) = 4$.

SOLUTION   We know that $y(t) = ce^{-2t}$ for some number c. Since $y(t) = ce^{-2t}$, we have in particular that $y(0) = ce^{-2 \cdot 0} = c$. Also, we are given that $y(0) = 4$. Hence, $c = 4$, and thus

$$y(t) = 4e^{-2t}$$

Notice that we have exponential decay. If $y(t)$ is some population of objects at time t, then that population is decreasing. Note also that $y(0)$ is the population at time 0 (that is, the initial population); hence, in this problem, the initial population is four objects.

EXAMPLE 2   Assume that a population grows so that its rate of growth is directly proportional to its size, and assume that its initial size is 1,000 objects, and in 10 days, it has 2,000 objects. Find the number of objects in an arbitrary time t.

SOLUTION   We are dealing with a problem in exponential growth (since the number of objects has *increased* in 10 days). Thus, if we denote by $y(t)$ the number of objects at time t (in days), then

$$y(t) = ce^{At}$$

The problem now is to determine c and A. Since the initial size of the population is 1,000, we have $y(0) = 1,000$. But $y(0) = ce^{A \cdot 0} = c$. Hence $c = 1,000$, and

$$y(t) = 1,000e^{At}$$

To find A, note that in 10 days we have 2,000 objects; hence,

$$2,000 = y(10) = 1,000e^{10A}$$

that is, $$2 = e^{10A}$$

Consequently, $$10A = \ln 2$$

hence $$A = \frac{\ln 2}{10}$$

The population at time t (days), therefore, is

$$y(t) = 1{,}000 \exp\left(\frac{t \ln 2}{10}\right)$$

EXAMPLE 3   Consider again the population given in Example 2. How fast is this population growing at the end of 6 days?

SOLUTION   The rate of change of the population at the end of 6 days is $y'(6)$. However, since we have exponential growth,

$$y'(t) = Ay(t)$$

$$= \frac{\ln 2}{10} y(t)$$

$$= \frac{\ln 2}{10} \cdot 1000 \exp\left(\frac{t \ln 2}{10}\right)$$

$$= 100 \ln 2 \exp\left(\frac{t \ln 2}{10}\right)$$

Hence, $$y'(6) = 100 \ln 2 \exp\left(\frac{6 \ln 2}{10}\right)$$

$$= 100 \ln 2 \exp\left(\tfrac{3}{5} \ln 2\right)$$

It is possible to find numerical estimates of this number, but we shall not concern ourselves with this problem here.

Problems   In problems 1–4, find $y(t)$.

1. $y'(t) = y(t)$ and $y(0) = 2$
2. $y'(t) = 2y(t)$ and $y(0) = 1$
3. $y'(t) = -y(t)$ and $y(0) = 5$
4. $y'(t) = -\frac{1}{2}y(t)$ and $y(0) = \frac{1}{3}$
5. For large samples of radium, the rate at which the radium decreases is known to be proportional to the amount of radium. If it takes 1 gram of radium 1,700 years to decrease to half its original size, find the amount of the original 1 gram of radium present after 2,000 years.
6. A population is known to grow exponentially. During the first 10 minutes of observation, the population grows from 1,000 objects to 5,000 objects. When will the population contain 6,000 objects? (Leave your answer in terms of the natural logarithm.) What will be the population at the end of 20 minutes? How fast is the population changing at the end of 20 minutes?
7. Repeat problem 6 if now the population grows from 1,000 to 4,000 in 10 minutes.

8. A population is known to grow exponentially. It is observed that the size of the population triples in the first 10 minutes of observation. The size of the population at the end of 20 minutes is found to be 9,000 objects. What is the size of the population at any time t? How fast is the size of the population changing?

9. A population is known to experience exponential decay. The initial population (that is, the population at time 0) is found to be 10,000 objects. At the end of 20 minutes, the population is found to be 5,000 objects. When will there be 1,000 objects in the population? When will there be 100 objects in the population? How fast is the size of the population declining when there are 100 objects in the population?

10. The experience of a certain company leads it to believe that the benefits in terms of increased sales from an advertising campaign undergo exponential decay. At the end of t days after the campaign has ended, the additional sales $f(t)$ is given by

$$f(t) = 1,000e^{-t/4}$$

How many extra sales will there be at the end of 8 days? How fast are the sales declining at that time?

11. The following technique is used to date once-living objects. A living organism is known to be composed of stable carbon C_1 and radioactive carbon C_2 in such a way that the ratio of the amount of C_2 to the amount of C_1 in the tissue is approximately constant, regardless of what part of the tissue is investigated. When the organism dies, C_2 experiences exponential decay such that in about 5,550 years only half as much C_2 is present.

a. Find the number of grams $y(t)$, of C_2, t years after the organism has died. (Use the fact that ln 2 is approximately equal to 0.693.)

b. A piece of ancient charcoal is found to have only 12 percent of its original C_2 content. How many years ago did the tree from which the charcoal came die?

6

MORE ON INTEGRATION

The main objective of this chapter is to indicate that the limit of a certain sum turns out to be an integral. This is the result that was promised when we first viewed the integral as an area, and, as was mentioned then, is the result that pays off in applications. For example, when the physicist investigates the problem of center of mass, or moment of inertia, or work (we shall *not* pursue examples of these problems), he finds that his physical model first leads to a certain sum (in each case, different in particulars but identical in general form), and then—and this is still part of his model—to a certain limit of this sum. The mathematical theory then steps in, and identifies this limit of a sum as an *integral*. Thus, the physical model is ultimately described by an integral, and the information about integrals becomes available to gain insight into the problem.

The chapter begins with new information about evaluating integrals. The techniques could have been introduced earlier in the book when the integral was first defined via area, but now, with exponential and logarithmic functions available, it is possible to give richer examples. We then view the integral as a limit of a sum. The immediate application of this is to the geometrical problems of area and volume. Again, geometrical examples precede examples from, say, biology because they are felt to be very intuitive and accessible to a more general audience.

The chapter concludes with some extensions of the integral concept (so-called improper integrals) and a brief discussion of differential equations.

SECTION 1

FURTHER TECHNIQUES FOR INTEGRATION: SUBSTITUTION AND INTEGRATION BY PARTS

We have found that it was possible to evaluate many derivatives explicitly without using the definition of the derivative as a limit. The fundamental theorem is not a complete analog for finding integrals; indeed, the theorem is useful only when we can explicitly exhibit an antiderivative for the continuous function whose definite integral we wish to evaluate. For instance, although it is routine to evaluate $D[(x + 1)\sqrt[3]{x^2 + 2x + 5}]$, we see that to evaluate $\int_0^1 (x + 1)\sqrt[3]{x^2 + 2x + 5}\, dx$ requires finding an antiderivative of $(x + 1)\cdot\sqrt[3]{x^2 + 2x + 5}$, and the fundamental theorem does not tell us how to do this. We shall exhibit in this section two new techniques for integrating more complicated functions; this will be a *partial* answer to the problem of evaluating integrals. The first result is: the **method of substitution** or **change of variables**.

Let $f(w)$ be a continuous function and let $g(x)$ be a function such that $g'(x)$ exists. Then, under the substitution $w = g(x)$,

$$\int f(w)\, dw = \int f[g(x)]g'(x)\, dx = \int (f \circ g)(x)g'(x)\, dx$$

The precise statement of the method of substitution suggests an informal procedure. The virtue of the procedure is that it is easy to remember and easy to work with. If we replace w by $w(x)$, we see that our result becomes

$$\int f(w)\, dw = \int f[w(x)]\frac{dw(x)}{dx}\, dx$$

that is, "w" is replaced by "$w(x)$" and "dw" is replaced by "$(dw/dx)\, dx$," or, for purposes of computation,

$$dw = \frac{dw}{dx}\, dx = w'(x)\, dx$$

The next examples illustrate the procedure.

EXAMPLE 1   Find $\int 2x(x^2 + 1)^{1/2}\, dx$.

SOLUTION   Let

$$w(x) = x^2 + 1$$

then

$$\frac{dw}{dx} = 2x$$

hence,

$$dw = w'(x)\, dx = 2x\, dx$$

Thus, replacing $2x\,dx$ by dw and $\sqrt{x^2 + 1}$ by $\sqrt{w}$, we obtain

$$\int \sqrt{x^2 + 1}\; 2x\,dx = \int \sqrt{w}\,dw$$
$$= \int w^{1/2}\,dw$$
$$= \tfrac{2}{3}w^{3/2} + c \qquad \text{(where } c \text{ is a real number)}$$
$$= \tfrac{2}{3}[w(x)]^{3/2} + c \qquad \text{(since ``} w = w(x) \text{'')}$$
$$= \tfrac{2}{3}(x^2 + 1)^{3/2} + c \qquad \text{(since } w(x) = x^2 + 1 \text{)}$$

Notice that what we have done amounts to handling dw/dx as if it were an ordinary fraction (that is, $dw \div dx$), since if we have fractions, then certainly $dw = (dw/dx) \cdot dx$. In fact, no harm is done if, in applying the rule of substitution, we treat dw/dx as $dw \div dx$. Suppose we redo our original problem as follows:

Let $w(x) = x^2 + 1$, then $dw/dx = 2x$, so that $dw = 2x\,dx$. Hence, $dx = dw/2x$, and replacing "dx" by "$dw/2x$,"

$$\int 2x(x^2 + 1)^{1/2}\,dx = \int 2x \cdot (w)^{1/2} \cdot \frac{dw}{2x} = \int (w)^{1/2}\,dw$$

which is exactly what we obtained before.

EXAMPLE 2   Find $\int (x + 1)\sqrt[3]{x^2 + 2x + 5}\,dx$.

Let $w(x) = x^2 + 2x + 5$. Then $dw/dx = 2(x + 1)$; hence $\tfrac{1}{2}dw = (x + 1)\,dx$. Substituting, we obtain

$$\int (x + 1)\sqrt[3]{x^2 + 2x + 1}\,dx = \tfrac{1}{2}\int w^{1/3}\,dw$$
$$= \tfrac{1}{2} \cdot \tfrac{3}{4} \cdot w^{4/3} + c$$
$$= \tfrac{3}{8}w^{4/3} + c$$
$$= \tfrac{3}{8}(x^2 + 2x + 5)^{4/3} + c$$

EXAMPLE 3   Evaluate

$$\int \frac{10x}{(x^2 + 1)^{1/2}}\,dx$$

SOLUTION   Let $w(x) = x^2 + 1$. Thus, $dw/dx = 2x$, or $2x\,dx = dw$. Hence,

$$\int \frac{10x}{(x^2 + 1)^{1/2}}\,dx = \int \frac{5}{(x^2 + 1)^{1/2}}\,2x\,dx = 5\int w^{-1/2}\,dw$$
$$= 10w^{1/2} + c$$
$$= 10(x^2 + 1)^{1/2} + c$$

The corresponding result for definite integrals is if $f \circ w$, w and w' are continuous on $[a,b]$, then

$$\int_a^b (f \circ w)(x)\, w'(x)\, dx = \int_{w(a)}^{w(b)} f(w)\, dw$$

that is, with $w = w(x)$, as x goes from a to b, w goes from $w(a)$ to $w(b)$. The next examples illustrate the application of this result.

EXAMPLE 4   Find $\int_0^1 2x(x + 1)^{1/2}\, dx$.

SOLUTION   We use the same substitutions as in Example 1. Thus, $w(x) = x^2 + 1$; hence, $dw = 2x\, dx$. Furthermore, $w(0) = 1$ and $w(1) = 2$. Hence,

$$\int_0^1 2x(x^2 + 1)^{1/2}\, dx = \int_{w(0)}^{w(1)} (w)^{1/2}\, dw$$

$$= \int_1^2 w^{1/2}\, dw$$

$$= \tfrac{2}{3} w^{3/2} \, \big|_1^2$$

$$= \tfrac{2}{3}(2^{3/2} - 1^{3/2})$$

$$= \tfrac{2}{3}(2\sqrt{2} - 1)$$

EXAMPLE 5   Find $\int_0^1 (x + 1)\sqrt[3]{x^2 + 2x + 5}\, dx$.

SOLUTION   We use the same substitutions as in Example 2: $w(x) = x^2 + 2x + 5$, and thus $dw = 2(x + 1)\, dx$. Note that $w(0) = 5$ and $w(1) = 8$. We now see that

$$\int_0^1 (x + 1)\sqrt[3]{x^2 + 2x + 5}\, dx = \tfrac{1}{2} \int_{w(0)}^{w(1)} w^{1/2}\, dw$$

$$= \tfrac{1}{2} \int_5^8 w^{1/2}\, dw$$

$$= \tfrac{3}{8} w^{4/3} \, \big|_5^8$$

$$= \tfrac{3}{8}(8^{4/3} - 5^{4/3})$$

$$= \tfrac{3}{8}(16 - 5\sqrt[3]{5})$$

(since $8^{4/3} = (8^{1/3})^4 = 2^4 = 16$, and $5^{4/3} = \sqrt[3]{5^4} = \sqrt[3]{5^3 \cdot 5} = 5\sqrt[3]{5}$).

EXAMPLE 6   Find $\int_1^2 x^2(x^3 + 1)^{20}\, dx$.

SOLUTION Let $w(x) = x^3 + 1$. Then

$$\frac{dw}{dx} = \underline{\hspace{2cm}}$$

hence,

$$\frac{1}{\boxed{}} dw = x^{\boxed{}} dx$$

Also, $\quad w(1) = \underline{\hspace{1.5cm}} \quad$ and $\quad w(2) = \underline{\hspace{1.5cm}}$

hence, $\quad \displaystyle\int_1^2 x^2(x^3 + 1)^{20} \, dx = \boxed{} \int_{\boxed{}}^{\boxed{}} w^{\boxed{}} \, dw$

The final technique we shall exhibit is **integration by parts.** Let $f(x), g(x), f'(x)$, and $g'(x)$ be continuous. Then

$$\int f(x) g'(x) \, dx = f(x) g(x) - \int g(x) f'(x) \, dx$$

or, for definite integrals,

$$\int_a^b f(x) g'(x) \, dx = f(x) g(x) \, \big|_a^b - \int_a^b g(x) f'(x) \, dx$$

To see this, we have from the product rule for derivatives

$$(f(x) g(x))' = f'(x) g(x) + f(x) g'(x)$$

Hence, $[f(x) g(x)]'$ is continuous, and $\int_a^b [f(x) g(x)]' \, dx$ exists and, by the fundamental theorem, equals $f(x) g(x) \, \big|_a^b$. Next observe that, since $\int [f(x) g(x)]' \, dx = f(x) g(x) + c$ (keep in mind that this is a statement about antiderivatives, and is true simply because the derivative of $f(x) g(x) + c$ is $[f(x) g(x)]'$),

$$f(x) g(x) + c = \int [f(x) g(x)]' \, dx$$
$$= \int \{f'(x) g(x) + f(x) g'(x)\} \, dx$$
$$= \int f'(x) g(x) \, dx + \int f(x) g'(x) \, dx$$

For the case of definite integrals, note that

$$\int_a^b f'(x) g(x) \, dx = \int f'(x) g(x) \, dx \, \big|_a^b$$

$$\int_a^b f(x) g'(x) \, dx = \int f(x) g'(x) \, dx \, \big|_a^b$$

Remark   The $g(x)$ that occurs on the right side of the integration by parts formula is *any* antiderivative of the $g'(x)$ that appears on the left side; therefore, the constant which arises when we antidifferentiate $g'(x)$ to get the $g(x)$ on the right side may be chosen to be any convenient value. It is usually chosen to be 0, although occasionally a different choice is efficacious.

EXAMPLE 7   Find $\int xe^x\,dx$.

SOLUTION   Let $f(x) = x$ and $g'(x) = e^x$. Then

$$\int xe^x\,dx = \int f(x)\,g'(x)\,dx$$

Clearly $f'(x) = 1$ and $g(x) = e^x$ [$g(x)$ is any antiderivative; therefore, we choose $c = 0$ in the general antiderivative $e^x + c$]. That is,

$$f(x) = x \qquad f'(x) = 1$$
$$g'(x) = e^x \qquad g(x) = e^x$$

Hence,

$$\int xe^x\,dx = xe^x - \int 1 \cdot e^x\,dx = xe^x - e^x + c$$

EXAMPLE 8   From Example 6,

$$\int_0^1 xe^x\,dx = \int xe^x\,dx \Big|_0^1 = (xe^x - e^x)\Big|_0^1 = 1$$

Another way to write this is

$$\int_0^1 xe^x\,dx = xe^x \Big|_0^1 - \int_0^1 1 \cdot e^x\,dx$$

$$= e - \left(e^x \Big|_0^1\right) = 1$$

EXAMPLE 9   Find $\int \ln x\,dx$.

SOLUTION   We have a choice to make: either let $f(x) = \ln x$ and $g'(x) = 1$ or let $f(x) = 1$ and $g'(x) = \ln x$. If we choose $g'(x) = \ln x$ and $f(x) = 1$, we must then find $g(x)$. To find $g(x)$, however, we must antidifferentiate $\ln x$; and to do this involves already knowing $\int \ln x\,dx$, which is, unfortunately, exactly our problem in this example.

Hence, let $f(x) = \ln x$ and $g'(x) = 1$. Then $f'(x) = 1/x$, and

$g(x) = x$. We now have

$$\int \ln x \, dx = x \ln x - \int x \cdot (1/x) \, dx$$
$$= x \ln x - \int dx = x \ln x - x + c$$

Answers Example 6: $\dfrac{dw}{dx} = 3x^2; \quad \tfrac{1}{3}dw = x^2 \, dx; \quad w(1) = 2; \quad w(2) = 9;$

$\tfrac{1}{3} \int_2^9 w^{20} \, dw$

Problems Use substitution to evaluate each of the following:

1. $\displaystyle\int \frac{x^2}{(x^3 + 2)^{1/2}} \, dx$

2. $\displaystyle\int \frac{x^2}{x^3 + 2} \, dx$

3. $\displaystyle\int \frac{x - 3}{\sqrt[3]{x^2 - 6x + 15}} \, dx$

4. $\displaystyle\int_0^1 \frac{x^5}{x^6 + 1} \, dx$

5. $\displaystyle\int_0^1 \frac{x^5}{(x^6 + 1)^2} \, dx$

6. $\displaystyle\int xe^{x^2} \, dx$

7. $\displaystyle\int_1^e \frac{\ln x}{x} \, dx$

8. $\displaystyle\int \frac{1}{x \ln x} \, dx$

9. $\displaystyle\int \frac{1}{2x + 3} \, dx$

10. $\displaystyle\int \frac{1}{5 - 4x} \, dx$

11. $\displaystyle\int \frac{x}{1 - x^2} \, dx$

12. $\displaystyle\int \frac{1}{(3x + 2)^2} \, dx$

13. $\displaystyle\int_1^e \frac{\ln (x^2)}{x} \, dx$

14. $\displaystyle\int \frac{\ln 5x}{x} \, dx$

15. $\displaystyle\int_1^e \frac{\ln \sqrt{x}}{x} \, dx$

16. $\displaystyle\int \frac{x + 1}{(8 - 2x - x^2)^{1/2}} \, dx$

17. $\displaystyle\int \frac{x \ln (1 + x^2)}{1 + x^2} \, dx$

18. $\displaystyle\int \frac{(\ln x)^4}{x} \, dx$

19. $\displaystyle\int \frac{e^{5x}}{1 + e^{5x}} \, dx$

20. $\displaystyle\int \frac{e^x - e^{-x}}{e^x + e^{-x}} \, dx$

21. $\displaystyle\int \frac{x}{x + 1} \, dx$

22. $\displaystyle\int \frac{x - 1}{x + 1} \, dx$

23. $\displaystyle\int x (1 + x)^{1/2} \, dx$

24. $\displaystyle\int x^2 (1 + x)^{1/2} \, dx$

Use integration by parts to compute each of the following:

25. $\displaystyle\int xe^{-x} \, dx$

26. $\displaystyle\int x \ln x \, dx$

27. $\displaystyle\int x^{10} \ln x \, dx$

28. $\displaystyle\int \frac{\ln x}{x^{10}} \, dx$

29. $\displaystyle\int \sqrt{x} \ln x \, dx$

30. $\displaystyle\int (\ln x)^2 \, dx$

31. $\int x^2 e^x \, dx$ (*Hint:* Let $f(x) = x$, $g'(x) = xe^x$. Consider Example 7.)

32. $\int \dfrac{x^3}{(1 + x^2)^{1/2}} \, dx$ (*Hint:* Let $f(x) = x^2$, $g'(x) = \dfrac{x}{(1 + x^2)^{1/2}}$. (Use substitution at the appropriate time.)

33. Let P be a polynominal function. Show that
$$\int e^x P(x) \, dx = e^x [P(x) - P'(x) + P''(x) - P'''(x) + \cdots]$$

34. Let f'' be continuous in $[a,b]$. Use integration by parts to show that
$$\int_a^b xf''(x) \, dx = bf'(b) - f(b) + f(a) - af'(a)$$

35. Assume that we wish to evaluate $\int (1/x) \, dx$ by integrating by parts. If we let $f(x) = (1/x)$ and $g'(x) = 1$, then $f'(x) = -1/x^2$ and $g(x) = x$, and we obtain
$$\int \frac{1}{x} \, dx = 1 + \int \frac{1}{x} \, dx$$

and thus, apparently, $1 = 0$. Explain this apparent paradox.

SECTION 2
TABLES OF INTEGRALS

In order to evaluate integrals, mathematicians have compiled tables of integrals. Many such tables appear in more general handbooks of mathematical tables where they occupy a fairly large portion of the book. There are also books, usually found in the reference section of a library (especially a technical library), whose sole content is the evaluation of integrals.

A short table of integrals appears on pages 359–70. Actually, it is a table of antiderivatives; to compute the integral, the appropriate limits of integration must be included. The table is set up according to the type if integrand being investigated. For completeness, the table includes functions that have not been studied in this book (for example, the trigonometric and hyperbolic functions and their inverses). For those of you who are familiar with these functions, the table provides additional information: curiously enough, the integration of some rational functions involves trigonometric and hyperbolic functions (see, for example, formulas 25, 29, and 31). To use the table, determine the category into which your integral fits, and look up that category in the table. At times, some ingenuity may be needed to change the term of an integral so that it is like one found in the table; substitution, integration by parts, and often simple algebra are used to make such an appropriate change.

EXAMPLE 1   Find $\int_0^1 x\sqrt{3 + 7x}\, dx$.

SOLUTION   We are dealing with an integrand involving "$a + bx$"; formulas (14)–(23) are of this type. Upon closer investigation we see that it is exactly like number 14, with $a = 3$ and $b = 7$ (and, of course, $u = x$). Thus,

$$\int_0^1 x(3 + 7x)^{1/2}\, dx = -\frac{2(6 - 21x)(3 + 7x)^{3/2}}{(15)(49)}\Big|_0^1$$

The remaining computations are left as an exercise.

EXAMPLE 2   Compute $\int_0^1 [e^x/(e^x + 2)]\, dx$.

SOLUTION   Formulas (125)–(133) cover exponential forms. Unfortunately, none of these fit the given problem. If we make the following change of variables

$$u = e^x + 2 \qquad du = e^x\, dx$$

the integral becomes

$$\int_3^{e+2} \frac{du}{u}$$

Of course, the antiderivative should be recognized as a logarithm function; to use the table, we see that formula (2) evaluates this integral. Again, the details are left as an exercise.

Problems   Compute each of the following integrals. Use the tables when necessary.

1. $\int \dfrac{dx}{x^2 - 5}$

2. $\int \dfrac{dx}{x^2 + 3x + 2}$

3. $\int \dfrac{dx}{(x + 1)(x^2 + 3x + 2)}$

4. $\int x^2(1 + x)^{1/2}\, dx$

5. $\int_1^2 \dfrac{dw}{w^2(4 - w^2)^{1/2}}$

6. $\int \dfrac{dx}{x^2 - 100}$

7. $\int \dfrac{x + 1}{x^2 - 100}\, dx$

8. $\int \dfrac{dx}{x(1 + x)^2}$

9. $\int \dfrac{dx}{x^3(1 + x)^2}$

10. $\int \dfrac{x}{(1 + x)^3}\, dx$

11. $\int \dfrac{x + x^2}{(1 + x)^3}\, dx$

12. $\int_1^2 \dfrac{z + 1}{z^2(2 + z)}\, dz$

SECTION 3

SIMPSON'S RULE

The problem of evaluating the definite integral relies on our cleverness at finding an antiderivative. There are times when all the tricks for integrating fail. What we shall investigate now is a numerical procedure for *estimating* integrals. It turns out that the procedure we shall outline, called **Simpson's rule**, gives a fairly accurate estimate for the integral.

Simpson's rule hinges on the fact that exactly one parabola can be drawn through three points (which do not lie on a common line) in the plane. What is done is to approximate the graph of a function f by a sequence of parabolic arcs, each drawn through three points on the graph. Then the area is found under each of the parabolic arcs, and these areas are added together. The first point used on the graph is $(a, f(a))$. In order to get groupings of three points on the graph, in addition to $(a, f(a))$, we need an *even* number of points on the graph of f (see Figure 6-1). The points are usually found by dividing the interval $[a,b]$ into n (with n an *even* integer) intervals of equal length, and taking the points on the graph over the subdivision points, as is done in Figure 6-1 (with $n = 6$). Notice that, since all the subintervals have equal length, they each have length $(b - a)/n$, which we will abbreviate by h (that is, $h = (b - a)/n$). The details of this construction are carried out in most calculus books oriented to physical science.* Here, we shall be content with the final result.

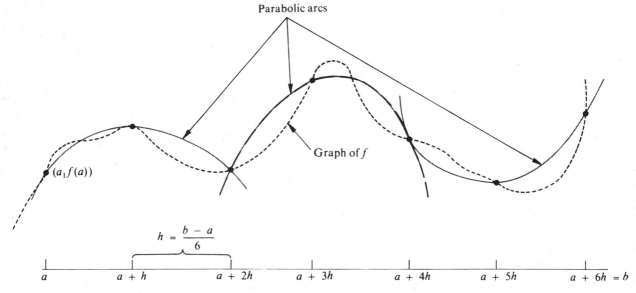

Parabolic arcs

$(a_1 f(a))$

Graph of f

$$h = \frac{b - a}{6}$$

a $a + h$ $a + 2h$ $a + 3h$ $a + 4h$ $a + 5h$ $a + 6h = b$

Figure 6-1

---

*See, for example, Johnson and Kiokemeister, *Calculus With Analytic Geometry*, 4th ed. (Boston: Allyn and Bacon, Inc., 1969).

Simpson's Rule. *Suppose that f is continuous on the interval [a,b]. Let n be an even integer, and let h = (b − a)/n. Then,*

$$\int_a^b f(x)\, dx \sim \frac{b-a}{3n}\, [f(a) + 4f(a+h)$$

$$+ 2f(a+2h) + 4f(a+3h) + 2f(a+4h)$$

$$+ 4f(a+5h) + \cdots + 2f(a+(n-2)h)$$

$$+ 4f(a+(n-1)h) + f(a+nh)]$$

Notice the pattern: $f(a)$ and $f(a+nh)$ (which is simply $f(b)$) have 1 as their coefficients. After the coefficient 1 for $f(a)$, the coefficients alternate 4, 2, 4, 2, 4, 2, etc., until $f(a+nh) = f(b)$, which again has the coefficient equal to 1.

EXAMPLE 1   Let $n = 8$. Estimate

$$\int_0^2 (1+x^3)^{1/2}\, dx$$

using Simpson's rule.

SOLUTION   Here, with $a = 0$, $b = 2$ and $n = 8$, we have $h = (b-a)/n = (2-0)/8 = \frac{1}{4}$. Thus,

$$a + h = \tfrac{1}{4}$$
$$a + 2h = \tfrac{1}{2}$$
$$a + 3h = \tfrac{3}{4}$$
$$a + 4h = 1$$
$$a + 5h = \tfrac{5}{4}$$
$$a + 6h = \tfrac{3}{2}$$
$$a + 7h = \tfrac{7}{4}$$
$$a + 8h = b = 2$$

In the accompanying table, we compute $f(a)$, $f(a+h)$, etc., to three-place accuracy.

x	0	$\frac{1}{4}$	$\frac{1}{2}$	$\frac{3}{4}$	1	$\frac{5}{4}$	$\frac{3}{2}$	$\frac{7}{4}$	2
$f(x)$	1.000	1.008	1.061	1.192	1.414	1.718	2.092	2.522	3.000

Hence,

$$\int_0^2 \sqrt{1+x^3}\, dx$$

$$\sim \tfrac{1}{12}[1.000 + 4(1.008) + 2(1.061) + 4(1.192) + 2(1.414)$$

$$+ 4(1.718) + 2(2.092) + 4(2.522) + 3.000]$$

$$= \tfrac{1}{12}[1.000 + 4(1.008 + 1.192 + 1.718 + 2.522) + 2(1.061$$

$$+ 1.414 + 2.092) + 3] = \tfrac{1}{12}[4.000 + 4(6.440) + 2(4.567)]$$

$$\sim 3.24$$

Use Simpson's Rule to approximate the following integrals.

1. $\displaystyle\int_0^1 \sqrt{1 - x^2}\, dx \qquad n = 4$

    (Use three-decimal-place accuracy.)

2. $\displaystyle\int_1^4 \frac{1}{x}\, dx \qquad n = 6$

    (*Note:* What you are approximating is ln 4.)

3. $\displaystyle\int_{-0.5}^{0.3} \frac{x}{1 + x}\, dx \qquad n = 8$

    (Use two-decimal-place accuracy.)

4. $\displaystyle\int_0^1 \frac{1}{1 + x^2}\, dx \qquad n = 4$

5. $\displaystyle\int_{-1}^3 \sqrt{4 + x^3}\, dx \qquad n = 4$

    (Use three-decimal-place accuracy.)

SECTION 4

THE INTEGRAL AS THE LIMIT OF A SUM. AREA BETWEEN CURVES

As a matter of convenience, the integral was introduced in our earlier work in terms of area under graphs. In applications, the integral usually arises only indirectly as an area. What often happens is that the description of many processes gives rise to the limit of a certain sum, and this limit of a sum turns out to be an integral. It is because of this that the physicist uses the integral in problems relating to such determinations as center of mass, moment of inertia, and work along a path. The next sections give geometrical examples of how the question of area and volume lead naturally to a certain limit of a sum which then turns out to be an integral. This interpretation of the integral in problems in economics and the life sciences is pursued in the "Supplementary Readings" at the end of this chapter.

Let us consider the graph of a function that is continuous on an interval $[a,b]$, or possibly piecewise continuous (in which case

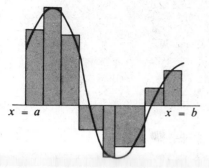

$x = a$ $x = b$

Figure 6-2

the graph can be constructed by pasting together side by side a finite number of graphs of functions which are continuous on closed intervals and do not blow up anywhere). A reasonable *approximation* to the signed area "under" the graph of f is a collection of rectangles whose bases are subintervals of the interval $[a,b]$ and whose heights are drawn from arbitrary points in each subinterval. See Figure 6-2.

Our intuition should lead us to believe that we get closer and closer approximations to the area if we take more and more rectangles (whose bases are getting smaller and smaller) and that the desired area should be, in some sense, the "limit of the sum" of all such rectangles.

To carry out this construction, divide the interval $[a,b]$ into n subintervals as follows: let $x_0 = a$, $x_n = b$, and insert the points $x_1, x_2, \ldots, x_{n-1}$ arbitrarily between $x_0 = a$ and $x_n = b$, where

$$a = x_0 < x_1 < x_2 < \cdots < x_n = b$$

Thus, we have constructed the (closed) subintervals

$$[x_0, x_1], [x_1, x_2], [x_2, x_3], \ldots, [x_{n-1}, x_n]$$

These will serve as the bases of our rectangles. Let

$$\Delta x_k = \text{length of } k\text{th subinterval} = x_k - x_{k-1}$$

that is,

$$\Delta x_1 = x_1 - x_0, \Delta x_2 = x_2 - x_1, \text{etc.}$$

Thus, Δx_k is the length of the base of the kth rectangle so constructed.

We next turn to the construction of the heights of the rectangles. Pick *any* point in the first subinterval $[x_0, x_1]$; call it x_1'. We shall use $f(x_1')$ as the height of our first rectangle. Hence, the *area* of the first rectangle is

$$f(x_1')\Delta x_1$$

Next pick any point x_2' in the second subinterval $[x_1, x_2]$ and use $f(x_2')$ as the height of the second rectangle, whose area is, therefore,

$$f(x_2')\Delta x_2$$

Continue in this way (see Figure 6-3). The sum of the areas of these rectangles is

$$f(x_1')\Delta x_1 + f(x_2')\Delta x_2 + \cdots + f(x_n')\Delta x_n$$

or, equivalently,

$$\sum_{k=1}^{n} f(x_k')\Delta x_k$$

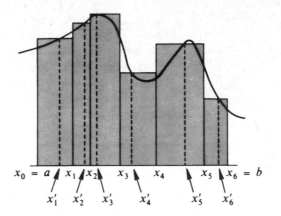

$$x_0 = a \quad x_1 \quad x_2 \quad x_3 \quad x_4 \quad x_5 \quad x_6 = b$$

$$x_1' \quad x_2' \quad x_3' \quad x_4' \quad x_5' \quad x_6'$$

Figure 6-3

This area is the desired approximation to the area under the graph. Now let δ be the largest of the numbers Δx_1, Δx_2, ..., Δx_n (that is, δ is the length of the *largest* subinterval). We get more and more rectangles (which, if all goes well, will be better and better approximations to the area under the graph) by taking n larger and larger and δ smaller and smaller. In fact,*

If f is **piecewise continuous in** $[a,b]$, **then as** $\delta \to 0$ **(in which case,** $n \to \infty$ **)**

$$\sum_{k=1}^{n} f(x_k')\Delta x_k \to \int_a^b f(x)\, dx$$

An elementary application of this procedure is to find the area between two curves. To simplify matters, we shall use as our curves graphs of functions, say f and g, that are continuous on some interval $[a,b]$ (see Figure 6-4). Divide the interval $[a,b]$ into

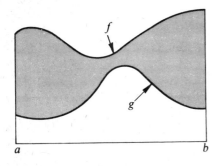

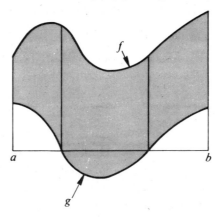

Figure 6-4

---

*This construction of area by rectangles is often used to *define* the integral.

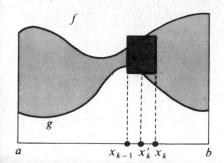

Figure 6-5

subintervals, and let the kth subinterval be $[x_{k-1},x_k]$. Pick any point x'_k in this subinterval and construct the rectangle in Figure 6-5. If the graph of f is above the graph of g in $[x_{k-1},x_k]$ (that is, if $f(x) \geq g(x)$ for $x \in [x_{k-1},x_k]$), then the area of the rectangle is

$$[f(x'_k) - g(x'_k)] \, \Delta x_k$$

If the graph of g is above that of f in $[x_{k-1},x_k]$, then the area is

$$[g(x'_k) - f(x'_k)]\Delta x_k$$

Both cases can be combined by taking as the area

$$|f(x'_k) - g(x'_k)| \, \Delta x_k$$

The sum of these areas is

$$\sum_{k=1}^{n} |f(x'_k) - g(x'_k)| \, \Delta x_k$$

and *the area between the curves is*

$$\lim_{\substack{n \to \infty \\ (\delta \to 0)}} \sum_{k=1}^{n} |f(x'_k) - g(x'_k)| \, \Delta x_k = \int_a^b |f(x) - g(x)| \, dx$$

EXAMPLE 1 Let $f(x) = x$ and $g(x) = x^2$. Find the area bounded by the graphs of f and g.

SOLUTION We begin by drawing a graph of the desired region. To find where the graphs intersect, we see that

$$f(x) = g(x) \qquad \text{if and only if} \qquad x = x^2$$

That is, if $x^2 - x = x(x - 1) = 0$. Hence, the graphs intersect at points whose first coordinates are 0 and 1. The line and parabola are graphed in Figure 6-6. From Figure 6-5 it is apparent that for $x \in [0,1]$, $f(x) \geq g(x)$, and thus the graph of f is above the graph of g on $(0,1)$, and the graphs intersect at 0 and 1.

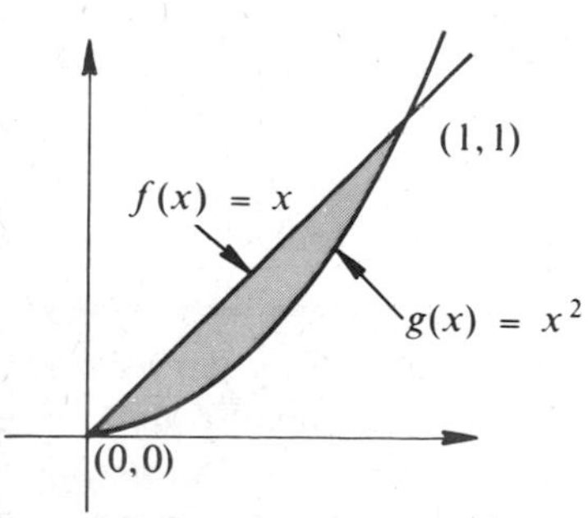

Hence, the area between the graphs of f and g is

$$\int_0^1 (f(x) - g(x)) \, dx = \int_0^1 (x - x^2) \, dx = \frac{1}{2} x^2 - \frac{1}{3} x^3 \Big|_0^1 = \frac{1}{6}$$

Figure 6-6

EXAMPLE 2 Let $f(x) = x^2 - 2$ and $g(x) = -x$. Find the area bounded by the parabola that is the graph of f, and the straight line that is the graph of g.

SOLUTION   The graphs are in Figure 6-7. To see where the graphs intersect, note that $f(x) = g(x)$ if and only if

$$x^2 - 2 = -x$$
$$x^2 + x - 2 = 0$$
$$(x - 1)(x + 2) = 0$$

Hence, the first coordinates of the intersection points are 1 and -2. It is clear from the geometry that the line given by $y = -x$ is above the parabola given by $y = x^2 - 2$ if x is in $[-2,1]$. Hence, the desired area is

$$\int_{-2}^{1} (g(x) - f(x))\, dx = \int_{-2}^{1} (-x - x^2 + 2)\, dx = \tfrac{9}{2}$$

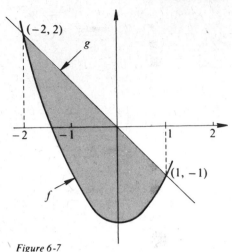

Figure 6-7

Problems   In problems 1–10, find the area bounded by the graphs of the following functions:

1. $f(x) = 2 - x^2$ and $g(x) = -x$
2. $f(x) = x^2$ and $g(x) = x^3$
3. $f(x) = x$ and $g(x) = \sqrt{x}$
4. $f(x) = x^2$ and $g(x) = \sqrt[3]{x}$
5. $f(x) = x$ and $g(x) = x^3$
6. $f(x) = x^2 - 4$ and $g(x) = 2$
7. $f(x) = x^4 - 3x^2$ and $g(x) = 6x^2$
8. $f(x) = x^3 - x^2$ and $g(x) = 6x$
9. $f(x) = 2x^3$ and $g(x) = 8x$
10. $f(x) = |x|$ and $g(x) = x^2$ (*Hint:* Consider symmetry.)
11. Find the area bounded by the curve $(x)^{1/2} + (y)^{1/2} = 1$ and the coordinate axes.
12. Find the area between the part of the graph of $f(x) = [x]$ and $g(x) = \tfrac{4}{9}x^2 - 2$ that is in the right half-plane.

SECTION 5

VOLUMES OF REVOLUTION

If we take a region in the xy-plane that is enclosed by curves and revolve this region about a straight line not passing through the region, the volume so generated is called a **volume of revolution**. Specifically, we shall take the region enclosed by the graph of some function f, continuous and nonnegative on $[a,b]$, the **x**-axis, and the lines $x = a$ and $x = b$, and we shall describe two types of volumes of revolution and methods for obtaining the volumes so generated: the "disk method" if the region is revolved about the **x**-axis, and the "method of cylindrical shells" if the region is of the form $0 \le a \le x \le b$, and is revolved about the **y**-axis.

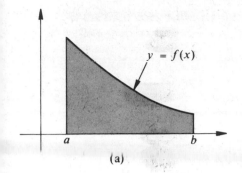

$y = f(x)$

a b

(a)

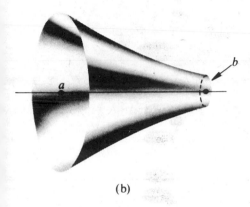

b

a

(b)

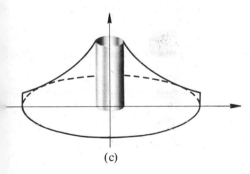

(c)

Figure 6-8

In Figure 6-8 the region sketched in (a) is revolved about the **x**-axis to get the volume described in (b); then the region in (a) is revolved about the **y**-axis to get the region sketched in (c).

First, we consider the problem of revolving about the **x**-axis and, in so doing, shall describe the **disk method**. We approach the problem in the same spirit as in the problem of finding the area bounded by curves. Divide the interval from $x = a$ to $x = b$ into N parts, each of length Δx_i. Let x_i' be any point in the ith subinterval. On the first subinterval, construct a rectangle of base Δx_1 and height $f(x_1')$; on the second subinterval, construct a rectangle of base Δx_2 and height $f(x_2')$; and so on. The kth rectangle will have base Δx_k and height $f(x_k')$ (see Figure 6-9). Now revolve about the **x**-axis the regions enclosed by these rectangles. We obtain a set of consecutive solid right circular cylinders, or disks, as shown in Figure 6-10(a), whose total volume approximates the volume we want.

If we isolate the kth disk, as shown in Figure 6-10(b), we see that it has radius $f(x_k')$ and height Δx_k; hence, volume $\pi \cdot (radius)^2 \cdot height = \pi[f(x_k')]^2 \Delta x_k$. The total volume of all the disks is

$$\pi \sum_{k=1}^{N} [f(x_k')]^2 \Delta x_k$$

Since f is continuous on $[a,b]$, so is f^2. Taking limits, we see that, as $N \to \infty$ (and $\delta \to 0$),

$$\pi \lim_{N \to \infty} \sum_{k=1}^{N} [f(x_k')]^2 \Delta x_k = \pi \int_a^b [f(x)]^2 dx$$

so we take the volume of revolution obtained by revolving about the **x**-axis the region bounded by the graph of the continuous

Figure 6-9

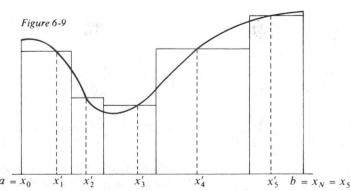

$a = x_0$ x_1' x_2' x_3' x_4' x_5' $b = x_N = x_5$

nonnegative function f, the **x**-axis, and the lines $x = a$ and $x = b$ to be this limit:

(6.1)
$$V = \pi \int_a^b [f(x)]^2 \, dx$$

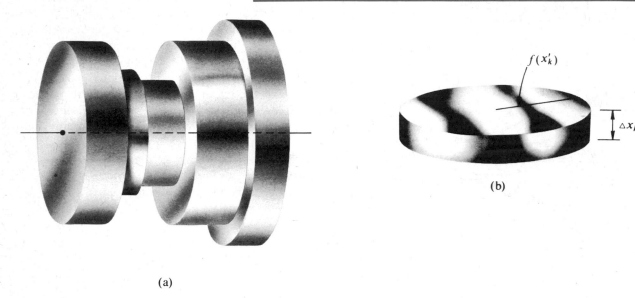

(b)

(a)

Figure 6-10

EXAMPLE 1   Let R and H be fixed positive numbers. Use the disk method to find the volume generated by revolving about the **x**-axis the region bounded by the graph of $f(x) = Rx/H$, the **x**-axis, and the lines $x = 0$ and $x = H$ (see Figure 6-11). Clearly, $[f(x)]^2 = R^2 x^2 / H^2$ is continuous on $[0, H]$ (in fact, it is continuous everywhere). The desired volume is

$$V = \pi \int_0^H \frac{R^2 x^2}{H^2} \, dx = \frac{\pi R^2}{H^2} \int_0^H x^2 \, dx$$

Applying the fundamental theorem, this becomes

$$V = \frac{\pi R^2}{H^2} \frac{x^3}{3} \Big|_0^H = \frac{\pi R^2}{H^2} \cdot \frac{H^3}{3} = \frac{1}{3} \pi R^2 H$$

If we examine the geometrical object obtained by revolving this triangle, we see that it is a right circular cone of radius R and height H.

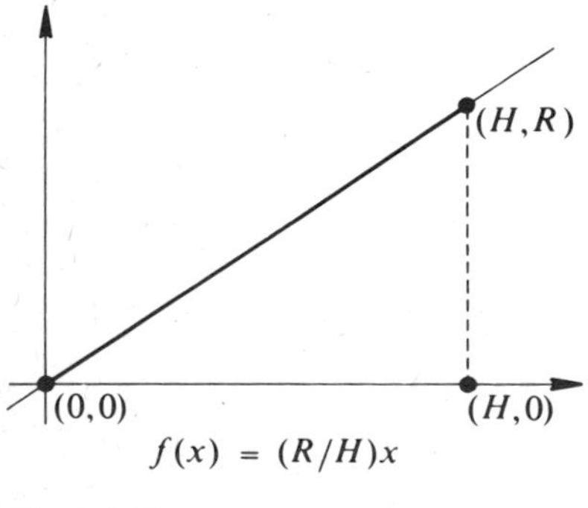

$f(x) = (R/H)x$

Figure 6-11

EXAMPLE 2

Find the volume generated by revolving about the **x**-axis the area bounded by the graphs of $f(x) = x$ and $g(x) = x^2$ (see Figure 6-12). The reader can easily verify that $f(x) = g(x)$ at $x = 0$ and $x = 1$; hence, the graphs intersect at $(0,0)$ and $(1,1)$. If we call V_2 the volume obtained by revolving about the **x**-axis the area under $f(x) = x$, from $x = 0$ to $x = 1$, and V_1 the volume obtained by revolving about the **x**-axis the area under $g(x) = x^2$, from $x = 0$ to $x = 1$, the desired volume V is given by $V = V_2 - V_1$. Using the disk method, we obtain

$$V = V_2 - V_1 = \pi \int_0^1 x^2\, dx - \pi \int_0^1 x^4\, dx$$

$$= \frac{\pi x^3}{3}\Big|_0^1 - \frac{\pi x^5}{5}\Big|_0^1 = \frac{2\pi}{15}$$

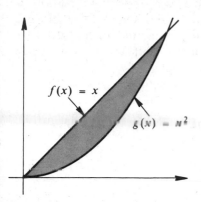

Figure 6-12

We observe, in passing, that we might have written

$$V = V_2 - V_1 = \pi \int_0^1 [f(x)]^2\, dx - \pi \int_0^1 [g(x)]^2\, dx$$

$$= \pi \int_0^1 \left\{ [f(x)]^2 - [g(x)]^2 \right\} dx$$

EXAMPLE 3

Find the volume obtained by revolving about the **x**-axis the region bounded by the graphs of $f(x) = x^2 + 1$, $g(x) = x^2$, $x - 0$ and $x = 1$.

SOLUTION

See Figure 6-13. Since $f(x) = x^2 + 1$ and $g(x) = x^2$, we see that

$$[f(x)]^2 = \underline{\hspace{3cm}}$$

and

$$[g(x)]^2 = \underline{\hspace{3cm}}$$

Call V_2 and V_1 the volumes obtained by revolving the graphs of f and g (over $[0,1]$), respectively, about the **x**-axis. Then, $V_2 - V_1$ is the desired volume, and

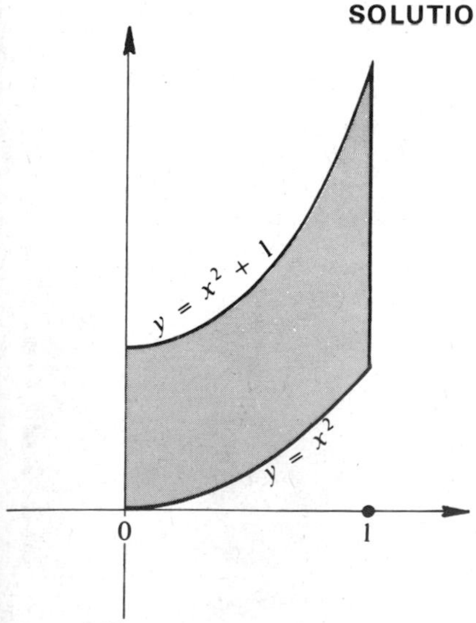

$$V_2 - V_1 = \pi \int_{\square}^{\square} [(\qquad) - (\qquad)]\, dx$$

$$= \pi \int_{\square}^{\square} \underline{\hspace{3cm}}\, dx$$

$$= \pi \left(\underline{\hspace{2cm}} \Big|_{\square}^{\square} \right)$$

$$= \underline{\hspace{3cm}}$$

Figure 6-13

Again, consider a function f that is continuous and nonnegative on $[a,b]$. We wish to take the area bounded by the graph of f, the **x**-axis, and the lines $x = a$ and $x = b$, and revolve this area about the **y**-axis, thus generating a volume; the method this time is called the **method of cylindrical shells**. To insure that our region does not cross the **y**-axis, we assume that $a \geq 0$.

Again, divide the interval from $x = a$ to $x = b$ into N parts, each of length Δx_i. Construct the same rectangles as were constructed for the disk method, but this time revolve them about the **y**-axis. This time we obtain a set of concentric cylindrical shells (a typical shell is drawn in Figure 6-14) whose total volume approximates the volume of revolution we want. Our plan is to take a limit of the volumes of these shells.

We next find the volume of a typical shell. In general, if we have two concentric cylinders of the same height h, the inside cylinder having radius r and the outside cylinder having radius R, the volume of the shell is just the difference in volumes of the cylinders; that is, $\pi R^2 h - \pi r^2 h = \pi h(R^2 - r^2)$ (see Figure 6-15). Comparing this with Figure 6-14, we see that

$$h = f(x_k')$$
$$R = x_k$$
$$r = x_{k-1}$$

Hence, the volume of the shell is

$$
\begin{aligned}
\pi h(R^2 - r^2) &= \pi f(x_k')(x_k^2 - x_{k-1}^2) \\
&= \pi f(x_k')(x_k + x_{k-1}) \cdot (x_k - x_{k-1}) \\
&= \pi f(x_k')(x_k + x_{k-1}) \cdot \Delta x_k
\end{aligned}
$$

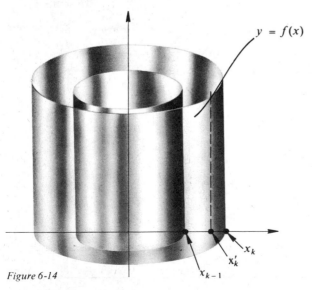

$y = f(x)$

x_k

x_k'

x_{k-1}

Figure 6-14

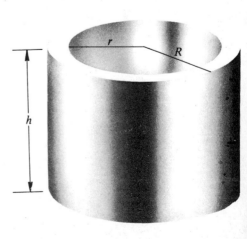

r

R

h

Figure 6-15

The sum of all such volumes (which is the approximation to the volume of revolution) is

$$\pi \sum_{k=1}^{N} f(x'_k)(x_k + x_{k-1})\Delta x_k$$

Let $N \to \infty$ (and $\delta \to 0$); we obtain

$$Volume = \pi \int_a^b f(x) \cdot 2x \, dx$$

That is,

(6.2) $$Volume = 2\pi \int_a^b xf(x) \, dx$$

EXAMPLE 4 Use the method of cylindrical shells to find the volume of the solid obtained by revolving about the **y**-axis the region bounded by the graph of $f(x) = x^2$, the **x**-axis and the lines $x = 0$ and $x = 1$. The desired volume is given by

$$2\pi \int_0^1 x \cdot x^2 \, dx = 2\pi \left. \frac{x^4}{4} \right|_0^1 = \frac{\pi}{2}$$

EXAMPLE 5 Find the volume obtained by revolving about the **y**-axis the region bounded by the graphs of $f(x) = x$, and $g(x) = x^2$.

SOLUTION As in Example 2, the graphs intersect at $(0,0)$ and $(1,1)$, and on $[0,1]$, $f(x) \geq g(x)$. If V_2 is the volume obtained by revolving about the **y**-axis the area under $y = f(x)$ from $x = 0$ to $x = 1$, and V_1 is the volume obtained by revolving about the **y**-axis the area under $y = g(x)$ from $x = 0$ to $x = 1$, then the desired volume is $V_2 - V_1$, which is

$$2\pi \int_0^1 x \cdot x \, dx - 2\pi \int_0^1 x \cdot x^2 \, dx = 2\pi (\tfrac{1}{3} - \tfrac{1}{4}) = \frac{\pi}{6}$$

In passing, we might observe that $V_2 - V_1$ can also be written as

$$2\pi \int_0^1 xf(x) \, dx - 2\pi \int_0^1 xg(x) \, dx$$

$$= 2\pi \int_0^1 x\{f(x) - g(x)\} \, dx$$

EXAMPLE 6 Consider the region given in Example 3 (that is, between the graphs of $f(x) = x^2 + 1$ and $g(x) = x^2$, with $0 \leq x \leq 1$); now revolve this region about the y-axis.

Complete: In $[0, 1]$, the inequality between f and g is

$$f(x) \boxed{} g(x)$$

Hence, the volume is

$$2\pi \int_{\Box}^{\Box} x(\underline{} - \underline{})\, dx = 2\pi \int_{\Box}^{\Box} \underline{}\, dx$$

$$= \underline{}$$

In conclusion, we should remark that in both the disk method and the method of cylindrical shells, we have tacitly assumed that our "approximations to the true volume" found by constructing disks or cylindrical shells actually do approach the true volume as the number of subdivisions becomes arbitrarily large; that is, we have tacitly assumed that the integrals in (6.1) and (6.2) give the correct volumes. Since we have no definition of the "true" volume, we have nothing with which to check our constructions. We simply assert that the volumes obtained are the correct volumes. The definition of volume involves the concept of *double integral* and is discussed later.

Answers   Example 3:

$$[f(x)]^2 = x^4 + 2x^2 + 1 \quad \text{and} \quad [g(x)]^2 = x^4$$

$$V_2 - V_1 = \pi \int_0^1 [(x^4 + 2x^2 + 1) - (x^4)]\, dx$$

$$= \pi \int_0^1 (2x^2 + 1)\, dx$$

$$= \pi \left(\frac{2}{3} x^3 + x \, \Big|_0^1 \right)$$

$$= \frac{5\pi}{3}$$

Example 6: $f(x) \geq g(x)$

$$2\pi \int_0^1 x(x^2 + 1 - x^2)\, dx = 2\pi \int_0^1 x\, dx = \pi$$

Problems   **1.**   Find the volume of revolution obtained by revolving about the **y**-axis the region bounded by the graph of $f(x) = -(H/R)(x - R)$ and the coordinate axes. Interpret your result geometrically (H and R are positive numbers).

2.   A semicircle of radius R and center $(0,0)$ has equation

$y = (R^2 - x^2)^{1/2}$. Obtain the volume of a sphere by revolving this semicircle about the **x**-axis.

3. Use the disk method to find the volumes obtained by revolving about the **x**-axis the region bounded by the graphs of
    a. $f(x) = (x)^{1/2}$, the **x**-axis, $x = 0$, and $x = 1$
    b. $f(x) = x + (x)^{1/2}$, the **x**-axis, $x = 0$, and $x = 1$
    c. $f(x) = (x)^{1/2}(x^2 + 1)^{1/4}$, the **x**-axis, $x = 0$, and $x = 1$
    d. $f(x) = |x^3|$, the **x**-axis, $x = -1$, and $x = 1$
    e. $f(x) = xe^{(x)^{1/2}}$, the **x**-axis, $x = 0$, and $x = 1$
    f. $f(x) = x(x^3 + 5)^\pi$, the **x**-axis, $x = 0$, and $x = 1$

4. Use the method of cylindrical shells to find the volume obtained by revolving about the **y**-axis the region bounded by the graphs of
    a. $f(x) = (x)^{1/2}$, the **x**-axis, $x = 0$, and $x = 1$
    b. $f(x) = x + (x)^{1/2}$, the **x**-axis, $x = 0$, and $x = 1$
    c. $f(x) = (16 - x^2)^{1/2}$, the **x**-axis, $x = 0$, and $x = 4$
    d. $f(x) = x(x^3 + 5)^\pi$, the **x**-axis, $x = 0$, and $x = 1$
    e. $f(x) = 1/(x^2 + 3)^{1/2}$, the **x**-axis, $x = 1$, and $x = (6)^{1/2}$
    f. $f(x) = xe^x$, the **x**-axis, $x = \ln 2$, and $x = \ln 3$

5. Find the volume obtained by revolving the area bounded by each of the following curves about the prescribed axis:
    a. $f(x) = x^2, g(x) = x^3$, each axis
    b. $f(x) = x^2, g(x) = 3$, the **x**-axis
    c. $f(x) = 4 - x^2, g(x) = 0$, the **x**-axis
    d. $f(x) = 4 - x^2, g(x) = -1, x \geq 0$, the **y**-axis
    e. $f(x) = x^{1/3}, g(x) = x^2$, each axis
    f. $f(x) = \frac{1}{8}(12x - x^3), g(x) = 2$, and $x = 0$, each axis.
        (*Hint:* Note that $f(x) = g(x)$ when $x = 2$)

6. Find the volume obtained by revolving the region bounded by the graph of $f(x) = 1 - x^2$ and the **x**-axis about the line $y = -1$.

SECTION 6

IMPROPER INTEGRALS

So far, we have studied the integral of continuous (or piecewise continuous) functions on closed intervals $[a,b]$. If there is some point in $[a,b]$ at which a function blows up, then our previous construction of the integral (either as an area or the limit of a sum) runs into trouble. For example, what meaning can we give the area under the graph if the graph blows up? One of the things we shall do in this section is to use the limit concept to obtain a meaning for the integral from a to b of f for *some* functions f (but, as we shall see, not all) that blow up on $[a,b]$. For convenience, we shall say that f has an **unbounded discontinuity** at c if f is *continuous near c, but*

$$\lim_{x \to c} f(x) = \pm \infty$$

We shall list a few functions of this type in the next example.

EXAMPLE 1   Consider $F(x) = x^{-(1/2)} = 1/(x)^{1/2}$, $0 < x \le 1$; $G(x) = 1/x$, $0 < x \le 1$; $H(x) = (x - 1)^{-(2/3)}$, $0 \le x < 1$. Observe that F and G are continuous at each point of $(0,1]$, but $1/(x)^{1/2}$ and $1/x$ become arbitrarily large as x tends to 0, with $x > 0$. Similarly, $H(x)$ is continuous on $[0,1)$, but as $x \to 1$ $(x < 1)$, $H(x)$ becomes arbitrarily large. Thus, F and G have an unbounded discontinuity at 0, and H has one at 1.

What, if any, meaning can be given

$$\int_0^1 F, \int_0^1 G, \quad \text{and} \quad \int_0^1 H?$$

Let us, for illustration, consider $\int_0^1 F$.

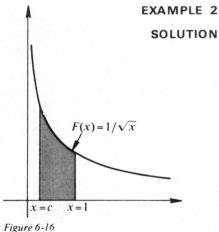

$F(x) = 1/\sqrt{x}$

$x = c$ $x = 1$

Figure 6-16

EXAMPLE 2   Let $F(x) = x^{-1/2}$, $0 < x \le 1$ (see Figure 6-16). Find $\int_0^1 F$.

SOLUTION   For any number c, with $0 < c < 1$, observe that F is continuous everywhere in the closed interval $[c,1]$. Hence, with respect to this interval, $\int_c^1 F$ exists, and, in fact, the fundamental theorem applies to this function in this interval. Consequently,

$$\int_c^1 x^{-(1/2)} \, dx = 2x^{1/2} \Big|_c^1 = 2 - 2(c)^{1/2}$$

Since $\lim_{c \to 0} [2 - 2(c)^{1/2}] = 2$, we say that the integral $\int_0^1 F$ exists *in the limiting sense;* we call the integral an **improper integral;** and we write

$$\int_c^1 \frac{1}{(x)^{1/2}} \, dx = 2$$

Generalizing, we have:

Let f be defined and continuous everywhere in $(a,b]$, and let f have a discontinuity at $x = a$ at which f is unbounded. Then we say that $\int_a^b f$ exists as an **improper integral** *and is equal to A if*

$$\lim_{\substack{c \to a \\ c > a}} \int_a^b f = A$$

and we write simply

$$\int_a^b f = \lim_{\substack{c \to a \\ c > a}} \int_c^b f$$

If the limit does not exist, we call $\int_a^b f$ **divergent**.

We shall write

$$\lim_{x \to a+} \; to \; mean \; \lim_{x \to a} with \; x > a$$

and

$$\lim_{x \to a-} \; to \; mean \; \lim_{x \to a} with \; x < a$$

Example 2 now can be rephrased to read:

$$\int_0^1 \frac{1}{\sqrt{x}} dx = \lim_{c \to 0+} \int_c^1 x^{-(1/2)} dx = \lim_{c \to 0+} \left(2x^{1/2} \Big|_c^1 \right)$$

$$= \lim_{c \to 0+} (2 - 2\sqrt{c}) = 2$$

EXAMPLE 3   We next investigate the improper integral (with $1 > c > 0$)

$$\int_0^1 \frac{1}{x} dx = \lim_{c \to 0+} \int_c^1 \frac{1}{x} dx = \lim_{c \to 0+} \left(\ln x \Big|_c^1 \right) = \lim_{c \to 0+} (-\ln c)$$

This limit does not exist. Hence, $\int_0^1 (1/x) \, dx$ is divergent.

We set a similar construction if f has an unbounded discontinuity at b. For this case, we have

*Let f be defined and continuous everywhere in [a,b), and let f have a discontinuity at $x = b$, at which f is unbounded. Then $\int_a^b f$ exists as an **improper integral** and is equal to B if*

$$\lim_{c \to b-} \int_a^c f = B$$

and we write

$$\int_a^b f = \lim_{c \to b-} \int_a^c f$$

If the limit does not exist, we say the improper integral **diverges**.

EXAMPLE 4   Investigate the improper integral

$$\int_0^1 (x - 1)^{-2/3} \, dx$$

SOLUTION   To emphasize some of our previous remarks, the function $G(x) = (x - 1)^{-(2/3)}$ is continuous on each interval $0 \le x < c$ for c fixed and $c < 1$. Hence, we can apply the fundamental theorem to obtain

$$\int_0^c (x-1)^{-(2/3)}\,dx = 3(x-1)^{1/3}\Big|_0^c = 3(c-1)^{1/3} + 3$$

and we see that

$$\int_0^1 (x-1)^{-(2/3)}\,dx = \lim_{c\to 1-}\int_0^c (x-1)^{-(2/3)}\,dx$$

$$= \lim_{c\to 1-}[3(c-1)^{1/3}+3] = 3$$

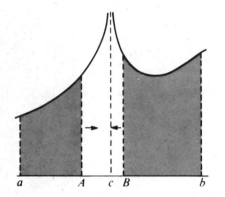

Figure 6-17

Another possibility is that the function is unbounded at a point c in the interior of the interval; that is, at a point c for which $a < c < b$. For this, we have

Let f be defined and continuous at each point of the closed interval $[a,b]$ except at $x = c$, with $a < c < b$, where f is unbounded (see Figure 6-17). Then we say that $\int_a^b f$ exists as an improper integral if

(6.3) $\displaystyle\lim_{A\to c-}\int_a^A f$ exists

(6.4) $\displaystyle\lim_{B\to c+}\int_B^b f$ exists

and we define

$$\int_a^b f = \lim_{A\to c-}\int_a^A f + \lim_{B\to c+}\int_B^b f$$

EXAMPLE 5   Let $f(x) = x^{-(2/3)}$, $-1 \le x \le 1$ with $x \ne 0$ (see Figure 6-18). Clearly f is continuous at each x in $[-1, 1]$ except at $x = 0$, where it is unbounded. Investigate

$$\int_{-1}^1 f = \int_{-1}^1 x^{-2/3}\,dx$$

SOLUTION   Since f is defined and continuous at all points x in $[-1, A] \cup [B, 1]$, $A < 0 < B$, the definite integrals exist over these intervals and the fundamental theorem applies over these intervals. Hence,

$$\int_{-1}^A x^{-(2/3)}\,dx = 3x^{1/3}\Big|_{-1}^A = 3A^{1/3} + 3, \qquad A < 0$$

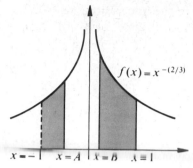

$f(x) = x^{-(2/3)}$

$x = -1$ $x = A$ $x = B$ $x = 1$

Figure 6-18

and
$$\int_B^1 x^{-(2/3)}\, dx = 3x^{1/3}\Big|_B^1 = 3 - 3B^{1/3}, \qquad B > 0$$

therefore,
$$\int_{-1}^1 x^{-(2/3)}\, dx = \lim_{A \to 0-} (3A^{1/3} + 3) + \lim_{B \to 0+} (3 - 3B^{1/3}) = 6$$

It is possible, of course, to have an improper integral of a so-called mixed type; that is, one with discontinuities at more than one point, with the function unbounded at each of the discontinuities. For such problems, the integral must exist in the sense of each definition that applies. We shall illustrate.

EXAMPLE 6   We wish to evaluate $\int_0^1 f$, where

$$f(x) = \begin{cases} x^{-(1/2)}, & 0 < x < \tfrac{1}{2} \\ (1 - x)^{-(1/2)}, & \tfrac{1}{2} \le x < 1 \end{cases}$$

SOLUTION   Observe that this function *is* continuous at $x = \tfrac{1}{2}$. It has an unbounded discontinuity at $x = 0$ and $x = 1$, the endpoints of the interval (see Figure 6-19).

We take any convenient point of continuity of f that is between the two "bad" points, 0 and 1. In the manner in which our particular function is defined, $x = \tfrac{1}{2}$ is a reasonable point to choose. Then,

$$\int_0^1 f = \int_0^{1/2} x^{-(1/2)}\, dx + \int_{1/2}^1 (1 - x)^{-(1/2)}\, dx$$

We see that for $c > 0$,

$$\int_0^{1/2} x^{-(1/2)}\, dx = \lim_{c \to 0+} \int_c^{1/2} x^{-(1/2)}\, dx$$

$$= \lim_{c \to 0+} \left(2x^{1/2}\, \Big|_c^{1/2}\right)$$

$$= \lim_{c \to 0+} \left(\frac{2}{(2)^{1/2}} - 2(c)^{1/2}\right) = \frac{2}{(2)^{1/2}}$$

With $c < 1$,

$$\int_{1/2}^1 (1 - x)^{-(1/2)}\, dx = \lim_{c \to 1-} \int_{1/2}^c (1 - x)^{-(1/2)}\, dx$$

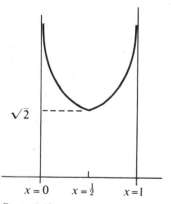

$\sqrt{2}$

$x = 0$ $x = \tfrac{1}{2}$ $x = 1$

Figure 6-19

Applying the rule of substitution to the integral on the right, we let $g(x) = 1 - x$, so that $dg = -dx$; also, $g(\frac{1}{2}) = \frac{1}{2}$, $g(c) = 1 - c$; thus,

$$\int_{1/2}^{c} (1 - x)^{-(1/2)}\, dx = -\int_{1/2}^{1-c} g^{-(1/2)}\, dg$$

$$= -2g^{1/2}\, \big|_{1/2}^{1-c} = -2(1 - c)^{1/2} + \frac{2}{(2)^{1/2}}$$

Hence, with $c < 1$,

$$\int_{1/2}^{1} (1 - x)^{-(1/2)}\, dx = \lim_{c \to 1-} \left[-2(1 - c)^{1/2} + \frac{2}{(2)^{1/2}} \right] = \frac{2}{(2)^{1/2}}$$

Consequently,

$$\int_{0}^{1} f = \frac{2}{(2)^{1/2}} + \frac{2}{(2)^{1/2}} = \frac{4}{(2)^{1/2}}$$

EXAMPLE 7   Let

$$g(x) = \begin{cases} x^{-1}, & 0 < x \le 1 \\ (2 - x)^{-1/2}, & 1 \le x < 2 \end{cases}$$

Investigate $\int_{0}^{2} g$.

SOLUTION   Verify that g is continuous everywhere except at $x = 0$ and $x = 2$. Using $x = 1$ as a convenient point between the problem points 0 and 2, we have

$$\int_{0}^{2} g = \int_{0}^{1} g + \int_{1}^{2} g$$

$$= \int_{0}^{1} x^{-1}\, dx + \int_{1}^{2} (2 - x)^{-(1/2)}\, dx$$

provided each of the integrals

$$\int_{0}^{1} x^{-1}\, dx, \qquad \int_{1}^{2} (2 - x)^{-(1/2)}\, dx$$

exists. But, from Example 3,

$$\int_{0}^{1} x^{-1}\, dx = \int_{0}^{1} \frac{1}{x}\, dx$$

does not exist, and so $\int_{0}^{2} g$ does not exist (that is, is divergent).

Problems In problems 1–14, compute the integral (if it exists).

1. $\displaystyle\int_0^1 \frac{1}{x^{1/5}}\,dx$

2. $\displaystyle\int_0^4 \frac{1}{x(x)^{1/2}}\,dx$

3. $\displaystyle\int_{-1}^1 \frac{1}{x^{1/5}}\,dx$

4. $\displaystyle\int_0^1 \frac{1}{x^{0.99}}\,dx$

5. $\displaystyle\int_0^1 \frac{1}{r^{1.1}}\,dx$

6. $\displaystyle\int_0^1 \frac{1}{(1-x)^{1/2}}\,dx$

7. $\displaystyle\int_0^4 \frac{1}{\sqrt[3]{4-x}}\,dx$

8. $\displaystyle\int_0^1 \ln x\,dx$

9. $\displaystyle\int_0^1 x \ln x\,dx$

(*Hint*· Assume $\lim_{x\to 0} x \ln x = 0$.)

10. $\displaystyle\int_{-1}^1 \frac{1}{\sqrt{|x|}}\,dx$

11. $\displaystyle\int_0^1 \frac{x}{(1-x^2)^{1/2}}\,dx$

12. $\displaystyle\int_1^2 \frac{x}{(x^2-1)^{1/2}}\,dx$

13. $\displaystyle\int_0^1 \frac{x}{1-x^2}\,dx$

14. $\displaystyle\int_0^1 \exp\left(-\tfrac{1}{2}\ln x\right)dx$

15. For what values of p does $\int_0^1 (1/x^p)\,dx$ exist? Evaluate the integral for the appropriate p.

16. Discuss the following calculation:

$$\int_{-1}^1 \frac{1}{x^2}\,dx = \int_{-1}^1 x^{-2}\,dx = \frac{-1}{x}\bigg|_{-1}^1 = -[1-(-1)] = -2$$

SECTION 7

SOME OTHER TYPES OF IMPROPER INTEGRALS

In the previous section, the integrals were improper because of difficulties in the function. We next consider integrals which are improper because of the limits of integration; that is, because of the region over which integration takes place. Specifically, we shall define integrals of the form

$$\int_a^\infty f(x)\,dx, \qquad \int_{-\infty}^b f(x)\,dx, \qquad \int_{-\infty}^\infty f(x)\,dx$$

Such integrals arise in probability and statistics, and in physics. If f has an integral (proper or improper) over *every* interval $[a, T]$, then we shall take

$$\int_a^\infty f(x)\,dx = \lim_{T\to\infty} \int_a^T f(x)\,dx$$

if the limit on the right exists. If the limit on the right does not exist, then $\int_a^\infty f(x)\, dx$ does not exist, and is said to be **divergent**; similarly,

$$\int_{-\infty}^{b} f(x)\, dx \;=\; \lim_{s \to -\infty} \int_{s}^{b} f(x)\, dx$$

if the limit on the right exists. If the limit on the right does not exist, then $\int_{-\infty}^{b} f(x)\, dx$ does not exist, and is said to be divergent.

EXAMPLE 1   Investigate $\int_1^\infty (1/x^2)\, dx$.

SOLUTION
$$\int_1^\infty \frac{1}{x^2}\, dx \;=\; \lim_{T \to \infty} \int_1^T x^{-2}\, dx$$

The function f given by $f(x) = x^{-2}$ is continuous on $[1, T]$ for every $T > 1$. Hence, on the interval $[1, T]$, the fundamental theorem applies, and

$$\int_1^T x^{-2}\, dx \;=\; -\frac{1}{x}\Big|_1^T \;=\; -\frac{1}{T} + 1$$

Consequently,

$$\int_1^\infty \frac{1}{x^2}\, dx \;=\; \lim_{T \to \infty} \left(-\frac{1}{T} + 1\right) = 1$$

EXAMPLE 2   Investigate $\int_1^\infty (1/x)\, dx$.

SOLUTION   Let $f(x) = 1/x$. Then,

$$\int_1^\infty f(x)\, dx \;=\; \lim_{T \to \infty} \int_1^T \frac{1}{x}\, dx$$

The function f is continuous on $[1, T]$ for every $T > 1$. Hence, the fundamental theorem applies, and

$$\int_1^T \frac{1}{x}\, dx \;=\; \ln x\, \Big|_1^T \;=\; \ln T - \ln 1 \;=\; \ln T$$

Note that $\lim_{T \to \infty} \ln T$ does not exist. Hence,

$$\int_1^\infty \frac{1}{x}\, dx \;=\; \lim_{T \to \infty} \ln T$$

does not exist; that is,

$$\int_1^\infty f$$

is divergent.

EXAMPLE 3 Define the function $G(x)$ by

$$G(x) = \begin{cases} x^{-1/2}, & 0 < x \le 1 \\ x^{-2}, & 1 < x \end{cases}$$

Investigate $\int_0^\infty G(x)\, dx$.

SOLUTION The integral is of a mixed type because of the upper limit and because G has an unbounded discontinuity at $x = 0$. We write

$$\int_0^\infty G(x)\, dx = \int_0^1 G(x)\, dx + \int_1^\infty G(x)\, dx$$

$$= \int_0^1 x^{-1/2}\, dx + \int_1^\infty \frac{1}{x^2}\, dx$$

provided each integral on the right exists. By previous examples (see Example 2 of section 6 and Example 1 of this section),

$$\int_0^\infty G(x)\, dx = 3$$

EXAMPLE 4 Find $\int_{-\infty}^0 e^x\, dx$.

SOLUTION We must consider

$$\lim_{s \to -\infty} \int_s^0 e^x\, dx$$

The function given by e^x is continuous on $[s,0]$ for every $s < 0$. Hence, the fundamental theorem applies, and

$$\int_s^0 e^x\, dx = e^x \big|_s^0 = 1 - e^s$$

Consequently,

$$\int_{-\infty}^0 e^x\, dx = \lim_{s \to -\infty} (1 - e^s) = 1$$

since $\lim_{s \to -\infty} e^s = 0$. (Why?)

Assume that $\int_a^\infty f(x)\,dx$ and $\int_{-\infty}^a f(x)\,dx$ each exists. Then

$$\int_{-\infty}^\infty f(x)\,dx = \int_{-\infty}^a f(x)\,dx + \int_a^\infty f(x)\,dx$$

If either of the integrals on the right is divergent, then so is the integral on the left.

Remark The value of a is arbitrary, and chosen for convenience. Often, 0 is chosen instead.

EXAMPLE 5 Investigate

$$\int_{-\infty}^\infty e^{-|x|}\,dx$$

SOLUTION First, observe that

$$
\begin{aligned}
e^{-|x|} &= e^{-x} && \text{if} && x \geq 0 \\
&= e^x && \text{if} && x < 0
\end{aligned}
$$

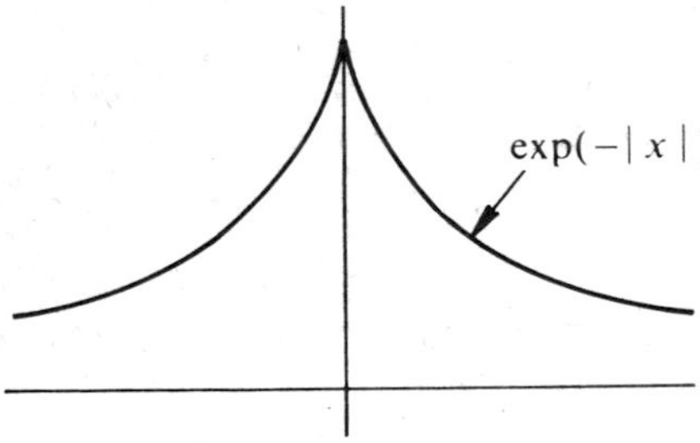

$\exp(-|x|)$

and the function is continuous everywhere (see Figure 6-20). Hence,

$$\int_{-\infty}^\infty e^{-|x|}\,dx = \lim_{s \to -\infty} \int_s^0 e^x\,dx + \lim_{T \to \infty} \int_0^T e^{-x}\,dx$$

Figure 6-20

By Example 4,

$$\int_{-\infty}^0 e^x\,dx = 1$$

Also,

$$\lim_{T \to \infty} \int_0^T e^{-x}\,dx = \lim_{T \to \infty} \left(-e^{-x}\,\Big|_0^T\right) = \lim_{T \to \infty} \left(-e^{-T} + 1\right) = 1$$

Thus,

$$\int_{-\infty}^\infty e^{-|x|}\,dx = 2$$

EXAMPLE 6 In so-called continuous models in probability, it is possible to associate the events (that is, the outcomes) of interest in an experiment with the real line. For example, suppose our experiment is to light at the same time a large number of light bulbs; and the particular event of interest is *a bulb burns out at the end of t hours.* We can identify this event with the number t itself.* If we now

*Probabilists call the function which makes this identification a **random variable**.

look at the number 7.1 on the line, we associate this with the event "a bulb burns out at the end of 7.1 hours."

Associated with our experiment and set of events is a function f with the following properties:

(6.5) $$f(x) \geq 0, \quad \text{all } x \text{ in } \mathbf{R}$$

(6.6) $$\int_{-\infty}^{\infty} f(x)\,dx = 1$$

Such a function f is called a **density function** and is used to measure the probability of an event. Specifically, if f is the density function associated with an experiment and its set of events (with the set of events interpreted as numbers on the line), then

$$\left(\begin{matrix}\textit{The probability that the outcome of} \\ \textit{the experiment is a number} \leq t\end{matrix}\right) = \int_{-\infty}^{t} f(x)\,dx$$

and

$$\left(\begin{matrix}\textit{The probability that the outcome of} \\ \textit{the experiment is a number in the} \\ \textit{interval } [a,b]\end{matrix}\right) = \int_{a}^{b} f(x)\,dx$$

For example, in our light-bulb experiment, given the appropriate f for this experiment,

$$\left(\begin{matrix}\textit{The probability that a bulb burns} \\ \textit{out in the first 7.1 hours}\end{matrix}\right) = \int_{-\infty}^{7.1} f(x)\,dx$$

or

$$\left(\begin{matrix}\textit{The probability that a bulb burns} \\ \textit{out between the 7.1st hour and the} \\ \textit{10th hour}\end{matrix}\right) = \int_{7.1}^{10} f(x)\,dx$$

Note that here, $\int_{-\infty}^{\infty} f(x)\,dx = 1$ simply says that with probability one, a bulb must burn out at some time.

One important density function which you may have encountered is the *normal density function N with mean u and standard deviation s > 0* given by

$$N(x) = \frac{1}{\sigma(2\pi)^{1/2}} \exp\left(\frac{-(x-u)^2}{2\sigma^2}\right)$$

Thus, $$\int_{-\infty}^{\infty} N(x)\,dx = 1$$

although the verification of this is beyond the scope of this book.

Problems   In problems 1–15, evaluate each of the integrals whenever they exist. Assume that you already know that $\lim_{x \to \infty} x^n e^{-x} = 0$ for all positive integers n.

1. $\displaystyle\int_1^\infty \frac{1}{x^3}\, dx$ 2. $\displaystyle\int_1^\infty \frac{1}{x^{1.01}}\, dx$

3. $\displaystyle\int_0^\infty xe^{-x}\, dx$ 4. $\displaystyle\int_0^\infty x^2 e^{-x}\, dx$

5. $\displaystyle\int_{-\infty}^0 xe^x\, dx$ 6. $\displaystyle\int_2^\infty \frac{x}{(x^2 - 1)^{10}}\, dx$

7. $\displaystyle\int_0^\infty \frac{1}{\sqrt{x}}\, dx$ 8. $\displaystyle\int_{-1}^\infty \frac{2x + 1}{(x^2 + x + 10)^\pi}\, dx$

9. $\displaystyle\int_{-\infty}^\infty x \exp(-|x|)\, dx$ 10. $\displaystyle\int_{-\infty}^\infty x \exp(-x^2)\, dx$

11. $\displaystyle\int_1^\infty \frac{1}{x^{0.99}}\, dx$ 12. $\displaystyle\int_{-\infty}^4 \frac{1}{(5 - x)^2}\, dx$

13. $\displaystyle\int_{-\infty}^0 \frac{1}{(2 - x)^{1/2}}\, dx$ 14. $\displaystyle\int_0^\infty e^{-x/2}\, dx$

15. $\displaystyle\int_0^\infty \frac{1}{(2x + 1)^2}\, dx$ (*Hint:* Use substitution.)

16. Let $f(x)$ be defined by

$$f(x) = \begin{cases} 1/x^2, & x \geq 1 \\ 1, & -1 \leq x < 1 \\ e^{x+1}, & x < -1 \end{cases}$$

Sketch the graph. Then, if it exists, evaluate

$$\int_{-\infty}^{+\infty} f(x)\, dx$$

17. For what numbers p does $\int_1^\infty (1/x^p)\, dx$ exist? Evaluate the integral for those p.

18. We have already seen that $\int_1^\infty (1/x)\, dx$ is divergent. Show that $\int_a^\infty (1/x)\, dx$ is divergent no matter what real number $a > 0$ is chosen. What happens if $a = 0$? If $a < 0$?

19. Show that

$$\int_2^\infty \frac{1}{x \ln x} \, dx$$

is divergent, but

$$\int_2^\infty \frac{1}{x(\ln x)^2} \, dx$$

exists, and find its value. Then find the values of p such that

$$\int_2^\infty \frac{1}{x(\ln x)^p} \, dx$$

exists, and the value of the integral for these p.

20. A painter with a good mathematical background is given the task of painting the region bounded by the graph of $f(x) = (1/x)$ and the x-axis, with the additional requirement that $x \geq 1$. Since $\int_1^\infty (1/x)\,dx$ is divergent, the painter realizes the task is futile, and he decides to revolve the graph about the x-axis, taking as his volume $\pi \int_1^\infty [f(x)]^2 dx$. Show that this "volume" is finite. Hence, all our painter has to do is pour the paint into the container formed by revolving $f(x) = (1/x)$, $x \geq 1$, about the x-axis, and then take a cross section; and, in so doing, he has painted the required region (or has he?).

21. Let g be the function given by

$$g(x) = \begin{cases} (1/a)\,e^{-x/a}, & x > 0 \ (a > 0 \text{ and fixed}) \\ 0, & x \leq 0 \end{cases}$$

It is easy to see that $g(x) \geq 0$, all x. Show that g is a density function by verifying that

$$\int_{-\infty}^\infty g(x)\,dx = 1$$

(*Remark:* g is said to determine an **exponential distribution**.)

22. For a constant $a > 0$, define

$$h(x) = \begin{cases} ax^{a-1}, & 0 < x < 1 \\ 0, & \text{elsewhere} \end{cases}$$

Show that h is a density function.

SECTION 8
A DIFFERENTIAL EQUATION OF INTEREST. SEPARABLE VARIABLES

Historically, one of the most powerful mathematical tools used by the physicist in his efforts to describe physical phenomena has been that of differential equations; that is, equations in which derivatives occur. Vast numbers of technical articles and texts at all levels have been devoted exclusively to a systematic investigation of such equations. We shall confine ourselves now to the investigation of a single such equation; namely, with $f(x)$ and $g(y)$ continuous functions of x and y, respectively, we consider

$$(6.7) \qquad g(y)\,\frac{dy}{dx} = f(x)$$

A solution to such an equation is a differentiable function u which satisfies (6.7); that is, a differentiable function u for which

$$(6.8) \qquad g[u(x)]\,D\,u(x) = f(x)$$

We wish to determine this function u. Assume that (6.7) has $u(x)$ as its solution, and thus (6.8) holds. If we formally antidifferentiate both sides of (6.8), we have

$$\int g[u(x)]u'(x)\,dx = \int f(x)\,dx + c$$

where c is the arbitrary constant arising through antidifferentiation. Applying substitution, we have with $y = u(x)$

$$\int g[u(x)]u'(x)\,dx = \int g(y)\,dy$$

Since $f(x)$ is continuous, $\int f(x)\,dx$ exists; namely, $F(x) = \int_a^x f$ is an antiderivative of $f(x)$. Hence, the solution to (6.7) must have the form

$$(6.9) \qquad \int g(y)\,dy = \int f(x)\,dx + c$$

The analysis first given suggests the following informal technique:

$$g(y)\,\frac{dy}{dx} = f(x)$$

$$g(y)\,dy = f(x)\,dx$$

(Note that the variables have now been "separated," since everything involving x is on one side and everything involving y is on the other.)

Hence, taking antiderivatives,

$$\int g(y)\,dy = \int f(x)\,dx + c$$

EXAMPLE 1   Solve the differential equation

$$\frac{dy}{dx} = \frac{x}{y}$$

SOLUTION   Since

$$\frac{dy}{dx} = \frac{x}{y}$$

we have $y \, dy = x \, dx$

and $\int y \, dy = \int x \, dx$

Thus, $\dfrac{y^2}{2} = \dfrac{x^2}{2} + c$

Multiplying by 2 and letting $k = 2c$, we have

$$y^2 = x^2 + k$$

Of course, this yields two solutions, $y = (x^2 + k)^{1/2}$ and $y = -(x^2 + k)^{1/2}$. Each of these functions is defined when $x^2 + k \geq 0$ and differentiable when $x^2 + k > 0$. Hence, the specific nature of the solutions depends on the value of the constant k. Some graphs for various values of k are drawn in Figure 6-21.

If we differentiate implicitly with respect to x the equation $y^2 = x^2 + k$, we obtain $2y \, D \, y = 2x$, or $dy/dx = x/y$, as expected.

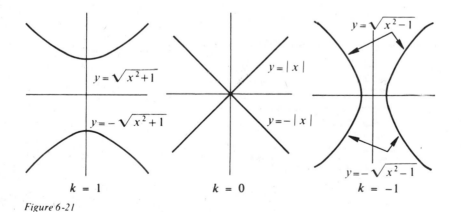

$y = \sqrt{x^2+1}$

$y = -\sqrt{x^2+1}$

$k = 1$

$y = |x|$

$y = -|x|$

$k = 0$

$y = \sqrt{x^2-1}$

$y = -\sqrt{x^2-1}$

$k = -1$

Figure 6-21

EXAMPLE 2   Find y for which

$$y'(x) = x \sqrt{y}$$

given that $y = 1$ when $x = 0$.

SOLUTION   Rewriting this equation, we obtain

$$\frac{dy}{dx} = x \sqrt{y}$$

so that

$$y^{-1/2} dy = x \, dx$$

and thus

$$\int y^{-1/2} dy = \int x \, dx + c, \qquad y > 0$$

or,

$$2(y)^{1/2} = \frac{x^2}{2} + c, \qquad y > 0$$

Since $y = 1$ when $x = 0$, we have $2(1)^{1/2} = 0^2/2 + c$; hence, $c = 2$, and our final solution is

$$2(y)^{1/2} = \frac{x^2}{2} + 2, \qquad y > 0$$

EXAMPLE 3   Find y if

$$D y = \frac{(y^2 + 1)^{1/2}}{10 y}$$

SOLUTION

$$\frac{dy}{dx} = \frac{(y^2 + 1)^{1/2}}{10 y}$$

$$\frac{10 y}{(y^2 + 1)^{1/2}} \, dy = dx$$

Hence,

$$\int \frac{10 y}{(y^2 + 1)^{1/2}} \, dy = \int 1 \, dx + c = x + c$$

By using substitution, the integral on the left is $10(y^2 + 1)^{1/2}$ (we omit the constant which appears, since it already appears on the right side of our solution). Hence, the solution is implicitly given by

$$10(y^2 + 1)^{1/2} = x + c$$

EXAMPLE 4   Let A be any number. Solve

$$y' = A y$$

That is, find a function y which is directly proportional to its derivative.*

SOLUTION

$$\frac{dy}{dx} = A y$$

---

*This is simply the problem of exponential growth or decay, now viewed as a special case of a separable variables differential equation.

Thus, for $y \neq 0$,

$$\frac{dy}{y} = A \, dx$$

$$\int \frac{1}{y} \, dy = \int A \, dx + k$$

Consequently, $\ln |y| = Ax + k$, and

$$\exp (\ln |y|) = \exp (Ax + k) = e^{Ax} \cdot e^{k}$$

Since $\exp (\ln |y|) = |y|$, this simplifies to

$$|y| = e^{k} \cdot e^{Ax}$$

No matter what value is chosen for k, we see that $e^{k} > 0$. Let $c = e^{k}$; then $c > 0$. Also, no matter what A is, or what value of x we choose, $e^{Ax} > 0$. Hence, $ce^{Ax} > 0$. Thus, $y = \pm ce^{Ax}$. Summarizing, *the solutions to the differential equation $y' = Ay$ are*

$$y = ce^{Ax}$$

where c is any real number.

EXAMPLE 5   In an early work in quantitative theory in psychology, Gustav Fechner in his *Elemente der Psychophysik* (printed in 1860) attempted to find a mathematical relationship between the measure of a stimulus R (to use our familiar English abbreviations instead of Fechner's symbols) and the resulting sensation S. If R is the original stimulus on an organism, then Fechner called dR a small change in the stimulus.* Thus, the relative increase in R is dR/R. Similarly, denote a small change in stimulus S by dS. Fechner based his mathematical formulas on two assumptions:

1. dS is constant so long as the relative increase in R (that is, dR/R) is constant (this is called **Weber's Law**); and
2. dS and dR vary proportionately so long as they are small.

This led him to the formula (that is, the differential equation)

$$dS = c \, \frac{dR}{R}$$

---

*The use of the derivative-like symbol dR to denote a small change, or increment, in R is common in application. It comes from the fact that df/dx is the limit of $(f(x + h) - f(x))/h$, and that for h a small number, this difference quotient represents a small change in f divided by a small change in x. Thus, df/dx is interpreted as a small change in f over a small change in x.

where c is a constant that depends on the units chosen for R and S. The solution to this equation is

$$S = c \ln R + k$$

for a certain constant k. Assume that when the sensation is 0, the stimulus is r. We then obtain

$$0 = c \ln r + k$$

That is,
$$k = -c \ln r$$

Thus,
$$S = c \ln R - c \ln r$$
$$= c (\ln R - \ln r)$$
$$= c \ln \frac{R}{r}$$

Problems Solve each of the following differential equations:

1. $y' = xy$

2. $Dy = -\dfrac{y}{x}$

3. $Dy = (x/y)^{1/2}$

    if $y = 4$ when $x = 0$

4. $y' = y^2 \left(x + \dfrac{1}{x} \right)$

5. $\left(\dfrac{dy}{dx} \right) = x$

    if $y = 1$ when $x = 1$

6. $Dy = \sqrt{y}\, e^x$

7. $Dy = \dfrac{yx^2}{(x^3 + 1)^{1/2}}$

8. $Dy = \dfrac{(\log x)(y^2 - 1)^{1/2}}{xy}$

9. $yDy = xe^x + 1$

10. $Dy = \dfrac{3x^2 + 2x}{(\log y)(x^3 + x^2 + 1)^{1/5}}$

11. $Dy = \dfrac{y}{x}$

    if $y = 5$ when $x = 1$

12. $Dy = \dfrac{y}{(x + 1)}$

13. $Dy = \dfrac{1}{xy(y + 2)^{1/2}}$

14. $Dy = xy(\ln x)(\ln y)$

15. The following is a theory of nerve excitation evinced by H. A. Blair (see *J. of General Physiology*, **15** (1932) 7–9). If a current V is applied to a nerve, ionization occurs; this, in turn, leads to excitation. If ε (epsilon) denotes the concentration of the ion causing excitation in excess of the normal concentration, then ε is a measure of the excitation. It is assumed that the time rate of change of ε increases proportionally with the voltage and decreases proportionally with ε; that is, $d\varepsilon/dt = aV - b\varepsilon$ where $a > 0$ and $b > 0$ are pro-

portionality constants. Determine ε as a function of time t, $\varepsilon(t)$, assuming constant current V.

16. Let ε and V be as in problem 15, and assume again that

$$\frac{d\varepsilon}{dt} = aV - b\varepsilon$$

Find ε if $a = 1, b = 1, V(t) = 2$ for all t and $\varepsilon(0) = 0$.

17. The following is a model proposed by L. F. Richardson to describe the spread of war fever (see "War-Moods: I," *Psychometrica*, 1948). Let $y(t)$ represent the proportion of the population advocating war at time t. Then for some constant c, *the term* $y(t)$ satisfies the differential equation $(dy/dt) = cy(1 - y)$ where $y(0) = 1/(e + 1)$. Find $y(t)$. (*Hint:* Observe that $1/[y(1 - y)] = [1/y] + [1/(1 - y)]$.)

18. The following probabilistic learning model is due to Louis L. Thurstone (*Journal of General Psychology*, 1930). A learner has a goal, and a series of acts he may initiate toward the attainment of the goal. Some of these acts may be thought of as successful if they lead to the desired goal, failure if they do not. Let

s = total number of successful acts initiated by the learner
w = total number of acts which are failures

The differential equation relating s to w is

$$\frac{ds}{dw} = -\frac{s}{w}$$

Show that for some constant m, that $sw = m$; that is, show that the product of the total number of failures and the effective number of successes that the learner can initiate is constant.

Supplementary Readings:

APPLICATIONS OF THE INTEGRAL TO THE BIOLOGICAL SCIENCES AND ECONOMICS

The subsequent sections are taken from the life sciences and economics. They afford the interested reader further opportunities to see how integration can be applied to such disciplines. They also indicate how many problems in application give rise to limits of sums that are integrals.

SECTION 1

SURVIVAL FUNCTIONS AND RENEWAL THEORY

We shall consider a population of some sort for which we know that there is some function f to be determined for which $f(x)$ is the number of objects in the population at time x. We are, therefore, thinking of populations that change with time, such as a population of bacteria in a culture, or the population of vacuum tubes in a television set. Although we do not know $f(x)$ for general x, we do know the population at time $x = 0$; that is, the initial population, which we designate as $f(0)$. Our objective is to determine $f(x)$ at any time x.

Suppose we look at our population at time $x > 0$. First, we ask how many of the original population survived at time x. This value can often be expressed as a proportion of the initial population; for example, perhaps the number that survive at some instant is $\frac{1}{2} \cdot f(0)$ or $\frac{3}{4} \cdot f(0)$. We can often determine, perhaps empirically, a **survival function** s with the property that at time x, $f(0) \cdot s(x)$ is the population which has survived. Typically, $s(0) = 1$ (that is, at time 0, our population is just $f(0)$), and $0 \leq s(x) \leq 1$. Very often, $s(x)$ is **S.D.** and $\lim_{x \to \infty} s(x) = 0$.

We now consider a situation in which new objects are added to the population; when this happens, we shall say that the population is *renewed*, and call r the **renewal function**. We turn to an investigation of this function.

Let $A > 0$ be some time of interest, and consider the closed time interval $[0, A]$. Divide this interval into N parts by means of the subdivisions $0 = x_0 < x_1 < x_2 < \cdots < x_{N-1} < x_N = A$. We shall suppose that at time x_k, $r(x_k)\Delta x_k$ new objects are *added* to the population. Let s be the survival function for the original population, and suppose that the new population after objects are added is subject to the same survival function s.

At time x_1, $r(x_1)\Delta x_1$ objects are added to the population; thinking of this as a new initial population, we find that at $A - x_1$ time units later (that is, when we are at time A),

$$s(A - x_1) \cdot r(x_1) \cdot \Delta x_1$$

of the population that was added at time x_1 has survived. At time $x = x_2$, we add $r(x_2)\Delta x_2$ objects, and of this added population, $A - x_2$ time units later (that is, again when we are at time A), the number surviving is

$$s(A - x_2) \cdot r(x_2)\, \Delta x_2$$

In general, then, at time $x = x_k$, we add $r(x_k)\,\Delta x_k$ to the population, and the number of this added population that survive $A - x_k$ time units later is

$$s(A - x_k) \cdot r(x_k)\, \Delta x_k$$

Hence, the *total* population $f(A)$ at time $x = A$, due to the initial population $f(0)$ and the N additions to the population at times $x_1, \ldots, x_N$, is given by

$$(6.10) \qquad f(A) = f(0)\,s(A) + \sum_{k=1}^{N} s(A - x_k)\, r(x_k)\, \Delta x_k$$

Assuming that s and r are continuous, we see that as we take the limit as $N \to \infty$

$$(6.11) \qquad f(A) = f(0)\,s(A) + \int_0^A s(A - x)\, r(x)\, dx$$

This can be interpreted as meaning that *we are continuously adding to the population*. By definition of limit, this also means that for N sufficiently large, (6.11) is a "good" approximation to (6.10).

If we think of A as variable, and replace A by t, we obtain

$$f(t) = f(0)\,s(t) + \int_0^t s(t - x)\, r(x)\, dx$$

Problems

1. Suppose we have an initial population of 10,000 objects, and we know that the survival function is given by $s(x) = e^{-x}$ and the renewal function is $r(x) = k$ for some constant k, with x measured in hours. We assume continuous renewals. What is the size of the population at the end of 12 hours?

2. Repeat problem 1 with $r(x) = x^2$.

3. Repeat problem 1 for arbitrary time $t = A$ and $k = 1$.

4. Show that if the survival function s is given by $s(x) = \exp(-kx/m)$ and $f(0) = m$, then the population size has the constant value $f(x) = f(0) = m$ when the renewal function has the constant value k. Of course, k and m are positive constants, and we are assuming continuous renewals.

SECTION 2

AN APPLICATION FROM ECONOMICS: CONSUMER SURPLUS

Let p be the price function for a commodity, expressed as a function of demand x. Suppose that whatever conditions govern the market determine the market demand to be $x = A$ with the corresponding price $p(A)$. If we are at a point x at which $p(x) > p(A)$, then the consumer has gained, since the market price is fixed at $p(A)$. That is, since the consumer is willing to pay $p(x)$ but only has to pay the smaller price $p(A)$, it is a gain for the consumer. A measure of the consumer gain is often taken to be the area between the graphs of $y = p(x)$ and $y = p(A)$ in the region in which $p(x) > p(A)$. (Of course, $y = p(A)$ is a constant function whose graph is a straight line parallel to the x-axis.)

Since $p(x)$ is assumed to be **S.D.** (this only assumes that as demand x increases, price $p(x)$ decreases), we have $p(x) > p(A)$ on $0 \le x < A$ (see Figure 6-22). The area between the graphs $y = p(x)$ and $y = p(A)$ in the region in which $p(x) > p(A)$; that is, $0 \le x \le A$, is given by

$$\int_0^A \{p(x) - p(A)\}\, dx$$

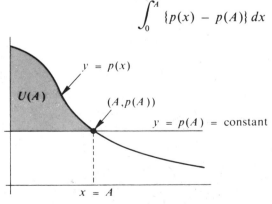

Figure 6-22

Therefore, we define the **consumer surplus** $U(A)$ related to the demand $x = A$ by

$$U(A) = \int_0^A \{p(x) - p(A)\}\, dx$$

Problems

1. Assume that price as a function of demand is given by $p(x) = e^{-x}$, and the demand is fixed at $x = \ln 10$. Determine the consumer surplus.

2. Repeat problem 1 with price now given by $p(x) = \ldots$

3. It is known that for a certain commodity, the price is constant at $1/(2 \ln 2)$ when demand x is such that $0 \le x \le 2$, and then for demand $x > 2$, the price behaves like $1/(x \ln x)$; that is,

$$p(x) = \begin{cases} 1/(2 \ln 2), & 0 \le x \le 2 \\ 1/(x \ln x), & x > 2 \end{cases}$$

Verify first that $p(x)$ is never increasing, and that for $x > 2$, $p(x)$ is **S.D.** Then find $U(x)$ if demand is fixed at $x = A = 5$.

SECTION 3

SOME APPLICATIONS OF DIFFERENTIAL EQUATIONS

The problems we shall consider are taken mainly from the life sciences.*

EXAMPLE 1

PARASITES

We consider the situation in which a population of parasites lives on a host population. Each population exerts an influence on the other, in some cases aiding and in other cases impeding the growth of that population. Hence, the time rate of change of the population of parasites at any given time depends both on the population of parasites and on the host population at that time; because of the interaction of the populations, the same statement can be made about the time rate of change of the host population.

If $x(t)$ denotes the size of the population of parasites at time t, and $y(t)$ denotes the size of the host population, with $x \ge 0$ and $y \ge 0$, then each of dx/dt and dy/dt depends on both x and y. The reader may recall that the total number of ways of

---

*The examples in this section make use of the variables separable differential equation discussed in Chapter 6, section 8.

pairing m distinct objects with n other distinct objects, distinct also from the first m, is $m \cdot n$; hence, x parasites can be paired with y hosts in xy distinct ways. This total pairing often contributes to dx/dt and dy/dt.

It has been determined that, in many cases, the differential equations which describe the populations are

$$(6.12) \qquad \frac{dx}{dt} = Ax + Bxy \qquad \text{and} \qquad \frac{dy}{dt} = Ry + Sxy$$

where A, B, R, and S are appropriate constants, $B \neq 0$, $S \neq 0$, and $x \geq 0$, $y \geq 0$.

We wish to find a relationship between x and y. Suppose the relationship is given by $y(x)$; our problem is to find $y(x)$. By the chain rule,

$$\frac{dy[x(t)]}{dt} = \frac{dy(x)}{dx} \cdot \frac{dx(t)}{dt}$$

Hence, if $dx/dt \neq 0$,

$$(6.13) \qquad \frac{dy}{dx} = \frac{(dy/dt)}{(dx/dt)}$$

and consequently, substituting Eq. (6.12) into Eq. (6.13), we have (thinking of y in terms of x)

$$\frac{dy}{dx} = \frac{Ry + Sxy}{Ax + Bxy} = \frac{y(R + Sx)}{x(A + By)}$$

Therefore,

$$\left(\frac{A}{y} + B \right) \frac{dy}{dx} = \left(\frac{R}{x} + S \right)$$

Appealing to our previous work on separable variables differential equations, with $g(y) = (A/y) + B$ and $f(x) = (R/x) + S$, we obtain

$$\int \left(\frac{A}{y} + B \right) dy = \int \left(\frac{R}{x} + S \right) dx + c$$

$$A \ln |y| + By = R \ln |x| + Sx + c$$

This can be written as

$$\ln \frac{(y^A)}{(x^R)} = Sx - By + c$$

which implies that

$$\frac{y^A}{x^R} = \exp (Sx - By + c)$$

or finally, with $k = e^c$,

$$y^A \exp(By) = kx^R \exp(Sx)$$

which is the desired relationship.

EXAMPLE 2

THE DIFFUSION EQUATION

We shall consider a cell which produces some substance at its center. For simplicity, we shall assume that the rate at which this substance is produced inside the cell is q gm/cu cm/sec, and that this rate is independent of the concentration of the substance produced in the cell. The external environment in which the cell is normally found contains some normal concentration of this substance produced by the cell, call this concentration u gm/cu cm. We assume that u is smaller than the concentration in the cell. What happens is threefold. First, the substance produced diffuses from the central portion of the cell toward the cell's periphery; then the substance passes through the wall of the cell; and finally the substance interacts with the external environment. We shall investigate the equations which describe this diffusion process.

First, we make some assumptions about the geometry of the cell. We suppose the cell is more or less an ellipsoid; that is, shaped something like a football. Figure 6-23 represents a cross section with approximate length $2A$ and approximate width $2B$. We shall refer to PR and QS as the "ends" of the cell, and PQ and RS as its "sides." Figure 6-23(b) represents the geometrical idealization of the cross section of the cell, an ellipse whose equation is

(6.14)
$$\frac{x^2}{A^2} + \frac{y^2}{B^2} = 1$$

The idealization of the cell itself as a three-dimensional entity is the ellipsoid obtained by revolving the upper half of the ellipse of Figure 6-23(b) about the **x**-axis, as in Figure 6-23(c) the total volume of the disk can then be found by the *disk method*.

Figure 6-23

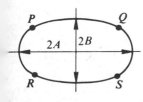
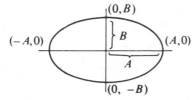
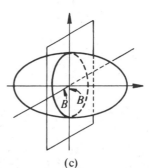

(a)   (b)   (c)

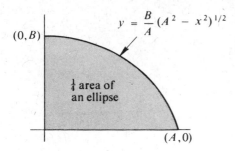

$$y = \frac{B}{A}(A^2 - x^2)^{1/2}$$

(0,B)

¼ area of an ellipse

(A,0)

Figure 6-24

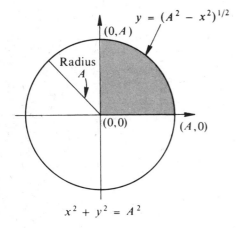

$$y = (A^2 - x^2)^{1/2}$$

(0,A)

Radius A

(0,0)   (A,0)

$$x^2 + y^2 = A^2$$

Figure 6-25

We now compute the area of the ellipse in Figure 6-23(b). By symmetry, it is clearly enough to find the area of the portion of the ellipse in the region bounded by the ellipse and the half-lines $x \geq 0$ and $y \geq 0$ (see Figure 6-24), and then multiply this result by 4. Solving (6.14) for y, we obtain as the equation of the upper portion of the ellipse $y = (B/A)(A^2 - x^2)^{1/2}$, and hence the area under the graph of the ellipse from $x = 0$ to $x = A$ is given by

(6.15) $\frac{1}{4}$(area of ellipse) $= \frac{B}{A} \int_0^A (A^2 - x^2)^{1/2}\, dx$

If we analyze the relation $y = (A^2 - x^2)^{1/2}$, we see that it is exactly the equation of the top semicircle of the circle $x^2 + y^2 = A$ (see Figure 6-25), hence, $\int_0^A (A^2 - x^2)^{1/2}\, dx$ is the area of one-fourth a circle of radius A; that is,

$$\int_0^A (A^2 - x^2)^{1/2}\, dx = \tfrac{1}{4}\pi A^2$$

Substituting this into (6.15) we see that (6.15) becomes $(B/A) \cdot (1/4)\pi A^2 = (1/4)\pi AB$. Since this is one-fourth the area of the ellipse, the *total area of the ellipse is* πAB.

The area of the *cross section* of the cell pictured in Figure 6-23(c) is exactly the area of a circle of radius B, and, hence, is πB^2. Because each "end" of the cell has approximately the area of the circle pictured in Figure 6-23(c),

(6.16)   Total surface area of ends of cell $\sim 2\pi B^2$

($\sim$ denotes approximately)

Each of the "sides" of the cell has a surface area approximately equal to the area of the ellipse of Figure 6-23(c), since there are four "sides,"

(6.17)   Total surface area of sides $\sim 4\pi AB$

If we revolve the region bounded by the graph of $y = (B/A)(A^2 - x)^{1/2}$ and the x-axis about the x-axis, we find by the disk method that the volume of the cell is

(6.18) $V = \frac{\pi B^2}{A^2} \int_{-A}^A (A^2 - x^2)\, dx$

and by symmetry,

$$V = \frac{2\pi B^2}{A^2} \int_0^A (A^2 - x^2)\, dx = \frac{4}{3}\pi AB^2$$

We now have all the background information necessary so that we may turn our attention to the diffusion process. We first consider the motion in the cell of the substance produced by the cell from the region of greater concentration in the center of the cell to the region of sparser concentration at the "ends" and "sides" of the cell. Call c the average concentration of the substance in the center of the cell, and let c_1 and c_2 be the concentrations at the "ends" and "sides" of the cell, respectively. By what we have already said, $c > c_1$ and $c > c_2$.

If we move from the center of the cell to either end, the concentration drops from c to c_1 as we move over roughly A units; thus, the average drop in concentration in one direction is $(c - c_1)/A$. Since there are two ends, the average drop in concentration per unit length in the direction corresponding to the length of the cell is $2(c - c_1)/A$. Similarly, the average drop in concentration per unit length in the direction of the width of the cell is $2(c - c_2)/B$. This implies that the average *flow* of the substance in the direction of the length and width of the cell is

$$(6.19) \quad 2M \frac{c - c_1}{A} \text{ gm/sq cm/sec}, \qquad 2M \frac{c - c_2}{B} \text{ gm/sq cm/sec}$$

respectively, where M is a constant called the **diffusion constant for the substance within the cell**.

We look next at the external environment in which the cell is suspended. As previously noted, we suppose that the environment normally contains a concentration u of the substance produced in the cell. The presence of the cell causes a greater concentration of this substance in the vicinity of the cell. Let u_1 be the concentration *in the environment* corresponding to the "ends" of the cell and u_2 be the concentration *in the environment* corresponding to the "sides" of the cell. We assume that the concentrations u_1 and u_2 are significant only in that part of the environment that is *near* the cell wall; call P the distance from the cell throughout which u_1 and u_2 are significant (see Figure 6-26).

The flow through the cell wall is proportional to the difference

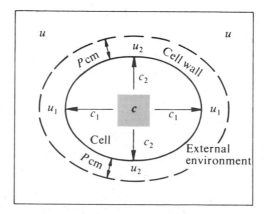

Figure 6-26

in concentrations of the substance within the cell and in the environment; and for a constant of proportionality we shall call L (the **permeability of the cell**), the flow through the "ends" and "sides" of the cell is given, respectively (see Figure 6-26), by

(6.20) $L(c_1 - u_1)$ and $L(c_2 - u_2)$ gm/sq cm/sec

Finally, we consider the situation in the environment that is within P-units of the cell. In this region, the drops in the concentration in the directions of the length and width of the cell are, respectively, $(u_1 - u)/P$ and $(u_2 - u)/P$. The equations analogous to (6.19) for the environment, that is, the equations describing the flow in the environment, are

(6.21) $\dfrac{N(u_1 - u)}{P}$ gm/sq cm/sec, $\dfrac{N(u_2 - u)}{P}$ gm/sq cm/sec

in the appropriate directions, where N is a constant called the **diffusion constant for the environment**.

The flow per square centimeter within the cell toward the cell wall, given by (6.19), must equal the flow per square centimeter through the cell wall given by (6.22), which in turn must equal the flow per square centimeter in the environment outside the cell wall, given by (6.21). Equating these expressions yields

(6.22) $2M\,(c - c_1) = AL\,(c_1 - u_1)$

(6.23) $2M\,(c - c_2) = BL\,(c_2 - u_2)$

(6.24) $\dfrac{2M}{A}\,(c - c_1) = \dfrac{N}{P}\,(u_1 - u)$

(6.25) $\dfrac{2M}{B}\,(c - c_2) = \dfrac{N}{P}\,(u_2 - u)$

The total flow through the "ends" and "sides" is the flow per square centimeter per second, multiplied by the total number of square centimeters; that is, it is flow per square centimeter per second multiplied by surface area. The total surface area of the "ends" if the cell, given by (6.16), is $2\pi B^2$, and the flow per square centimeter per second, given by (6.20) is $L(c_1 - u_1)$; thus, the *total flow through the "ends" is*

(6.26) $L(c_1 - u_1) \cdot 2\pi B^2$ gm/sec

In a similar way, appealing to (6.17) and (6.20), we see that the *total flow through the "sides" is*

(6.27) $L(c_2 - u_2) \cdot 4\pi AB$ gm/sec

Substituting from (6.22), (6.26) becomes

(6.28) $$\frac{4\pi MB^2(c - c_1)}{A} \text{ gm/sec}$$

and substituting from (6.23), (6.27) becomes

(6.29) $$8\pi MA(c - c_2) \text{ gm/sec}$$

Since the substance within the cell is produced on the average at q gm/cu cm/sec (see the first paragraph of the discussion), and the total volume of the cell, given by (6.18) is $(\frac{4}{3})\pi AB^2$, it follows that the *total amount of substance produced in the cell per second* is

(6.30) $$(\tfrac{4}{3})\pi AD^2q \text{ gm/sec}$$

The dimension of volume is cubic centimeters and the dimension of density is grams per cubic centimeter; therefore, this product gives the total amount of substance in the cell, which is $(\frac{4}{3})\pi AB^2c$ gm, and hence, the *rate of change* of the *total* amount of the substance is given by the time derivative

$$\frac{4}{3}\pi AB^2 \frac{dc}{dt} \text{ gm/sec}$$

and this must equal the total amount produced by the cell per second minus the total amounts leaving the cell per second. These amounts are given by (6.28), (6.29) and (6.30). Consequently,

$$\frac{4}{3}\pi AB^2 \frac{dc}{dt} = \frac{4}{3}\pi AB^2 q - \left[\frac{4\pi MB^2(c - c_1)}{A} + 8\pi MA(c - c_2)\right]$$

Multiplying this expression by $3/(4\pi AB^2)$ and factoring out $3M$, this expression becomes

(6.31) $$\frac{dc}{dt} = q - 3M\left(\frac{c - c_1}{A^2} + 2\frac{c - c_2}{B^2}\right)$$

and this is the differential equation which describes the diffusion process.

Equation (6.31) can be modified at the expense of introducing some rather lengthy computations. Without explicitly carrying out the details, we indicate what can be done. From (6.22) and (6.24), it is possible to eliminate u_1 and to express c_1 in terms of M, N, P, L, c, A, and u. From (6.23) and (6.25) we can eliminate u_2 and express c_2 in terms of M, N, P, L, c, B, and u. These expressions for c_1 and c_2 give us new expressions for $c - c_1$ and $c - c_2$ which can be substituted into (6.31) yielding a form equivalent to (6.31) namely,

(6.32) $$\frac{dc}{dt} = q - \frac{(c - u)}{w}$$

where w is the rather cumbersome constant

(6.33)

$$w = \frac{AB(2MN + 2PML + ALN)(2MN + 2PML + BLN)}{2(2MN + 2PML + ALN)A + (2MN + 2PML + BLN)B}$$

We shall finally solve (6.32). We put it into the form:

(6.34)
$$\frac{w}{(qw + u) - c}\frac{dc}{dt} = 1$$

where the expressions w, $(qw + u)$ are independent of time t; hence, constant. This variable separable differential equation has solution of the form

(6.35)
$$w \int \frac{1}{(qw + u) - c} dc = \int 1 \, dt - k$$

Setting $z = (qw + u) - c$, and, hence, $dz/dc = -1$, the integral on the left in (6.35) becomes $-w\int (1/z) \, dz$, so that (6.35) becomes

(6.36)
$$\int \frac{1}{z} dz = \frac{-1}{w} \int dt + k$$

This is exactly the form that appeared in the solution of the equations of exponential growth and decay. Hence, the solution to (6.36) is

(6.37)
$$z = a \exp(-t/w)$$

where a is an arbitrary constant. We replace z by $(qw + u) - c$ in (6.37), and finally obtain as the solution to (6.32)

(6.38)
$$c = u + qw - a \exp(-t/w)$$

where a is determined by the initial value of c, c is the average concentration in the center of the cell, q is the average rate at which the substance is produced within the cell, w is a constant which depends on the geometry and physics of the cell and its environment; and t is time.

EXAMPLE 3   In the first supplementary section, a result on elasticity of demand was mentioned without verification: If elasticity of demand N is always 1 for all prices p, then the product $p\mathbf{x}(p)$ is constant (we are using the notation of that section, with $\mathbf{x}(p)$ the demand corresponding to price p). Recall that $N(p) = 1$ for all p in some interval simply means

$$\left[\frac{-p}{\mathbf{x}(p)}\right]\frac{d\mathbf{x}(p)}{dp} = 1$$

for those prices p. Rewriting this differential equation, we obtain

$$\frac{1}{\mathbf{x}(p)} \frac{d\mathbf{x}(p)}{dp} = -\frac{1}{p}$$

and thus, from Equations (6.8) and (6.9) in section 8 on differential equations in this chapter,

$$\int \frac{1}{x} \, dx = -\int \frac{1}{p} \, dp$$

that is,

$$\ln x = -\ln p + c$$

where c is a constant. Rewriting, we obtain

$$\ln x + \ln p = c$$

Hence,

$$\ln (xp) = c$$

from which we obtain

$$px = e^c$$

Since c is a constant, so is e^c; thus, we see that px (that is, $p\mathbf{x}(p)$) is constant, which is what we wanted to show.

7

TAYLOR POLYNOMIALS AND SERIES

This brief chapter is concerned with the problem of approximating functions by polynomials. The virtue of such approximation is that polynomial functions are relatively easy to work with.

TAYLOR POLYNOMIALS

For a polynomial function P, it is relatively easy to compute the numbers $P(x)$ for given numbers x; in fact, it is relatively simple to program a computer to perform the necessary calculations. What we shall show in this section is that it is possible to approximate certain nonpolynomial functions by polynomial functions and to predict the error in these approximations. In such cases, we shall be able to approximate the numerical value of the nonpolynomial function by evaluating the appropriate polynomial function. For example, our procedure will give approximations to numbers like ln 1.1, and (in the next section), $\sqrt{e}$.

Thus, for a given f and a positive integer n, the problem is to find a polynomial function of degree n (or what amounts to the same thing, the coefficients $a_0, a_1, \ldots, a_n$) and an "error function" R_n for which

$$f(x) = a_0 + a_1 x + a_2 x^2$$
$$+ \cdots + a_n x^n + R_n(x), \qquad \text{all } x$$

Note that

$$R_n(x) = f(x) - (a_0 + a_1 x + a_2 x^2 + \cdots + a_n x^n)$$

that is, $R_n(x)$ is the difference between the value of $f(x)$ and the value of the polynomial approximation $a_0 + \cdots + a_n x^n$ and, hence, measures the error in the approximation, as desired.

Assume, first, that f is itself a polynomial function of degree n; thus, there are real numbers $a_0, \ldots, a_n$ for which

$$f(x) = a_0 + a_1 x + a_2 x^2 + \cdots + a_n x^n, \qquad \text{all } x$$

and $R_n(x) = 0$ for all x. Setting $x = 0$, we see that $f(0) = a_0$; that is, a_0 is the value of f at 0. Next, observe that since

$$f'(x) = a_1 + 2a_2 x + 3a_3 x^2 + 4a_4 x^3 + \cdots + na_n x^{n-1}$$

we see that $f'(0) = a_1$. Thus, a_0 and a_1, the first coefficients of f, are given by $f(0)$ and $f'(0)$ respectively. Continuing, we have

$$f''(x) = 2a_2 + 2 \cdot 3a_3 x + 3 \cdot 4a_4 x^2 + \cdots + n(n-1)a_n x^{n-2}$$
$$f'''(x) = 2 \cdot 3a_3 + 2 \cdot 3 \cdot 4a_4 x + \cdots + n(n-1)(n-2)a_n x^{n-3}$$

and, consequently, $f''(0) = 2a_2$, and $f'''(0) = 2 \cdot 3a_3 = 6a_3$. Thus,

$$a_2 = \frac{f''(0)}{2}$$

$$a_3 = \frac{f'''(0)}{2 \cdot 3}$$

and, in general, recalling that $k! = 1 \cdot 2 \cdot 3 \cdots k$,

$$a_k = \frac{f^{(k)}(0)}{k!}$$

Hence, the coefficients of f involve $f(0), f'(0), \ldots, f^{(n)}(0)$; specifically,

$$f(x) = a_0 + a_1 x + a_2 x^2 + \cdots + a_n x^n$$

$$= f(0) + f'(0)x + \frac{f''(0)}{2!}x^2 + \frac{f'''(0)}{3!}x^3 + \cdots + \frac{f^{(n)}(0)}{n!}x^n$$

EXAMPLE 1 Let $f(x) = 2 + 3x - x^2 + 4x^3$, a polynomial of degree four. We see that $f'(x) = 3 - 2x + 12x^2$, $f''(x) = -2 + 24x$, $f'''(x) = 24$, hence $f'(0) = 3$, $f''(0) = -2$, and $f'''(0) = 24$. Thus,

$$a_0 = f(0) = 2$$

$$a_1 = f'(0) = 3$$

$$a_2 = \frac{f''(0)}{2!} = \frac{-2}{2} = -1$$

$$a_3 = \frac{f'''(0)}{3!} = \frac{24}{6} = 4$$

which are, of course, the coefficients we began with.

The following, called **Taylor's Theorem**, generalizes this result to certain nonpolynomial functions; that is, for certain nonpolynomial functions, it gives a polynomial function approximation together with the error in making the approximation.

Let f be a function with the property that it and its first $(n + 1)$ derivatives are defined throughout some closed interval $[a, b]$ contain-

ing the origin (that is, $f(x)$, $f'(x)$,...,$f^{(n+1)}(x)$ exist for all x in some closed interval $[a,b]$ containing 0). Then there exists a number w between 0 and x such that

$$f(x) = f(0) + f'(0)x + \frac{f''(0)}{2!}x^2$$
$$+ \cdots + \frac{f^{(n)}(0)}{n!}x^n + \frac{f^{(n+1)}(w)}{(n+1)!}x^{n+1}$$

for all x in $[a,b]$.

Here, the remainder $R_n(x)$ for the given x is given by

$$R_n(x) = \frac{f^{(n+1)}(w)}{(n+1)!}x^{n+1}$$

for *some w* between 0 and x. Observe that, in general, we do not have an explicit value for the error since we do not know, in general, which value w between 0 and x to choose; all we know is that there *is* such a w. The polynomial $f(0) + f'(0)x + \cdots + f^{(n)}(0)/n!$ is called the **Taylor polynomial** (of degree n) for f. In the next examples we will find Taylor polynomials.

EXAMPLE 2   Let $f(x) = \ln(x+1)$. Find the Taylor polynomial of degree three for f, and then use this polynomial to approximate $\ln 1.1$.

SOLUTION   With $f(x) = \ln(x+1)$, we see that

$$f(0) + f'(0)x + \frac{f''(0)}{2!}x^2 + \frac{f'''(0)}{3!}x^3$$

is the required Taylor polynomial for f. We see that for x near 0,

$$f'(x) = \frac{1}{x+1}$$

$$f''(x) = -\frac{1}{(x+1)^2}$$

$$f'''(x) = \frac{2}{(x+1)^3}$$

hence $f(0) = \ln 1 = 0$, $f'(0) = 1$, $f''(0) = -1$, and $f'''(0) = 2$. Therefore, the Taylor polynomial of degree three for f is

$$x - \tfrac{1}{2}x^2 + \tfrac{1}{3}x^3$$

and thus $\ln(x+1) \sim x - \tfrac{1}{2}x^2 + \tfrac{1}{3}x^3$

By using this,

$$\ln 1.1 = \ln (0.1 + 1) \sim 0.1 - \tfrac{1}{2}(0.1)^2 + \tfrac{1}{3}(0.1)^3 = \tfrac{59}{1,000}$$

Since $\ln 1 = 0$, we expect $\ln 1.1$ to be near 0, so the approximation seems reasonable.

EXAMPLE 3   Let exp $(x) = e^x$, and let n be any positive integer.   Find the Taylor polynomial of degree n for exp

SOLUTION   We see that

$$f'(x) = f''(x) = f'''(x) = \cdots = f^{(n)}(x) = e^x$$

Hence, $f'(0) = f''(0) = f'''(0) = \cdots = f^{(n)}(0) = 1$

Consequently, the Taylor polynomial of degree n for exp is

$$1 + x + \frac{x^2}{2!} + \frac{x^3}{3!} + \cdots + \frac{x^n}{n!}$$

That is, $e^x \sim 1 + x + \dfrac{x^2}{2!} + \dfrac{x^3}{3!} + \cdots + \dfrac{x^n}{n!}$

EXAMPLE 4   Let $f(x) = xe^x$. Find the Taylor polynomial for f of degree n (where n is any positive integer).

SOLUTION   Complete.

$$f(x) = e^x \cdot x$$
$$f'(x) = e^x \cdot (\qquad)$$
$$f''(x) = e^x \cdot (\qquad)$$

Hence, $f^{(n)}(x) = e^x \cdot (\qquad)$

Thus,

$$f'(0) = \underline{\qquad}, \quad f''(0) = \underline{\qquad}, \quad f'''(0) = \underline{\qquad}$$
$$f^{(n)}(0) = \underline{\qquad}$$

The Taylor polynomial is

$$\underline{\qquad} + \underline{\qquad} x + \underline{\qquad} x^2 + \underline{\qquad} x^3$$
$$+ \cdots + \underline{\qquad} x^n$$

EXAMPLE 5   Discuss the error involved in replacing exp in $[0,1]$ by its nth-degree Taylor polynomial.

SOLUTION   Refer to Example 3. The remainder (which, of course, measures the error) is

$$R_n(x) = \frac{f^{(n+1)}(w)}{(n+1)!} x^{n+1}$$

$$= \frac{e^w}{(n+1)!} x^{n+1}$$

where w is some number between 0 and x. Since we are only looking at the interval $[0,1]$, and since exp is **S.I.** (thus, $0 \le w \le 1$ implies $1 = e^0 < e^w < e^1 = e$), we see that in $[0,1]$

$$R_n(x) = \frac{e^w x^{n+1}}{(n+1)!} < \frac{e \cdot 1^{n+1}}{(n+1)!}$$

That is, for x in $[0,1]$,

$$R_n(x) < \frac{e}{(n+1)!} < \frac{3}{(n+1)!}$$

(since $e < 3$), and thus the error never exceeds $3/(n+1)!$.

Remark   In general, the error that arises when a function f is replaced by its Taylor polynomial gets larger and larger when $f(x)$ is approximated for x further and further from 0.

EXAMPLE 6   Let $f(x) = 1/(x+1)$. Find the Taylor polynomial of degree four for f.

SOLUTION   For x near 0,

$$f'(x) = -\frac{1}{(x+1)^2}$$

$$f''(x) = \underline{\qquad}$$

$$f'''(x) = \underline{\qquad}$$

$$f^{(4)}(x) = \underline{\qquad}$$

Hence, $f(0) = 1$, $f'(0) = \underline{\qquad}$, $f''(0) = \underline{\qquad}$, $f'''(0) = \underline{\qquad}$, and $f^{(4)}(0) = \underline{\qquad}$.
The required polynomial is

$$\underline{\qquad} + \underline{\qquad} x + \underline{\qquad} x^2 + \underline{\qquad} x^3 + \underline{\qquad} x^4$$

Answers   Example 4: $x + 1$, $x + 2$, $x + n$; 1, 2, 3, n; 0, 1, 1, $\frac{1}{2}$, $1/(n-1)!$.
Another way to do Example 4 is to note from Example 3 that

$$e^x \sim 1 + x + \frac{x^2}{2!} + \frac{x^3}{3!} + \cdots + \frac{x^{n-1}}{(n-1)!} + \frac{x^n}{n!}$$

Hence, $xe^x \sim x + x^2 + \dfrac{x^3}{2!} + \dfrac{x^4}{3!} + \cdots + \dfrac{x^n}{(n-1)!}$

(dropping the last term from e^x, since all we want is a polynomial of degree n).

Example 6: $f''(x) = 2/(x+1)^3$; $f'''(x) = -6/(x+1)^4$; $f^{(4)}(x) = 24/(x+1)^5$; $f'(0) = -1$, $f''(0) = 2$, $f'''(0) = -6$, $f^{(4)}(x) = 24$. Taylor polynomial: $1 - x + x^2 - x^3 + x^4$.

Problems

1. Let $f(x) = (x+1)^{1/2}$. Find the Taylor polynomial of degree three for f. Use this to approximate $(1.2)^{1/2}$.

2. Let $f(x) = 1/(x^2 + 1)$. Find the Taylor polynomial of degree three for f.

3. Let $f(x) = \ln(x+1)$. Find the Taylor polynomial of degree five for f. Use this Taylor polynomial to approximate $\ln 1.1$. Compare your result with Example 2.

4. Let $f(x) = x^2 e^x$. Find the Taylor polynomial of degree five for f. What is the Taylor polynomial for f of degree n (where n is any positive integer)?

5. Let $f(x) = x \ln(x+1)$. Find the Taylor polynomial of degree three for f.

6. Use the Taylor polynomial of degree five for exp to find an approximation to the number e.

7. Use the Taylor polynomial of degree three for exp to approximate (a) $\sqrt[2]{e}$ and (b) $\sqrt[3]{e}$.

SECTION 2

A MORE GENERAL FORM OF TAYLOR'S THEOREM

In order to apply Taylor's theorem to a function f, it must be the case that f and certain of its derivatives are defined at, and near, 0. If $f(x) = 1/x$, we see f fails to meet this condition. Thus, f, for example, does not have a Taylor polynomial. Another difficulty with Taylor polynomials is that if we want to estimate, say, ln 101, the error in replacing ln 101 by a Taylor polynomial of small degree, evaluated at 101, is so large that the estimate is not useful. A partial remedy to such problems is the following generalization of Taylor's theorem.

Let f be a function with the property that it and its first $(n+1)$ derivatives are defined throughout some closed interval $[a,b]$ containing c. Then there is a number w between c and x for which

$$f(x) = f(c) + f'(c)(x-c) + \frac{f''(c)}{2!}(x-c)^2$$

$$+ \cdots + \frac{f^{(n)}(c)}{n!}(x-c)^n + \frac{f^{(n+1)}(w)}{(n+1)!}(x-c)^{n+1}$$

for $x \in [a,b]$.

The polynomial so obtained is called the **Taylor polynomial about c for f**, and is a polynomial in powers of $(x - c)$. Note that the Taylor polynomial about 0 for f is just the Taylor polynomial discussed in the last section; the Taylor polynomial for f about 0 is sometimes called the **Maclaurin polynomial** for f.

EXAMPLE 1   Let $f(x) = (x)^{1/2}$. Since $f'(x) = 1/[2(x)^{1/2}]$, we see that $f'(0)$ does not exist; thus, f has no Maclaurin polynomial (that is, in terms of our new definitions, f has no Taylor polynomial about 0). However, f and all its derivatives are defined for all positive x. Hence, we can find a Taylor expression for f about any positive x. For convenience and ease of computation, we choose $c = 1$ and, say, $n = 3$. The problem then is to find a Taylor polynomial about 1 of degree three for f.

SOLUTION   The desired polynomial (with remainder) is

$$f(1) + f'(1)(x - 1) + \frac{f''(1)}{2!}(x - 1)^2 + \frac{f'''(1)}{3!}(x - 1)^3$$

$$+ \frac{f^{(4)}(w)}{4!}(x - 1)^4$$

for x in any closed interval which contains 1 but excludes 0. Here, w is between 1 and x. Hence, we can use any closed interval $[a,b]$ containing 1 with $0 < a < b$. Thus, we need the following:

$$f'(x) = \tfrac{1}{2}(x)^{-1/2}, \quad f''(x) = -\tfrac{1}{4}x^{-3/2}, \quad f'''(x) = \tfrac{3}{8}x^{-5/2},$$

$$f^{(4)}(x) = -\tfrac{15}{16}x^{-7/2}$$

Hence, $f'(1) = \tfrac{1}{2}, \quad f''(1) = -\tfrac{1}{4}, \quad f'''(1) = \tfrac{3}{8}$

and $f^{(4)}(w) = \left(-\tfrac{15}{16}\right)w^{-7/2}$

Since we have to compute powers and roots to evaluate these derivatives, the choice of $c = 1$ was evidently a good one. Hence, for $n = 3$, we obtain, for $0 < a < b$, and for x in $[a,b]$,

$$(x)^{1/2} = 1 + \tfrac{1}{2}\cdot(x - 1) - \tfrac{1}{8}(x - 1)^2 + \tfrac{1}{16}(x - 1)^3$$

$$-\frac{15}{16\cdot 4!}\frac{(x - 1)^4}{w^{7/2}}$$

where w is some number between 1 and x.

EXAMPLE 2   Let $f(x) = \ln x$. Find the Taylor polynomial about 10 of degree two for f, and use this to approximate $\ln 10.2$ (given that $\ln 10 \sim 2.3$).

SOLUTION   Here, $n = 2$ and $c = 10$. Thus,

$$\ln x = f(x) \sim f(10) + f'(10)(x - 10) + \frac{f''(10)}{2!}(x - 10)^2$$

Since

$$f(x) = \ln x$$

we have

$$f'(x) = \frac{1}{x} \qquad f''(x) = -\frac{1}{x^2}$$

and thus $f(10) = \ln 10$

$$f'(10) = \frac{1}{10}$$

$$f''(10) = -\frac{1}{100}, \qquad \text{so that } \frac{f''(10)}{2!} = -\frac{1}{200}$$

Hence, $$\ln x \sim \ln 10 + \frac{(x - 10)}{10} - \frac{(x - 10)^2}{200}$$

Consequently,

$$\ln 10.2 \sim \ln 10 + \frac{0.2}{10} - \frac{(0.2)^3}{200} = \ln 10 + \frac{1}{50} - \frac{1}{5,000} \sim 2.3198$$

EXAMPLE 3   Let $f(x) = 1/x$. Find the Taylor polynomial about 2 of degree four for f.

SOLUTION   The required derivatives of f are $f'(x) = -1/x^2$, $f''(x) = 2/x^3$, $f'''(x) = -6/x^4, f^{(4)}(x) = 24/x^5$. Complete the following:

$$f(2) = \underline{\quad}, \qquad f'(2) = \underline{\quad}, \qquad f''(2) = \underline{\quad}$$
$$f'''(2) = \underline{\quad}, \qquad f^{(4)}(2) = \underline{\quad}$$

Hence,

$$\frac{f''(2)}{2!} = \underline{\quad}, \qquad \frac{f'''(2)}{3!} = \underline{\quad}, \qquad \frac{f^{(4)}(2)}{4!} = \underline{\quad}$$

Thus, the required polynomial is

$$\underline{\quad} + \underline{\quad}(x - \underline{\quad}) + \underline{\quad}(x - \underline{\quad})^2$$
$$+ \underline{\quad}(x - \underline{\quad})^3 + \underline{\quad}(x - \underline{\quad})^4$$

Answer   Example 3: $f(2) = \frac{1}{2}$, $f'(2) = -\frac{1}{4}$, $f''(2) = \frac{1}{4}$, $f'''(2) = -\frac{3}{8}$,

$f^{(4)}(2) = \frac{3}{4}, \frac{f''(2)}{2!} = \frac{1}{8}, \frac{f'''(2)}{3!} = -\frac{1}{16}, \frac{f^{(4)}(2)}{4!} = \frac{1}{32}$,

polynomial $= \frac{1}{2} - \frac{1}{4}(x - 2) + \frac{1}{8}(x - 2)^2 - \frac{1}{16}(x - 2)^3 + \frac{1}{32}(x - 2)^4$

Problems In problems 1–13, find the Taylor polynomial of the given degree n and about the given point c.

1. $f(x) = e^{-x}$, $n = 5, c = 0$

2. $f(x) = \dfrac{1}{(1-x)^2}$, $n = 4, c = 0$

3. $f(x) = x^5$, $n = 6, c = 0$, and $c = 1$

4. $f(x) = e^{x^2}$, $n = 3, c = 0$, and $c = 1$

5. $f(x) = (x)^{1/2}$ $n = 4, c = 1$

6. $f(x) = \dfrac{1}{(x)^{1/2}}$, $n = 3, c = 1$

7. $f(x) = \dfrac{1}{(1-x)^{1/2}}$, $n = 4, c = 0$

8. $f(x) = \ln x$, $n = 5, c = 1$

9. $f(x) = e^x$, $n = 4, c = -1$

10. $f(x) = \dfrac{1}{x+1}$, $n = 6, c = -2$

11. $f(x) = \ln(x+5)$, $n = 6, c = 0$
    (Maclaurin polynomial)

12. $f(x) = (1+x)^{1/2}$, $n = 5, c = 0$

13. $f(x) = e^{\sqrt{x}}$, $n = 2, c = 1$

14. Let $f(x) = (x)^{1/2}$. Use the third-degree Taylor polynomial about 100 for f to approximate $(101)^{1/2}$.

SECTION 3

TAYLOR SERIES

In the previous section, we found that it is possible to approximate certain functions by polynomial functions. We now investigate conditions under which certain functions f can be expressed as the infinite series

$$f(x) = f(0) + f'(0)x + \frac{f''(0)}{2!}x^2 + \frac{f'''(0)}{3!}x^3 + \cdots$$

for x near 0. This series is called the **Maclaurin series of f at x**. We shall require that f and all its derivatives exist near 0. Since

$$f(x) = f(0) + f'(0)x + \frac{f''(0)}{2!}x^2 + \cdots + \frac{f^{(n)}(0)}{n!}x^n + R_n(x)$$

we see that

$$\mathbf{f(x) = f(0) + f'(0)x + f''(0)x^2 + \cdots} \qquad \textbf{if and only if,}$$

$$\lim_{n \to \infty} R_n(x) = 0$$

Note that with the sigma notations the Maclaurin series for f at x is*

---

*Here, $f^{(0)}(x) = f(x)$, and, as before, $0! = 1$.

$$f(x) = \sum_{k=0}^{\infty} \frac{f^{(k)}(0)}{k!} x^k$$

We shall have need of the following result (which we shall not verify): for *any* x,

$$\lim_{n \to \infty} \frac{x^n}{n!} = 0$$

With this, we can show that for *any* x

$$e^x = \sum_{k=0}^{\infty} \frac{x^k}{k!} = 1 + x + \frac{x^2}{2!} + \frac{x^3}{3!} +$$

We must show that the remainder tends to zero. To see this, note that if $f(x) = e^x$, then

$$R_n(x) \equiv \frac{f^{(n+1)}(w)}{(n+1)!} x^{n+1} = \frac{e^w}{(n+1)!} x^{n+1} \qquad (w \text{ between } 0 \text{ and } x)$$

and for any x,

$$\lim_{n \to \infty} e^w \frac{x^{n+1}}{(n+1)!} = e^w \lim_{n \to \infty} \frac{x^{n+1}}{(n+1)!} = 0$$

We shall now consider functions which have Taylor series representations *near* 0; that is, a function f for which

$$f(x) = \sum_{k=0}^{\infty} \frac{f^{(k)}(0)}{k!} x^k$$

for all x in some open interval (a,b) containing 0. It turns out that for such f, *all derivatives* of f exist in (a,b). Thus, if

$$f(x) = \sum_{k=0}^{\infty} \frac{f^{(k)}(0)}{k!} x^k, \qquad x \text{ in } (a,b)$$

it makes sense to talk about

$$Df(x) = D \sum_{k=0}^{\infty} \frac{f^{(k)}(0)}{k!} x^k, \qquad x \text{ in } (a,b)$$

$$D^2 f(x) = D^2 \sum_{k=0}^{\infty} \frac{f^{(k)}(0)}{k!} x^k, \qquad x \text{ in } (a,b)$$

and so on. A remarkable fact is that within (a,b), Df can be found by differentiating the series for f, term-by-term; and $D^2 f$ can be found by differentiating the series derived for f', term-by-term, etc. For example,

$$e^x = 1 + x + \frac{x^2}{2!} + \frac{x^3}{3!} + \frac{x^4}{4!} + \cdots \qquad \text{all } x \text{ in } \mathbf{R}$$

hence for all x in $\mathbf{R}$,

$$De^x = D\left(1 + x + \frac{x^2}{2!} + \frac{x^3}{3!} + \frac{x^4}{4!} + \cdots\right)$$

$$= D1 + Dx + D\frac{x^2}{2!} + D\frac{x^3}{3!} + D\frac{x^4}{4!} + \cdots$$

$$= 1 + \frac{2x}{2!} + \frac{3x^2}{3!} + \frac{4x^3}{4!} + \cdots$$

$$= 1 + x + \frac{x^2}{2!} + \frac{x^3}{3!} + \cdots$$

$$= e^x$$

giving us another way to see that $De^x = e^x$.

A similar result is that f can be integrated term-by-term on any closed interval inside (a,b).

EXAMPLE 1 Recall from our work on geometric series

$$\sum_{n=0}^{\infty} x^n = \frac{1}{1-x} \text{ if } |x| < 1 \qquad \text{(that is, if } x \text{ is in } (-1, 1))$$

This series is also the Taylor series for $f(x) = 1/(1 - x)$, $|x| < 1$ (you might check some terms to convince yourself of this). If we replace x by $-x$, we obtain

$$\sum_{n=0}^{\infty} (-x)^n = \frac{1}{1+x}, \qquad |-x| < 1$$

That is, for $|x| < 1$,

$$\sum_{n=0}^{\infty} (-1)^n x^n = 1 - x + x^2 - x^3 + \cdots = \frac{1}{1+x}$$

Hence, for $|x| < 1$,

$$\ln(1+x) = \int_0^x \frac{1}{1+t}\, dt = \int_0^x \left(\sum_{n=0}^{\infty} (-1)^n t^n\right) dt$$

$$= \sum_{n=0}^{\infty} (-1)^n \int_0^x t^n dt$$

$$= \sum_{n=0}^{\infty} (-1)^n \frac{x^{n+1}}{n+1}$$

Writing some of the terms, we obtain for $|x| < 1$

$$\ln(1+x) = x - \frac{x^2}{2} + \frac{x^3}{3} - \frac{x^4}{4} + \frac{x^5}{5} - \cdots$$

EXAMPLE 2 Use the first three terms in the Taylor series about 0 for ln $(1 + x)$ to approximate

$$\int_0^{1/2} \ln (x + 1) \, dx$$

SOLUTION From Example 1 we see that for $|x| < 1$,

$$\ln (x + 1) = x - \frac{x^2}{2} + \frac{x^3}{3} - \frac{x^4}{4} + \cdots$$

Hence, for $0 < x < 1$, using the first three terms in the Taylor series,

$$\int_0^x \ln (t + 1) \, dt \sim \int_0^x t \, dt - \tfrac{1}{2} \int_0^x t^2 \, dt + \tfrac{1}{3} \int_0^x t^3 \, dt$$

and consequently

$$\int_0^{1/2} \ln (t + 1) \, dt \sim \frac{t^2}{2} \Big|_0^{1/2} - \frac{t^3}{6} \Big|_0^{1/2} + \frac{t^4}{12} \Big|_0^{1/2} = \frac{21}{192}$$

---

1. Use the fact that, for $|x| < 1$,

$$\frac{1}{1 + x} = 1 - x + x^2 - x^3 + \cdots$$

to find a series for the function f where $f(x) = 1/(1 + x^2)$ for x in $(-1, 1)$. (*Hint:* Replace x by x^2 in the series for $1/(1 + x)$.)

2. Use the first four terms in the Taylor series about 1 for ln to estimate

$$\int_1^{3/2} \ln x \, dx$$

Use the first three terms in the series for exp to approximate

3. $\displaystyle\int_0^1 xe^x \, dx$

4. $\displaystyle\int_0^1 x^3 e^x \, dx$

For $|x| < 1$, we have shown that

$$\frac{1}{1 - x} = 1 + x + x^2 + x^3 + \cdots$$

(Recall the section on geometric series.) By differentiating, show that for $|x| < 1$,

5. $\dfrac{1}{(1-x)^2} = 1 + 2x + 3x^2 + 4x^3 + 5x^4 + \cdots$

6. $\dfrac{1}{(1-x)^3} = 1 + 3x + 6x^2 + 10x^3 + \cdots$

7. Those of you who have studied trigonometry are familiar with the definitions of **sin** and **cos** (perhaps in terms of right triangles, or the unit circle). Another way to define these functions is by the series

$$\sin x = x - \frac{x^3}{3!} + \frac{x^5}{5!} - \frac{x^7}{7!} + \frac{x^9}{9!} - \cdots \qquad \text{all } x$$

and

$$\cos x = 1 - \frac{x^2}{2!} + \frac{x^4}{4!} - \frac{x^6}{6!} + \cdots \qquad \text{all } x$$

Note, for example, that sin 0 = 0 and cos 0 = 1. Prove that

$$D \sin x = \cos x$$

and

$$D \cos x = -\sin x$$

8. Use the first four terms in the series for cos (see Problem 7) to approximate $\int_0^1 \cos x \, dx$.

9. Recall that the Taylor series for exp is

$$e^x = 1 + x + \frac{x^2}{2!} + \frac{x^3}{3!} + \cdots \qquad \text{all } x$$

   a. Use this to obtain a series for $f(x) = e^{-x^2}$. (*Hint:* Replace x by $-x^2$ in the series for exp.)
   b. Find the Taylor series for

$$\int_0^x e^{-t^2} \, dt$$

   c. Use the first four terms of the series obtained in (b) to approximate

$$\int_0^1 e^{-x^2} \, dx$$

10. For x in $(-1, 1)$, let

$$f(x) = x + \frac{x^2}{2} + \frac{x^3}{3} + \cdots$$

   Show that f is a logarithmic function. (*Hint:* First find f' and then retrieve f by integrating f'.)

8

MULTIVARIABLE CALCULUS

In this chapter, we investigate functions whose domains are sets in the plane (or in three-dimensional space), and the graphs of such functions. As in the case of functions with domains on the line, we begin by defining limit, and then we extend the concepts of derivative and integral.

The point to bear in mind is that in most applications, a quantity will really depend on more than one other quantity. In our one-variable models, we say that the price of a commodity depends on the demand for that commodity. But a more realistic setting for such a problem is to take into account the fact that the price of the quantity really depends on many *variables*, such as the cost of labor, maintenance and insurance, and competitor's price. In a sense, then, it is the multivariable problem that more accurately describes our "real world."

SECTION 1

FUNCTIONS OF MANY VARIABLES. GRAPHS.

So far we have considered functions whose domain and range are subsets of the real line, with the convention that the domain is a subset of the horizontal axis and the range is a subset of the vertical axis. A common jargon is to refer to such functions as "real valued functions of one real variable," the "real valued function" because the **range** is a subset of the real line, and "one real variable" because the **domain** is a subset of the real line. The plan is to extend the concept of functions so that their domain can be more general than the real line; however, the range shall continue to be a subset of the line. Thus, to say that f is a real valued function of one real variable means that to each $x \in$ (domain f) $\subset$ **R**, the f associates to (or assigns) exactly one number in the range, and we call this number $f(x)$.

In practice, some one quantity will often depend on more than one other quantity. For example, the price to be charged for the manufacture of some commodity might depend on the three "variables" of labor costs, advertising costs, and maintenance costs (in fact, we might need even more than three variables). To begin with, let us concentrate on functions of two variables.

We say that *f is a (real valued)* **function of two variables** *if f assigns exactly one real number z to each point (x,y) in some set D of points in the plane, and D is called the* **domain** *of f. We then write*

$z = f(x,y)$ for $(x,y) \in D$. Note that the domain D of a function of two variables is a subset of the plane.

EXAMPLE 1   According to the Spearman-Brown formula (see page 140), the reliability of a test that has been lengthened by a factor of n is a function of two variables, namely, n and the reliability r. Specifically, if we call this function f, we have

$$f(n,r) = \frac{nr}{1 + (n - 1)r}, \qquad n \in I^+, \qquad 0 \leq r \leq 1$$

EXAMPLE 2   Let $f(x,y) = x - y$, all x,y. Here, f associates the number $x - y$ to each point (x,y) in the plane. For example, $f(0,0) = 0 - 0 = 0$, $f(1,2) = 1 - 2 = -1$, $f(2,1) = 2 - 1 = 1$ [thus, $f(1,2) \neq f(2,1)$], $f(-1,\frac{1}{2}) = -1 - \frac{1}{2} = -\frac{3}{2}$, $f(s,s) = 0$, all s, $f(s,0) = s$, all s, and $f(0,s) = -s$, all s. The domain of f is the plane.

EXAMPLE 3   Let $f(x,y) = x/(x - y)$, all (x,y) such that $x \neq y$. For example, $f(1,2) = 1/(1 - 2) = -1$, $f(-1,2) = -1/(-1 - 2) = \frac{1}{3}$, $f(2,-1) = 2/(2 - (-1)) = \frac{2}{3}$; $f(0,s) = 0$, all $s \neq 0$; $f(s,0) = 1$, all $s \neq 0$; $f(s,\frac{1}{2}s) = 2$, all $s \neq 0$; $f(1 + s,1 - s) = (1 + s)/2s$, $s \neq 0$. The domain of f is all (x,y) such that $x \neq y$; that is, all points in the plane that are off the line $y = x$.

We can define functions of three, four, or more variables in a completely analogous way. For example, to say that f is a function of three variables means that the domain S of f is a subset of three-dimensional space, and that to each $(x,y,z) \in S$, we get a real number $f(x,y,z)$.

EXAMPLE 4   Let $f(x,y,z) = xy + yz + xz$ for all (x,y,z) in three-dimensional space. Then, for example,

$$f(0,0,0) = 0$$
$$f(0,y,0) = 0 \cdot y + y \cdot 0 + 0 \cdot 0 = 0 \qquad \text{for any } y$$
$$f(0,0,z) = 0 \cdot 0 + 0 \cdot z + 0 \cdot z = 0 \qquad \text{for any } z$$
$$f(-1,1,1) = \underline{\qquad}$$

For any t, $\qquad f(t,t,t) = \underline{\qquad}$

For any t, $\qquad f(t,2t,3t) = \underline{\qquad}$

For any t, $\qquad f(t,t^2,t^3) = \underline{\qquad}$

EXAMPLE 5   Let $f(x,y,z) = xy/z$ for all (x,y,z) such that $z \neq 0$; f is a function of three variables. For example, $f(1,1,1) = (1 \cdot 1)/1 = 1$, $f(1,2,-1) = 1 \cdot 2/(-1) = -2$, $f(2,2,5) = \frac{4}{5}$, $f(0,s,t) = 0$, all $t \neq 0$, all s; $f(s^2 - t^2, u, s - t) = u \cdot (s + t)$, all $s \neq t$, all u; $f(s,t,1) = st$, all s, all t.

EXAMPLE 6   Let r be a positive real number. Let $f(x,y) = x^2 + y^2 - r^2$, all (x,y) in the plane. Observe that the set of (x,y) such that $f(x,y) = 0$ is just the circle of radius r whose center is $(0,0)$. Also, for example, $f(0,0) = -r^2$, $f(1,1) = f(-1,-1) = 2 - r^2$, and $f(a,a) = f(-a,-a) = f(a,-a) = f(-a,a) = 2a^2 - r^2$ for all a. We return to this example later (see Example 12).

Remark   We shall adopt the following domain convention. *Unless otherwise stated, the (implicit) domain of any function is the set of all points p for which the functions make sense (that is, for which f(p) is a real number).*

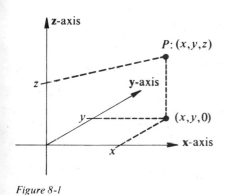

Figure 8-1

For the time being, to enable us to construct graphs, we shall restrict our attention to functions of two variables. To construct the graph of such a function, we shall need the following geometrical construction: given a coordinate plane, construct a line perpendicular to the plane and passing through $(0,0)$; lay units on this line in the same way as on the coordinate axes in the given coordinate plane. We refer to this perpendicular line as the **z-axis**, and the coordinate plane as a horizontal coordinate plane, or the **xy-plane**. This terminology is motivated by the construction in Figure 8-1. We choose a so-called right-hand orientation for three-dimensional space. That is, if the fingers of the right hand point from the **x**-axis to the **y**-axis, the thumb points in the direction of increasing z. We see from Figure 8-1 that there is a unique correspondence between points p in **xyz**-space (that is, three-dimensional space) and ordered triples of numbers (x,y,z). Also, the set of points $(x,y,0)$, all x, all y, is the **xy**-plane; hence, the "horizontal" coordinate plane is given by $z = 0$. Similarly, the **xz**-plane and **yz**-plane are given by $y = 0$ and $x = 0$, respectively. We define: (1.) *The **graph** of a function of two variables f is the set of all points $(x,y,f(x,y))$, all (x,y) in the domain of f.* (2.) *In general, a graph in three-dimensional space is any set of points (x,y,z).* We shall often refer to a continuous graph in three-dimensional space as a **surface**.

EXAMPLE 7   Let c be any real number. Sketch the surface S whose equation is $z = c$; that is, sketch the points (x, y, z) for which $z = c$.

SOLUTION   Since $z = c$, and x and y are arbitrary, we see that the surface is simply a translation of the horizontal coordinate plane by c units ("up" if $c > 0$ and "down" if $c < 0$). See Figure 8-2.

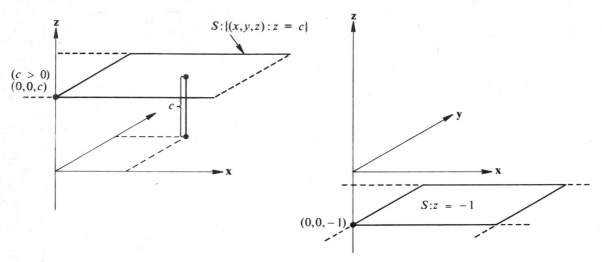

Figure 8-2

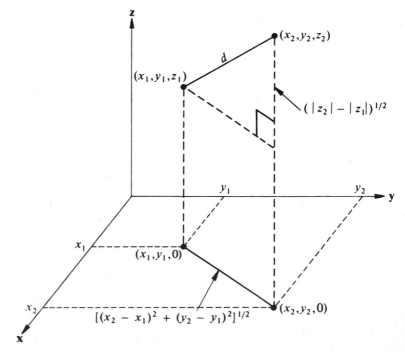

Figure 8-3

From Figure 8-3, we see (using the theorem of Pythagoras twice) that *the distance d between two points* (x_1, y_1, z_1) *and* (x_2, y_2, z_2) *is*

$$[(x_2 - x_1)^2 + (y_2 - y_1)^2 + (z_2 - z_1)^2]^{1/2}$$

EXAMPLE 8   The *sphere* with center (a, b, c) and radius r is given by

$$(x - a)^2 + (y - b)^2 + (z - c)^2 = r^2$$

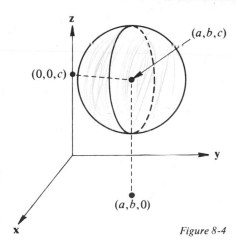

See Figure 8-4. Thus, $\{(x, y, z): x^2 + (y + 1)^2 + (z - \sqrt{2})^2 = 5\}$ is a sphere with center _____ and radius _____.

Figure 8-4

EXAMPLE 9   Sketch the surface S whose equation is $x^2 + z^2 = 1$.

SOLUTION   In the plane given by $y = 0$ (that is, the **xz**-plane), this is simply a circle of radius 1. Since y is any real number, we obtain the cylindrical surface sketched in Figure 8-5.

Figure 8-5

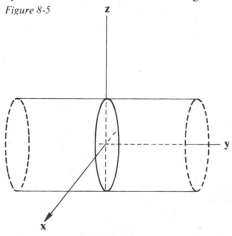

EXAMPLE 10   Let f be a function of *one variable*. The surface S associated with f (sketched in Figure 8-6) is the set of points $(x, f(x), z)$ with z arbitrary, for all x in the domain of f.

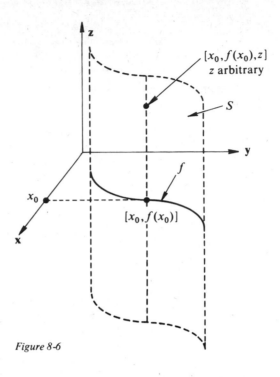

Figure 8-6

EXAMPLE 11   Sketch the graph S of the function f given by

$$f(x, y) = x^2 + y^2$$

SOLUTION   Let c be a real number. The intersection of this surface S with the plane whose equation is $z = c$ is

$$x^2 + y^2 = c$$

If $c < 0$, there are no points (x,y) for which $x^2 + y^2 = c$; if $c = 0$, only $(0,0)$ has the property; if $c > 0$, the intersection is the circle with center $(0,0,c)$ and radius $(c)^{1/2}$. As we take planes further above the **xy**-plane, c increases, and the radii of the circles get larger. If we look at the intersection of S and the plane $y = c$ (that is, the plane c units from the **xz**-plane), we obtain

$$z = x^2 + c^2$$

a *parabola*. The S is sketched in Figure 8-7.

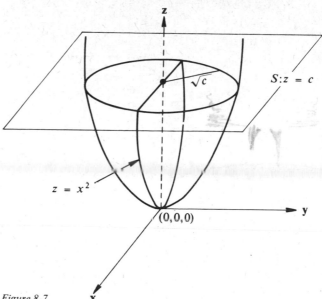

Figure 8-7

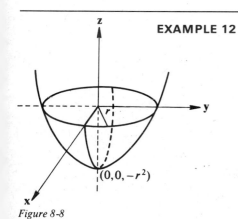

EXAMPLE 12 Let $r > 0$ be given. The graph S of $f(x, y) = x^2 + y^2 - r^2$, all x, y, is obtained from Figure 8-7 by displacing that graph r^2 units down (see Figure 8-8). Observe that the intersection of S and the **xy**-plane is the circle given by $x^2 + y^2 = r^2$.

Figure 8-8

For completeness, we graph the following equations. In each case, a, b, and c are nonzero real numbers. We leave the details as an exercise.

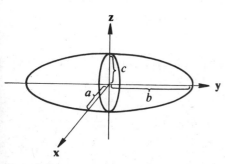

Ellipsoid $\dfrac{x^2}{a^2} + \dfrac{y^2}{b^2} + \dfrac{z^2}{c^2} = 1$

(See Figure 8-9.)

Saddle-shape $\dfrac{y^2}{b^2} - \dfrac{x^2}{a^2} = cz$

(See Figure 8-10 for $c > 0$.)

To help see that the surface is saddle-shaped, note that, since

Figure 8-9

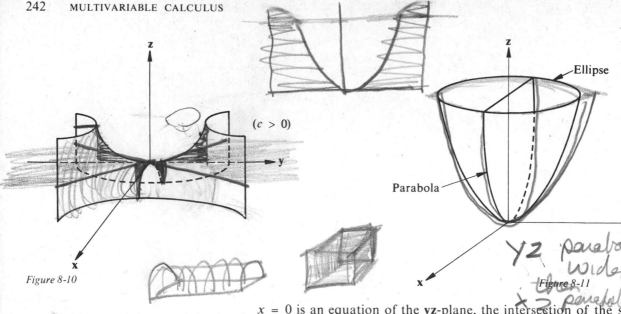

$(c > 0)$

Figure 8-10

Ellipse

Parabola

Figure 8-11

[handwritten: yz parabola wider than x3 parabola in this picture]

[handwritten: circle is just special form of ellipse]

$x = 0$ is an equation of the **yz**-plane, the intersection of the surface with the **yz**-plane is given by

$$\frac{y^2}{b^2} - \frac{0^2}{a^2} = cz$$

or, equivalently,

$$z = \frac{1}{b^2 c} y^2$$

and, since $c > 0$, we have $b^2 c > 0$; hence, this is the equation of a **concave up** parabola in the **yz**-plane. However, since $y = 0$ is an equation of the **xz**-plane, we see that the intersection of the surface with the **yz**-plane is given by

$$\frac{0^2}{b^2} - \frac{x^2}{a^2} = cz$$

thus,

$$z = -\frac{1}{a^2 c} x^2$$

Again, since $c > 0$, we see that this is the equation of a **concave down** parabola in the **xz**-plane. In the case $c < 0$, the concavity of the parabolas is simply interchanged.

Elliptic paraboloid $\qquad \dfrac{x^2}{a^2} + \dfrac{y^2}{b^2} = cz$

(See Figure 8-11.)

[handwritten: $a \neq b$]

Plane $\qquad ax + by + cz = d$

(See Figure 8-12.)

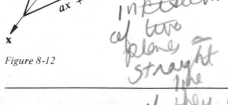

Figure 8-12

[handwritten labels in figure: $ax + cz = d$, $by + cz = d$, yz = plane, $ax + by = d$]

[handwritten: intersection of two planes = straight line if they intersect]

[handwritten: $ax + by = d$; $2x + 3y = 4$; $x = 0, y =$; $y = 0$, else x, y intersect on axis]

Answers   Example 4: $f(-1,1,1) = -1$; $f(t,t,t) = 3t^2$;
$f(t,2t,3t) = 11t^2$; $f(t,t^2,t^3) = t^3 + t^4 + t^5$
Example 8: Center $= [0,-1,(2)^{1/2}]$, radius $= (5)^{1/2}$

Problems

1. Let $f(x,y) = x^2y + xy^2$
   Find $f(0,0), f(-1,0), f(0,-1), f(1,1), f(2,4)$.
   If t is a real number, find $f(t,t), f(1 - t, t), f(t,t^2)$.

2. Let $f(x,y) = \left(1 - \dfrac{x}{y}\right)^2$

   Find $f(0,1), f(5,5), f(6,1), f(1,2)$.
   If t is a nonzero real number, find $f(t,t), f(5t,t), f(t,2t), f(1 + t,t)$.

3. Let $f(x,y,z) = x^2ye^{zx} + (x + y - z)^2$
   Find $f(0,0,0), f(1,-1,1), f(-1,1,-1)$.

   Find $\dfrac{d}{dx} f(x,x,x), \qquad \dfrac{d}{dy} f(1,y,1), \qquad \dfrac{d}{dz} f(1,1,z^2)$.

4. Let $w = f(x,y,z) = (x - 1)^2 + (y + 2)^2 + z^2$.   Describe
   the intersection of the graph of f and the following planes:
   a. $w = 0$
   b. $w = 4$
   c. $w = -4$

5. Find the distance between the following points:
   a. $(0,1,0)$ and $(1,2,-1)$
   b. $(1,2,-1)$ and $(2,-1,-1)$

   In problems 6–10, sketch the surface for the given equation.
6. $x = 5$
7. $z = 5$
8. $x^2 + y^2 = 4$, z in $\mathbf{R}$
9. $x^2 = y^2$, z in $\mathbf{R}$
10. $z = e^y$, x in $\mathbf{R}$

   In problems 11–17, sketch the graph of each of the following functions.
11. $f(x,y) = -4$
12. $f(x,y) = (1 - x^2 - y^2)^{1/2}$
13. $f(x,y) = x^2 + y^2 - 2$
14. $f(x,y) = x$ (y any real number)
15. $f(x,y) = y^2 + 1$ (x any real number)
16. $f(x,y) = y^2 - x^2$
17. $f(x,y) = \dfrac{x^2}{9} - \dfrac{y^2}{4}$

   In problems 18–27, sketch the surface whose equation is given.
18. $x + y + z = 1$
19. $x - y - z = 1$

20. $x + z = 1$

21. $y + z = 1$

22. $(x - 3)^2 + (y + 2)^2 + (z + 5)^2 = 1$

23. $(x + 1)^2 + (y - 2)^2 + (z - 3)^2 = 4$

24. $\dfrac{x^2}{9} + \dfrac{y^2}{4} + z^2 = 1$

25. $x^2 + 4y^2 + 16z^2 = 16$

26. $x^2 - 4y^2 = 4z$

27. $y^2 - 4x^2 = 4z$

SECTION 2

LIMITS AND CONTINUITY

The basic concept of limit for functions of two variables (that is, functions whose domains are sets in the plane) is essentially the same as for functions whose domains are sets on the real line. Recall $\lim_{x \to a} f(x) = L$ means that the numbers $f(x)$ can be made to stay as close to the number L as we please if the numbers x in the domain of f are close enough to a. Similarly,

$lim_{(x,y) \to (a,b)} f(x,y) = L$ *means that the numbers $f(x,y)$ can be made to stay as close to the number L as we please if the points (x,y) in the domain of f are close enough to the point (a,b); that is, $f(x,y)$ tends to L as (x,y) tends to (a,b)* **along any path** *in the domain of f.*

(See Figure 8-13.)   Continuity now becomes: f *is continuous at (a,b) means that $lim_{(x,y) \to (a,b)} f(x,y) = f(a,b)$.*

Figure 8-13

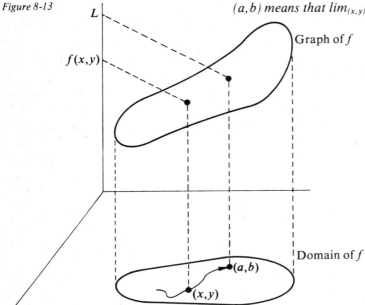

$f(x, y)$ tends to L as (x, y) tends to (a, b) along paths in the domain of f

EXAMPLE 1   Let $f(x,y) = x^2 + y^2$, all (x,y). Then

$$\lim_{(x,y)\to(0,0)} f(x,y) = \lim_{(x,y)\to(0,0)} (x^2 + y^2) = 0 = f(0,0)$$

$$\lim_{(x,y)\to(-1,1)} f(x,y) = 2 = f(-1,1)$$

and, in general,

$$\lim_{(x,y)\to(a,b)} f(x,y) = a^2 + b^2 = f(a,b)$$

Thus, f is continuous everywhere (in the plane).

EXAMPLE 2   Let $f(x,y) = xe^y$, all (x,y). Then,

$$\lim_{(x,y)\to(0,0)} f(x,y) = 0 \cdot e^0 = 0 \cdot 1 = 0$$

$$\lim_{(x,y)\to(1,0)} f(x,y) = \underline{\qquad}$$

For any a,

$$\lim_{(x,y)\to(a,0)} f(x,y) = \underline{\qquad}$$

and

$$\lim_{(x,y)\to(3,\ln 2)} f(x,y) = \underline{\qquad}$$

The set where f is continuous is, therefore, _____

EXAMPLE 3   Let $f(x,y) = \dfrac{x^2 - y^2}{x - y}$, all (x,y) such that $y \neq x$. Then,

$$f(x,y) = \frac{x^2 - y^2}{x - y} = \frac{(x + y)(x - y)}{x - y} = x + y \text{ if } x \neq y$$

Hence, if $a \neq b$, then

$$\lim_{(x,y)\to(a,b)} f(x,y) = a + b = f(a,b)$$

Thus, f is continuous everywhere except on the line given by $y = x$.

Note that f is not defined for any pair (x,y) with $x = y$; that is, for any pair (a,a), a in $\mathbf{R}$. Nevertheless, for any a in $\mathbf{R}$,

$$\lim_{(x,y)\to(a,a)} f(x,y)$$

exists, since

$$\lim_{(x,y)\to(a,a)} f(x,y) = \lim_{(x,y)\to(a,a)} \frac{x^2 - y^2}{x - y}$$

$$= \lim_{(x,y)\to(a,a)} \frac{(x + y)(x - y)}{x - y}$$

$$= \lim_{(x,y)\to(a,a)} (x + y) \quad (\text{since } (x,y) \neq (a,a))$$

$$= 2a$$

EXAMPLE 4   Complete:

$$\lim_{(x,y)\to(3,1)} \frac{x^2 - yx - 6y^2}{x - 3y} = \lim_{(x,y)\to(3,1)} \frac{(x - 3y)(\underline{\quad\quad})}{x - 3y}$$

$$= \underline{\hspace{4cm}}$$

EXAMPLE 5   Let f be given by

$$f(x,y) = \begin{cases} \dfrac{2xy}{x^2 + y^2}, & (x,y) \neq (0,0) \\ 0, & (x,y) = (0,0) \end{cases} \qquad [\text{that is, } f(0,0) = 0]$$

Show that f is *not* continuous at $(0,0)$.

SOLUTION   It is enough to find two distinct paths that terminate at $(0,0)$ and along which f has distinct limits. Let the first path be given by $x = 0$ (the path is the y-axis); that is, we let $(x,y) \to (0,0)$ along the y-axis. Since $f(0,y) = 0$, all y, we obtain *along this path*

$$\lim_{(x,y)\to(0,0)} f(x,y) = \lim_{(0,y)\to(0,0)} f(0,y) = 0$$

Let the second path be the straight line path given by $y = x$; we now let $(x,y) \to (0,0)$ along this straight line path. Since $f(x,x) = 1$ for $x \neq 0$, we obtain along this path

$$\lim_{(x,y)\to(0,0)} f(x,y) = \lim_{(x,x)\to(0,0)} f(x,x) = 1$$

Hence, f is discontinuous at $(0,0)$.

We conclude with the following:

If f and g are each continuous at (a,b), then so are $f + g$, $f \cdot g$, and $f - g$; if $g(a,b) \neq 0$, then f/g is also continuous at (a,b).

Answers   Example 2: 1; a; 6; the entire plane
Example 4: $x + 2y$; 5

Problems   1.   Let $f(x,y) = xy^2 + x^3y + 5$. Find

a. $\displaystyle\lim_{(x,y)\to(-1,-1)} f(x,y)$

b. $\displaystyle\lim_{(x,y)\to(-1,0)} f(x,y)$

c. $\displaystyle\lim_{(x,y)\to(0,0)} f(x,y)$

d. $\displaystyle\lim_{(x,y)\to(0,-1)} f(x,y)$

2. Let $f(x,y) = xe^y + x^2y^3 - \dfrac{y}{x}$. Find

a. $\displaystyle\lim_{(x,y)\to(1,0)} f(x,y)$

b. $\displaystyle\lim_{(x,y)\to(-1,0)} f(x,y)$

c. $\displaystyle\lim_{(x,y)\to(1,-1)} f(x,y)$

3. Find $\displaystyle\lim_{(x,y)\to(1,-1)} \dfrac{x^2 + 3xy + 2y^2}{x + y}$

4. Find $\displaystyle\lim_{(x,y)\to(1,1)} \dfrac{x^2 - y^2}{x - y}$

5. Find $\displaystyle\lim_{(x,y)\to(5,5)} \dfrac{x^4 - y^4}{x^2 - y^2}$

6. Find $\displaystyle\lim_{(x,y)\to(a,a)} \dfrac{x^4 - y^4}{x^2 - y^2}$

7. Let

$$f(x,y) = \frac{x^2 - 4y^2}{x - 2y} \qquad \text{for} \qquad x - 2y \neq 0$$

Let a be any real number. Find

$$\lim_{(x,y)\to(a,a/2)} f(x,y)$$

8. Let

$$f(x,y) = \frac{4x^2 - 9y^2}{2x - 3y} \qquad \text{for} \qquad 2x - 3y \neq 0$$

Let a be in $\mathbf{R}$. Find

$$\lim_{(x,y)\to(a/2,a/3)} f(x,y)$$

9. Let

$$f(x,y) = \frac{x^3 - y^3}{x - y}, \qquad x \neq y$$

Define f at each point (x,x) so that f will be continuous at all (x,y).

10. Let

$$f(x,y) = \frac{x^2 - y^2}{x^2 + y^2}, \qquad (x,y) \neq (0,0)$$

a. Show that $\lim_{(x,y)\to(0,0)} f(x,y)$ does not exist by considering the x-axis and y-axis.

b. Find the limit of f as $(x,y) \to (0,0)$ along the path given by $y = x$.
c. Find $\lim_{(x,y)\to(2,1)} f(x,y)$.

11. Let

$$f(x,y) = \frac{xy^2}{x^2 + y^4} \qquad (x,y) \neq (0,0)$$

Let L be the curve given by $x = y^2$. Find

$$\lim_{(x,y)\to(0,0)} f(x,y) \text{ for } (x,y) \text{ on } L$$

and

$$\lim_{(x,y)\to(0,0)} f(x,y) \text{ for } (x,y) \text{ on x-axis}$$

Does $\lim_{(x,y)\to(0,0)} f(x,y)$ exist?

12. Let

$$f(x,y) = \frac{xy^3}{x^2 + y^6}, \qquad (x,y) \neq (0,0)$$

Show that $\lim_{(x,y)\to(0,0)} f(x,y)$ does not exist.
(*Hint:* Consider the paths given by $x = 0$ and $y^3 = x$.)

13. Let

$$f(x,y) = \frac{3x^2y - x^7 + x^8(y^5 + 3)}{(x^2 + y^2)^2}, \qquad (x,y) \neq (0,0)$$

Show that

$$\lim_{(x,y)\to(0,0)} f(x,y)$$

does not exist. (*Hint:* Let one of your paths be the one given by $y = x$.)

14. Let

$$f(x,y) = \frac{x^4}{x^2 + y^2}, \qquad (x,y) \neq (0,0)$$

Show that

$$\lim_{(x,y)\to(0,0)} f(x,y) = 0$$

[*Hint:* Observe $f(x,y) \geq 0$, all $(x,y) \neq (0,0)$.]
Next, since $x^2 \leq x^2 + y^2$, all (x,y), we have

$$0 \leq f(x,y) \leq \frac{x^2 \cdot x^2}{x^2 + y^2} \leq x^2, \qquad \text{all } (x,y) \neq (0,0)$$

Now, let $(x,y) \to (0,0)$.]

SECTION 3

PARTIAL DERIVATIVES. IMPLICIT DIFFEREN- TIATION

We turn now to the question of the meaning of differentiation of a function of two variables. For convenience, let f be a function (of two variables) whose domain is the entire plane. Recall that for a function of *one* variable, say g, the number $g'(a)$ is the slope of the line tangent to the graph of g at $(a, g(a))$. The graph of f, our function of two variables, is a surface in three-dimensional space, call it S; and if we look at some point $[a, b, f(a, b)]$ on this surface S, we see that if S is smooth enough there may be infinitely many tangent lines that can be drawn to the surface at the point (in fact, under not too stringent requirements, we might expect that it is possible to draw a tangent *plane* to S at $(a, b, f(a, b))$, and this plane would contain infinitely many tangent lines to S). See Figure 8-14.

What we shall do is to single out *two* of the tangent lines to S at $(a, b, f(a, b))$, the one in the direction of the **x**-axis (that is, parallel to the **xz**-plane) and the other in the direction of the **y**-axis (that is, parallel to the **yz**-plane), and call their slopes the partial derivative of f with respect to x at (a, b) and the partial derivative with respect to y at (a, b), respectively (Figures 8-15(a) and (b)).

We shall denote the *partial derivative of f with respect to x at (a, b)* by

$$f_x(a, b), \qquad \frac{\partial f(a, b)}{\partial x}, \qquad \frac{\partial f}{\partial x}(a, b), \qquad or \qquad \frac{\partial}{\partial x} f(a, b)$$

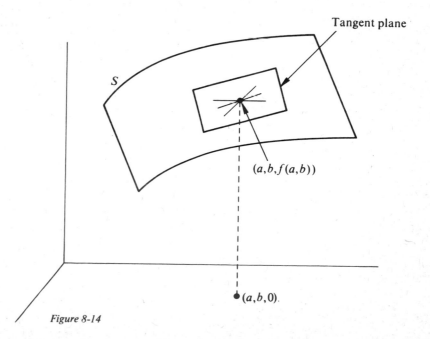

Figure 8-14

and *the partial derivative of f with respect to y at* (a,b) by

$$f_y(a,b), \qquad \frac{\partial f(a,b)}{\partial y}, \qquad \frac{\partial f}{\partial y}(a,b), \qquad or \qquad \frac{\partial}{\partial y} f(a,b)$$

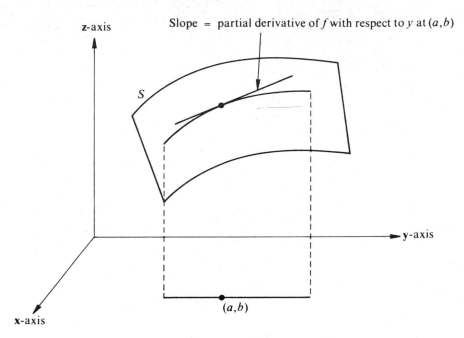

Slope = partial derivative of *f* with respect to *y* at (a,b)

Figure 8-15(a)

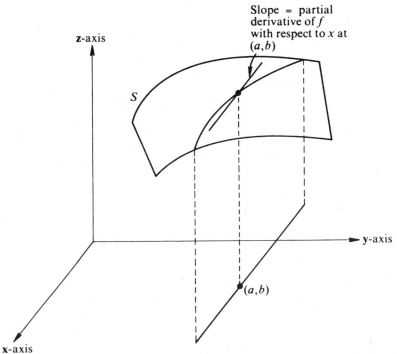

Slope = partial derivative of *f* with respect to *x* at (a,b)

Figure 8-15(b)

A nongeometrical application of partial derivatives can be found in economics. The **utility** of a commodity is the subjective benefit a consumer obtains from its possession, and that utility is often a function of the price of the commodity. Assume that we have two commodities, with prices x and y, respectively. If $u(x, y)$ is the utility related to the two commodities when they are respectively priced at x and y, then $\partial u/\partial x$ and $\partial u/\partial y$ are defined as the **marginal utilities** of the two commodities, respectively.

To compute, say, $f_x(a, b)$, note that we are considering the tangent line to the graph of f only in the direction of the x-axis; thus, to compute the slope, we consider the change in z only in the x-direction, holding y fixed. Similarly, to compute $f_y(a, b)$, we consider the change in z in the y direction, holding x fixed. In other words, to find f_x, we treat y as fixed and, thus, f as a function only of x; and to find f_y, we treat x as fixed and, thus, f as a function only of y.

EXAMPLE 1 Let $f(x, y) = xy^2 + y^4x^3 + x$, all (x, y). Find

$$f_x(-1, 2) \quad \text{and} \quad f_y(-1, 2)$$

SOLUTION To find $f_x(-1, 2)$, we treat y as fixed; that is, we treat y as a fixed number. Thus we can think of f as a function of x only. Hence, for *any* (x, y) in the plane,

$$f_x(x, y) = 1 \cdot y^2 + (3x^2)y^4 + 1 = y^2 + 3x^2y^4 + 1$$

and $\qquad\qquad\qquad f_x(-1, 2) = 53$

Similarly, to find f_y, we treat f as a function of y (with x fixed). Hence, for any (x, y),

$$f_y(x, y) = x(2y) + x^3(4y^3) + 0 = 2xy + 4x^3y^3$$

and $\qquad\qquad\qquad f_y(-1, 2) = -36$

EXAMPLE 2 Let $f(x, y) = x^3y^5 + 6(x)^{1/2}y$. Find $f_x(x, y)$ and $f_y(x, y)$ at any (x, y) with $x > 0$.

SOLUTION For any (x, y) with $x > 0$,

$$f_x(x, y) = 3x^2y^5 + 6 \cdot \frac{1}{2(x)^{1/2}} y = 3x^2y^5 + \frac{3y}{(x)^{1/2}}$$

$$f_y(x, y) = x^3(5y^4) + 6(x)^{1/2} \cdot 1 = 5x^3y^4 + 6(x)^{1/2}$$

It should be clear from the examples that if $\partial f/\partial x$, $\partial f/\partial y$, $\partial g/\partial x$, and if $\partial g/\partial y$ exist, then

$$\frac{\partial}{\partial x}(f+g) = \frac{\partial}{\partial x}f + \frac{\partial}{\partial x}g$$

$$\frac{\partial}{\partial y}(f+g) = \frac{\partial}{\partial y}f + \frac{\partial}{\partial y}g$$

The partial derivative concept extends in a natural way to functions of more than two variables. For example, if f is a function of three variables, then to find $f_x(x,y,z)$, treat y and z as fixed, and thus f as if it is a function of x alone, etc. Now, f_z (or $\partial f/\partial z$) is the partial derivative of f with respect to z.

EXAMPLE 3   Let $f(x,y,z) = xy^2z^3$. At any point (x,y,z), find f_x, f_y, and f_z.

SOLUTION
$$f_x(x,y,z) = 1 \cdot y^2 \cdot z^3 = y^2z^3$$
$$f_y(x,y,z) = x(2y)z^3 = 2xyz^3$$
$$f_z(x,y,z) = xy^2 \cdot 3z^2 = 3xy^2z^2$$

We can now define **higher order partial derivatives** as iterates of partial derivatives. We define the **mixed partials**

$$f_{xy} \equiv \frac{\partial^2 f}{\partial y \partial x} \equiv \frac{\partial}{\partial y}\left(\frac{\partial f}{\partial x}\right) \equiv (f_x)_y$$

$$f_{yx} \equiv \frac{\partial^2 f}{\partial x \partial y} \equiv \frac{\partial}{\partial x}\left(\frac{\partial f}{\partial y}\right) \equiv (f_x)_y$$

and the **second partials**

$$f_{xx} \equiv \frac{\partial^2 f}{\partial x^2} \equiv \frac{\partial}{\partial x}\left(\frac{\partial f}{\partial x}\right)$$

$$f_{yy} \equiv \frac{\partial^2 f}{\partial y^2} \equiv \frac{\partial}{\partial y}\left(\frac{\partial f}{\partial y}\right)$$

Analogous definitions can be made for functions of many variables. For example,

$$f_{zzz} \equiv \frac{\partial^3 f}{\partial z^3} \equiv \frac{\partial}{\partial z}\left[\frac{\partial}{\partial z}\left(\frac{\partial f}{\partial z}\right)\right] \equiv [(f_z)_z]_z$$

$$f_{xyz} \equiv \frac{\partial^2 f}{\partial x \partial y \partial z} \equiv \frac{\partial}{\partial z}\left[\frac{\partial}{\partial y}\left(\frac{\partial f}{\partial x}\right)\right] \equiv [(f_x)_y]_z$$

$$f_{zzxy} = \frac{\partial}{\partial y}\left[\frac{\partial}{\partial x}\left(\frac{\partial}{\partial x}\left\{\frac{\partial f}{\partial z}\right\}\right)\right] \equiv [(\{f_z\}_x)_x]_y$$

EXAMPLE 4   Let $f(x, y) = x^2ye^y$. Then,

$$f_x(x, y) = 2xye^y$$

$$f_y(x, y) = x^2[ye^y + e^y]$$

$$f_{xy}(x, y) = (f_x)_y = \frac{\partial}{\partial y}(2xye^y) = 2x[ye^y + y]$$

$$f_{yx}(x, y) = (f_y)_x = \frac{\partial}{\partial x}[x^2(ye^y + e^y)] = 2x[ye^y + y]$$

$$f_{xx}(x, y) = \frac{\partial}{\partial x}[2xye^y] = 2ye^y$$

$$f_{yy}(x, y) = \frac{\partial}{\partial y}[x^2(ye^y + e^y)] = x^2[ye^y + 2e^y]$$

EXAMPLE 5   Let $f(x, y) = (x + 1)/y$. Then,

$$f_x(x, y) = \frac{1}{y}$$

$$f_y(x, y) = -\frac{x + 1}{y^2}$$

$$f_{xy}(x, y) = \frac{\partial}{\partial y}[f_x] = -\frac{1}{y^2}$$

$$f_{yx}(x, y) = \frac{\partial}{\partial x}[f_y] = -\frac{1}{y^2}$$

$$f_{xx}(x, y) = \frac{\partial}{\partial x}[f_x] = 0$$

$$f_{yy}(x, y) = \frac{\partial}{\partial y}[f_y] = 2\left(\frac{x + 1}{y^3}\right)$$

$$f_{yyy}(x, y) = \frac{\partial}{\partial y}[f_{yy}] = -6\left(\frac{x + 1}{y^4}\right)$$

In both Example 4 and Example 5, it turned out that $f_{xy} = f_{yx}$. This is not always so, but the following can be proved:

Let f be defined in a region R of the plane, and let f_{xy} and f_{yx} be continuous everywhere in R. Then, $f_{xy} = f_{yx}$ everywhere in R.

The following example illustrates **implicit differentiation** for functions of more than one variable.

EXAMPLE 6   Consider all x, y, z for which $x^2/2 + y^2 + z^2/3 = 5$.   (a) Find $\partial z/\partial x$ and $\partial z/\partial y$; (b) find $z_x((2)^{1/2}, -1)$, if $z > 0$.

SOLUTION   (a) Rather than solve explicitly for z, we shall assume $z = z(x, y)$ and obtain

$$\frac{x^2}{2} + y^2 + \frac{[z(x, y)]^2}{3} = 5$$

Consequently,

$$\frac{\partial}{\partial x}\left[\frac{x^2}{2} + y^2 + \frac{[z(x, y)]^2}{3}\right] = \frac{\partial}{\partial x} 5 = 0$$

$$\frac{\partial}{\partial x}\frac{x^2}{2} + \frac{\partial}{\partial x} y^2 + \frac{\partial}{\partial x}\frac{[z(x, y)]^2}{3} = 0$$

$$x + \frac{2}{3} z \cdot \frac{\partial z(x, y)}{\partial x} = 0$$

Finally, $$\frac{\partial z(x, y)}{\partial x} = -\frac{3x}{2z}$$

at each point (x, y, z) on the surface whose equation is $x^2/2 + y^2 + z^2/3 = 5$.

To find $\partial z/\partial y$, we proceed analogously:

$$\frac{\partial}{\partial y}\left[\frac{x^2}{2} + y^2 + \frac{[z(x, y)]^2}{3}\right] = 0$$

$$2y + \frac{2}{3} z \cdot \frac{\partial z(x, y)}{\partial y} = 0$$

Finally, $$\frac{\partial z(x, y)}{\partial y} = -\frac{3y}{z}$$

(b) If $x = (2)^{1/2}$ and $y = -1$, then $[((2)^{1/2})^2/2] + (-1)^2 + z^2/3 = 5$, and thus $z^2 = 9$, so $z = \pm 3$. Since $z > 0$, z must be 3. We have already seen that $\partial z/\partial x = -(3x/2z)$, hence

$$\frac{\partial z}{\partial x}(\sqrt{2}, -1) = -\frac{3(2)^{1/2}}{2 \cdot 3} = -\frac{(2)^{1/2}}{2}$$

Problems   In problems 1–6, find f_x, f_y, and f_z.

1. $f(x, y, z) = xy^2 + yz^3 + xyz$
2. $f(x, y, z) = (x)^{1/2}ye^z$
3. $f(x, y, z) = (x + y^2)/z$
4. $f(x, y, z) = (xy + yz)/xz$
5. $f(x, y, z) = \ln(x + y^2 + z^3)$

6. $f(x,y,z) = \ln \sqrt[5]{x + y^2 + z^3}$

In problems 7–20, find f_x, f_y, f_{xx}, f_{yy}, and f_{xy}.

7. $f(x,y) = x^2y^3$

8. $f(x,y) = 1 + x^2y^3 - 3x^5y^3$

9. $f(x,y) = 1 + (xy)^{1/2}$

10. $f(x,y) = (2x + 3y)^{1/2}$

11. $f(x,y) = e^{2x-3y}$

12. $f(x,y) = e^{xy}$

13. $f(x,y) = \ln(2x + 3y)$

14. $f(x,y) = \ln[(x^2 + y^2)^{1/2}]$

15. $f(x,y) = \ln(xy)^{1/2}$

16. $f(x,y) = \ln(x^2y^4)$

17. $f(x,y) = xe^{xy} - y^2$

18. $f(x,y) = \exp(xy^2 + y)$

19. $f(x,y) = x/y + y/x$

20. $f(x,y) = (x^2 + y^2)/(xy)^2$

In problems 21–32, find $\partial z/\partial x$ and $\partial z/\partial y$.

21. $x^2 + y^2 + z^2 = 1$

22. $3x^2 + 4y^2 + 2z^2 = 5$

23. $x^2/4 + y^2 + z^2/3 = 2$

24. $x^2/9 - y^2/4 + z^2/2 = 1$

25. $3x^2y + y^3z - z^2x = 1$

26. $xy^2 + x^2y + xyz^2 = 5$

27. $x^{1/2} + y^{1/2} + z^{1/2} = 2$

28. $x^{1/3} - x^{1/2}y + yz = 1$

29. $e^{x^2y^3z} = 2x + 3y$

30. $e^{x+2y-z} = x^2 + y^3$

31. $\ln(1 + x^2 + y^2 + z^2) = 2x + y$

32. $\ln(xy + yz + xz) = 5$

Let f be a function of two variables, and suppose $\partial f/\partial x$ and $\partial f/\partial y$ are continuous in some open disk containing (a,b). Then the plane given by

$$z - f(a,b) = f_x(a,b)(x - a) + f_y(a,b)(y - b)$$

is called the **tangent plane** to the graph of f at $[a,b,f(a,b)]$. It can be shown that in a certain precise way, this plane is tangent to the surface given by f at the point on the surface $[a,b,f(a,b)]$.

33. Find the tangent plane to the paraboloid given by

$$f(x,y) = \frac{x^2}{24} + \frac{y^2}{16}$$

at the point $(3,2,\frac{5}{8})$ on the paraboloid.

34. Find the tangent plane to the paraboloid given by

$$f(x,y) = 4x^2 + y^2$$

at the point $(-1,2,8)$ on the paraboloid.

35. A function f is said to be *harmonic on a set S* (in the plane) if

$$f_{xx}(x, y) + f_{yy}(x, y) = 0$$

for all (x, y) in S.

Show that each of the following functions is harmonic on the given set S.

a. $f(x, y) = 3x^2y - y^3$
    $S =$ the plane
b. $f(x, y) = \ln(x^2 + y^2)$
    $S =$ the plane with $(0,0)$ removed

SECTION 4

THE CHAIN RULE

Suppose we are given $f(x, y)$, and we know that $x = u(t)$ and $y = v(t)$ for t in some set. We can then form the *composition function z* given by

$$z(t) = f(u(t), v(t))$$

That is, the function f depends on x and y, but x and y each depend on t, and thus f actually depends only on t (and $z(t)$ is this dependence). To emphasize the chain of dependencies, we shall write $x = \mathbf{x}(t)$, $y = \mathbf{y}(t)$, so that $f(x, y)$ becomes

(8.1) $$z(t) = f[\mathbf{x}(t), \mathbf{y}(t)]$$

EXAMPLE 1 Let $f(x, y) = xy^2 + x^3$, and let $\mathbf{x}(t) = 3t^2$ and $\mathbf{y}(t) = e^t$. Then,

$$f[\mathbf{x}(t), \mathbf{y}(t)] = \mathbf{x}(t)[\mathbf{y}(t)]^2 + [\mathbf{x}(t)]^3$$

$$= (3t^2)(e^t)^2 + (3t^2)^3$$

$$= 3t^2e^{2t} + 27t^6$$

Thus, $$z(t) = 3t^2e^{2t} + 27t^6$$

EXAMPLE 2 Let $f(x, y) = xe^y + y^2e^x$, and let $\mathbf{x}(t) = t + 1$ and $\mathbf{y}(t) = \ln t$. Then, for $t > 0$,

$$f[\mathbf{x}(t), \mathbf{y}(t)] = \mathbf{x}(t) \underline{\quad\quad} + [\mathbf{y}(t)]^2 \underline{\quad\quad}$$

$$= \underline{\quad\quad} + \underline{\quad\quad}$$

The following result in item 1 gives information about the continuity and differentiability of z. A form of the *chain rule* is shown in item 2.

1. *Let $\mathbf{x}$ and $\mathbf{y}$ be continuous for all t in some interval I, and let f be continuous on a region of the plane containing the*

*points [x(t), y(t)], all t in I. Then z defined by equation (8.1)
is continuous for all t in I.*

2. *Let x'(t) and y'(t) exist for all t in some interval I, and let
f be such that f_x and f_y are continuous in a region of the
plane containing [x(t), y(t)] for all t in I. Then, for all t in I,
z' exists, and*

$$z'(t) = f_x[x(t), y(t)] \cdot x'(t) + f_y[x(t), y(t)] \cdot y'(t)$$

or, equivalently,

(8.2)
$$\frac{dz}{dt} = \frac{\partial f}{\partial x}\frac{dx}{dt} + \frac{\partial f}{\partial y}\frac{dy}{dt}$$

EXAMPLE 3   Let $f(x, y) = x^2 + y^2$, all (x, y), $x(t) = 1/t$, $t \neq 0$, and $y(t) = t$,
all t. Observe that

$$z(t) = f(x(t), y(t)) = [x(t)]^2 + [y(t)]^2$$
$$= \frac{1}{t^2} + t^2, \qquad t \neq 0$$

Hence,
$$z'(t) = -\frac{2}{t^3} + 2t, \qquad t \neq 0$$

We obtain the same result using the chain rule as follows:
$f_x(x,y) = 2x$, $f_y(x,y) = 2y$; hence, $f_x[x(t),y(t)] = f_x(1/t,t) = 2/t$,
and $f_y[x(t), y(t)] = f_y(1/t,t) = 2t$. Also, $x'(t) = -1/t^2$, and
$y'(t) = 1$. Therefore, by using the chain rule,

$$z'(t) = f_x[x(t),y(t)]x'(t) + f_y[x(t),y(t)]y'(t)$$
$$= f_x(1/t,t) \cdot x'(t) + f_y(1/t,t)y'(t)$$
$$= \left(\frac{2}{t}\right)\left(-\frac{1}{t^2}\right) + (2t) \cdot 1$$
$$= -\frac{2}{t^3} + 2t \qquad t \neq 0$$

A slightly more general problem arises if x and y are functions
not of the single variable t but rather of two variables u, v. That
is, consider $f(x, y)$ where now $x = x(u,v)$ and $y = y(u,v)$; we
see that $f(x, y)$ becomes

$$f[x(u,v), y(u,v)]$$

and depends only on u, v.

EXAMPLE 4   Let $f(x, y) = 4x - y^2$, and let $\mathbf{x}(u,v) = uv^2$ and $\mathbf{y}(u,v) = u^3v$. Then,

$$f[\mathbf{x}(u,v),\, \mathbf{y}(u,v)] = 4\mathbf{x}(u,v) - [\mathbf{y}(u,v)]^2$$
$$= 4(uv^2) - (u^3v)^2$$
$$= 4\,uv^2 - u^6v^2$$

EXAMPLE 5   Let $f(x, y) = y^4(x^2 - 1)$, and let $\mathbf{x}(u,v) = u + v$ and $\mathbf{y}(u,v) = uv$. Then,

$$f[\mathbf{x}(u,v),\mathbf{y}(u,v)] = [\mathbf{y}(u,v)]^4 \underline{\hspace{2cm}}$$
$$= \underline{\hspace{1.5cm}} (u^2 + 2uv + v^2 - 1)$$

A generalization of the **chain rule** is given by the following:

The Chain Rule   *Let (u,v) be a point on which* $\mathbf{x}$ *and* $\mathbf{y}$ *are continuous and* $\partial \mathbf{x}/\partial u$, $\partial \mathbf{x}/\partial v$, $\partial \mathbf{y}/\partial u$, *and* $\partial \mathbf{y}/\partial v$ *each exist. Let f be such that f_x and f_y are continuous in some rectangular region containing the point* $[\mathbf{x}(u,v),$ $\mathbf{y}(u,v)]$. *Then,*

(8.3) $$\frac{\partial}{\partial u} f[\mathbf{x}(u,v),\, \mathbf{y}(u,v)] = f_x[\mathbf{x}(u,v),\, \mathbf{y}(u,v)]\mathbf{x}_u(u,v)$$
$$+ f_y[\mathbf{x}(u,v),\, \mathbf{y}(u,v)]\mathbf{y}_u(u,v)$$

(8.4) $$\frac{\partial}{\partial v} f[\mathbf{x}(u,v),\, \mathbf{y}(u,v)] = f_x[\mathbf{x}(u,v),\mathbf{y}(u,v)]\mathbf{x}_v(u,v)$$
$$+ f_y[\mathbf{x}(u,v),\mathbf{y}(u,v)]\mathbf{y}_v(u,v)$$

Equations (8.3) and (8.4) are sometimes written respectively as (8.3′) and (8.4′) below, a more succinct, less precise form, but one that is more easily remembered:

(8.3′) $$f_u = f_x \cdot \mathbf{x}_u + f_y \mathbf{y}_u$$

(8.4′) $$f_v = f_x \mathbf{x}_v + f_y \mathbf{y}_v$$

EXAMPLE 6   Let $f(x, y) = 4x - y^2$, $\mathbf{x}(u,v) = uv^2$ and $\mathbf{y}(u,v) = u^3v$. Find f_u {that is, $(\partial / \partial u) f[\mathbf{x}(u,v)]$} and f_v two ways.

SOLUTION   From Example 4, $f[\mathbf{x}(u,v),\mathbf{y}(u,v)] = 4uv^2 - u^6v^2$. Hence

(8.5) $$\frac{\partial}{\partial u} f[\mathbf{x}(u,v),\mathbf{y}(u,v)] = 4v^2 - 6u^5v^2$$

$$\frac{\partial}{\partial v} f[\mathbf{x}(u,v),\mathbf{y}(u,v)] = \underline{\hspace{1.5cm}}$$

A second way is to use the chain, obtaining

$$f_u = f_x x_u + f_y y_u$$
$$= 4v^2 + (-2y)(3u^2 v)$$
$$= 4v^2 - 6yu^2 v$$
$$= 4v^2 - 6(u^3 v) \cdot u^2 v \qquad \text{(since } y = u^3 v)$$
$$= 4v^2 - 6u^5 v^2$$

$$f_v = f_x x_v + f_y y_v$$
$$= 4 \underline{\quad\quad} - 2y \underline{\quad\quad}$$
$$= \underline{\quad\quad} - 2u^3 v \underline{\quad\quad}$$
$$= \underline{\quad\quad\quad\quad}$$

To this point, we have considered $f(x,y)$ where $x = \mathbf{x}(u,v)$ and $y = \mathbf{y}(u,v)$. Of course, there is nothing sacred about the particular choice of symbols for the variables. We might just as well have considered

$$f(u,v) \text{ where } u = \mathbf{u}(x,y) \qquad \text{and} \qquad v = \mathbf{v}(x,y)$$

and obtained $\qquad f[\mathbf{u}(x, y), \mathbf{v}(x, y)]$

which depends only on x, y. By the chain rule,

$$f_x = f_u \mathbf{u}_x + f_v \mathbf{v}_x$$
$$f_y = f_u \mathbf{u}_y + f_v \mathbf{v}_y$$

EXAMPLE 7 Let $f(u,v) = e^{(u/v)}$, $\mathbf{u}(x, y) = x^2 + y$ and $\mathbf{v}(x, y) = x - y$. A direct substitution yields

$$f[\mathbf{u}(x,y), \mathbf{v}(x,y)] = e^{(x^2 + y)/(x-y)}$$

from which f_x and f_y can be computed. However, we shall find f_x and f_y with the chain rule, as follows:

$$f_x[\mathbf{u}(x,y), \mathbf{v}(x,y)] = f_u[\mathbf{u}(x,y), \mathbf{v}(x,y)]\mathbf{u}_x(x,y)$$
$$+ f_v[\mathbf{u}(x,y), \mathbf{v}(x,y)] \cdot \mathbf{v}_x(x,y)$$
$$= e^{u/v}\left(\frac{1}{v}\right)2x + e^{u/v}\left(-\frac{u}{v^2}\right) \cdot 1$$
$$= \frac{\exp{(x^2 + y)/(x - y)}}{x - y}\left[2x - \frac{x^2 + y}{x - y}\right]$$

$$f_y = f_u \mathbf{u}_y + f_v \mathbf{v}_y$$
$$= e^{u/v}\left(\frac{1}{v}\right) \cdot 1 + e^{u/v}\left(-\frac{u}{v^2}\right)(-1)$$
$$= \frac{\exp{[(x^2 + y)/(x - y)]}}{x - y}\left[1 + \frac{x^2 + y}{x - y}\right]$$

EXAMPLE 8   If $z = f(x^2 + y^2)$, show that

$$x \frac{\partial f}{\partial y} - y \frac{\partial f}{\partial x} = 0$$

SOLUTION   Let $\mathbf{u}(x,y) = x^2 + y^2$. Then $f(x^2 + y^2) = f[\mathbf{u}(x,y)]$. Consequently,

$$f_x = f_u \cdot \mathbf{u}_x = f_u \cdot 2x$$
$$f_y = f_u \cdot \mathbf{u}_y = f_u \cdot 2y$$

Hence, $xf_y - yf_x = x \cdot f_u \cdot 2y - y f_u \cdot 2x = 0$

EXAMPLE 9   Let p be the price for some commodity, and let q be the quantity of the commodity produced. Economists talk about a *constant outlay* when for some positive constant c, we have $pq = c$. Assuming a constant outlay, and that the quantity q is a function of the price p, use implicit differentiation to find dq/dp.

SOLUTION   Let $q(p)$ be the quantity of the commodity produced when the commodity is offered at price p. Since $pq = c$, we see that $pq - c = 0$. Let $u(p,q) = pq - c$. Consequently,

$$u(p,q) = pq - c = 0$$

Thus, $$u(p,q(p)) = 0$$

And so, differentiating with respect to p, we see that

$$u_p + u_q q'(p) = 0$$

That is, $$q'(p) = -(u_p/u_q)$$

Notice that

$$u_p = \frac{\partial}{\partial p}(pq - c) = q$$

and $$u_q = p$$

Thus, $$q'(p) = -(q/p)$$

Answers   Example 2. $e^{y(t)}$ or $e^{\ln t}$; $e^{x(t)}$ or e^{t+1}; $(t + 1)e^{\ln t}$; $(\ln t)^2 e^{t+1}$ (the solution simplifies slightly to $t(t + 1) + (\ln t)^2 e^{t+1}$.)
Example 5. $[x(u,v)]^2 - 1$, or $(u + v)^2 - 1$; $(uv)^4$, or $u^4 v^4$
Example 6. $f_v = 8uv - 2u^6 v$;
$$f_v = f_x x_v + f_y y_v$$
$$= 4 \cdot 2uv - 2y \cdot u^3$$
$$= 8uv - 2(u^3 v) \cdot u^3$$
$$= 8uv - 2u^6 v$$

Problems   1.   Let $f(u,v) = 2uv + v^2$, and let $\mathbf{u}(t) = -3t^2$ and $v(t) = 1 + t^3$. Let
$$G(t) = f[\mathbf{u}(t), \mathbf{v}(t)]$$
a.   Find G explicitly, and then find G'.
b.   Use the chain rule to find G'.

2.   Repeat problem 1, if now $f(u,v) = -uv + v^3$, and $\mathbf{u}(t) = t^{1/3}$, and $\mathbf{v}(t) = (1 + t)^{1/3}$.

3.   Repeat problem 1 if now $f(u,v) = (1 + u^2 + v^2)^{1/2}$, and $\mathbf{u}(t) = 3t + 1$ and $v(t) = t - 1$.

4.   Repeat problem 1 if now $f(u,v) = (4 + v^2)u + e^u$, $\mathbf{u}(t) = \ln(1 + t^2)$, and $v(t) = 2e^{3t}$.

5.   Let $f(u,v) = uv + v^2$, and let $\mathbf{u}(x,y) = x + y$, and $\mathbf{v}(x,y) = x - y$. Let
$$g(x,y) = f[\mathbf{u}(x,y), \mathbf{v}(x,y)]$$
a.   Find g explicitly; then find $\partial g/\partial x$ and $\partial g/\partial y$.
b.   Use the chain rule to find $\partial g/\partial x$ and $\partial g/\partial y$.

6.   Let $f(u,v) = u^2 - v^2$, and let $\mathbf{u}(x,y) = x + 2y$ and $\mathbf{v}(x,y) = x - 2y$.
a.   Explicitly find $f[\mathbf{u}(x,y), \mathbf{v}(x,y)]$, and then find
$$\frac{\partial}{\partial x} f[\mathbf{u}(x,y), \mathbf{v}(x,y)] \quad \text{and} \quad \frac{\partial}{\partial y} f[\mathbf{u}(x,y), \mathbf{v}(x,y)]$$
b.   Use the chain rule to obtain the same partial derivatives.

7.   Repeat problem 6, but now let $f(u,v) = u^2 + v^2$.

8.   Let $f(u,v) = e^{u/v}$, and let $\mathbf{u}(x,y) = 2x - y$, and $\mathbf{v}(x,y) = x + 2y$.
Find $\partial f/\partial x$ and $\partial f/\partial y$.

In each of the following problems, assume that all the functions which arise have whatever derivatives are necessary for the problem to be meaningful.

9.   If $\mathbf{z}(x,y) = f(xy^2)$, show that $2xz_x - yz_y = 0$. (*Hint:* Let $\mathbf{u}(x,y) = xy^2$. Then, $z = f[\mathbf{u}(x,y)]$. Now use the chain rule to find z_x and z_y.)

10.   Let $\mathbf{z}(x,y) = x + f(x^2y^2)$. Show that $xz_x - yz_y = x$. (*Hint:* See problem 9.)

11.   Let $\mathbf{z}(x,y) = f(x - y, y - x)$. Show that $z_x + z_y = 0$. (*Hint:* Let $\mathbf{u}(x,y) = x - y$ and $\mathbf{v}(x,y) = y - x$.)

12.   Let $u(x,y)$ be the utility if two commodities are offered at prices x and y, respectively. Assume that there is some relationship between x and y for which u is constant; say that $y = g(x)$ is this relationship. The graph is g is said to be an **indifference curve**, since all prices (x, y) on this curve give rise to the same utility. Show that
$$g'(x) = -\frac{\text{marginal utility of } x}{\text{marginal utility of } y}$$

EXTREMA OF FUNCTIONS OF TWO VARIABLES

In an earlier chapter, we considered the problem of finding the highest or lowest points on the graph of a function. This was the problem of finding the extrema of a function (of one variable). The analogous problem for functions of two variables is to find the highest or lowest points on the surface, which is the graph of the function.

Let f be defined in a region R of the plane. Then, $f(a,b)$ *is the* **maximum of f in R** *if* $f(a,b) \geq f(x,y)$ *for all* (x,y) *in R.* *The* **minimum** *of f in R is defined similarly.* Let (a,b) be in the *interior** of the region R; to say that f has a **local** *(or* **relative***)* **maximum** *at* (a,b) means $f(a,b) \geq f(x,y)$ *for all* (x,y) *in R that are near* (a,b). A local minimum is defined similarly.

The problem of locating absolute extrema for functions of two variables is intrinsically more complicated than the analogous problem for functions of one variable. But we can conclude that **if f has a local maximum (or minimum) at (a,b) and if $f_x(a,b)$ and $f_y(a,b)$ exist, then $f_x(a,b) = 0 = f_y(a,b)$.** To see this, observe that if f has a local extremum at (a,b), then the tangent lines to the surface at $[a,b,f(a,b)]$ will be parallel to the **xy**-plane in the direction of both the **x**- and **y**-axes (in fact, in *all* directions). See Figure 8-16. Since $f_x(a,b)$ and $f_y(a,b)$ exist, we see that $f_x(a,b) = 0$ and $f_y(a,b) = 0$.

Unfortunately, $f_x(a,b) = 0 = f_y(a,b)$ does not insure the presence of a local extremum, as the next example illustrates.

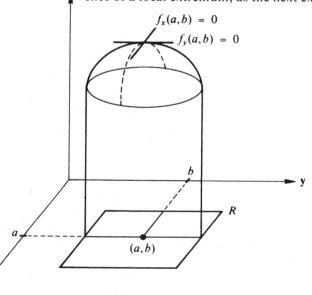

Figure 8-16

EXAMPLE 1 Let $f(x, y) = x^2 - y^2 + 1$. Then, $f_x(0,0) = 0 = f_y(0,0)$. However, we saw in section 1 that the surface for f is saddle-shaped, and has no extrema (see Figure 8-17).

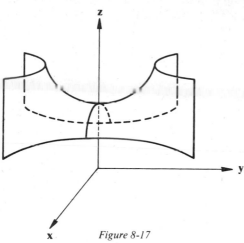

Figure 8-17

Nevertheless, for many problems, relatively simple considerations show that the given function has an extremum. The following is such an example.

EXAMPLE 2 Let $f(x, y) = 1 - x^2 - y^2$, all (x, y). Find the extrema of f.

SOLUTION Observe that $f_x(x, y) = -2x$ and $f_y(x, y) = -2y$; consequently, $f_x(x, y) = 0 = f_y(x, y)$ only at $(0,0)$. Hence, an extremum can occur only at $(0,0)$. Since $f(0,0) = 1$ and $1 > 1 - x^2 - y^2$ for all $(x, y) \neq (0,0)$, it follows that $f(0,0) = 1$ is the absolute maximum of f (see Figure 8-18).

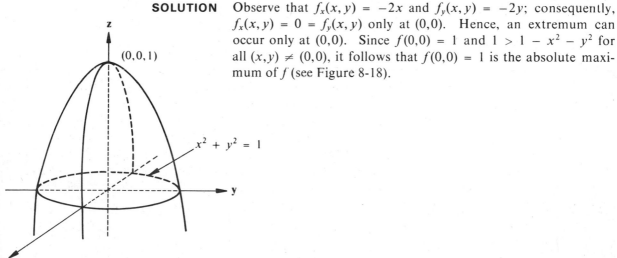

Figure 8-18

EXAMPLE 3   What are the dimensions of the rectangular box of fixed volume that has a minimum surface area?

SOLUTION   Let x, y, and z be the dimensions of the box, and let a^3 be the fixed volume. The problem then is to maximize the surface area

$$A = 2(xy + yz + xz) \qquad (x > 0, y > 0, z > 0)$$

subject to the constraint on volume that

$$xyz = a^3$$

From the constraint, we see that $z = a^3/xy$; hence, substituting into A, we see that we must minimize the function f where

$$A = f(x,y) = 2\left(xy + \frac{a^3}{x} + \frac{a^3}{y}\right)$$

We solve

$$f_x(x,y) = 2\left(y - \frac{a^3}{x^2}\right) = 0,$$

$$f_y(x,y) = 2\left(x - \frac{a^3}{y^2}\right) = 0$$

Hence,

$$x^2 y = a^3 \qquad \text{and} \qquad xy^2 = a^3$$

so that (after some algebra) $x = y = a$. Since $xyz = a^3$, we obtain $z = a$. Therefore, the cube is the rectangular box which has minimum surface area. To convince ourselves that we have minimized f, observe that for x or y near 0 (that is, near either coordinate axis in the plane), $f(x,y)$ is positive and large, whereas if x or y is large and positive then $f(x,y)$ is positive and large. Consequently, we can expect f to have a minimum for $x > 0$, $y > 0$.

In our earlier work on extrema, we discussed the use of the second derivative to determine whether or not a function (of one variable) had a local extremum. The following (which we shall not prove) is the analog of the second derivative test.

Test for local extrema.   *Let f be a function of two variables having continuous second partial derivatives in a rectangular region R (without boundary), and let (a,b) be a point in R at which*

$$f_x(a,b) = 0 \qquad \text{and} \qquad f_y(a,b) = 0$$

Define the function F by

$$F(x,y) = f_{xx}(x,y)\, f_{yy}(x,y) - [f_{xy}(x,y)]^2$$

1.   *If $F(a,b) > 0$ and if $f_{xx}(a,b) > 0$, then f has a local minimum at (a,b).*

2.   If $F(a,b) > 0$, and if $f_{xx}(a,b) < 0$, then f has a local max-
     imum at (a,b).
3.   If $F(a,b) < 0$, then f does not have a local extremum at
     (a,b).

EXAMPLE 4   In Example 1, with $f(x,y) = x^2 - y^2 + 1$, we saw that $f_x(0,0) = 0 = f_y(0,0)$, but the surface for f is saddle-shaped near $(0,0)$. Observe that $f_{xx}(x,y) = 2$, $f_{yy}(x,y) = -2$, and $f_{xy}(x,y) = 0$. Thus, with F defined above,

$$F(0,0) = f_{xx}(0,0)\, f_{yy}(0,0) - [f_{xy}(0,0)]^2 = -4$$

Hence, the test for local extrema shows that f has *no* local extremum at $(0,0)$.

EXAMPLE 5   As in Example 2, let $f(x,y) = 1 - x^2 - y^2$, and note that $f_x(x,y) = 0$ and $f_y(x,y) = 0$ if and only if $x = y = 0$. Note that $f_{xx}(x,y) = -2$, $f_{yy}(x,y) = -2$, and $f_{xy}(x,y) = 0$; hence, $F(0,0) = 4$. Thus, f has a local maximum at $(0,0)$, (a fact we already observed in Example 2).

EXAMPLE 6   Let $f(x,y) = 2x^3 - 6xy + y^3$. Find all the local extrema for f.

SOLUTION   We shall need the following information:

$$f_x(x,y) = 6x^2 - 6y = 6(x^2 - y)$$
$$f_y(x,y) = -6x + 3y^2 = 3(-2x + y^2)$$
$$f_{xx}(x,y) = 12x, \qquad f_{yy}(x,y) = 6y, \qquad f_{xy}(x,y) = -6$$

To find the possible locations of local extrema, we must find all (x,y) for which

$$f_x(x,y) = 6(x^2 - y) = 0$$

and $\qquad\qquad f_y(x,y) = 3(-2x + y^2) = 0$

Let (x,y) be such a point. Then, $x^2 - y = 0$ if and only if $y = x^2$; hence,

$$0 = -2x + y^2 = -2x + (x^2)^2 = -2x + x^4$$
$$= x(-2 + x^3)$$

Consequently,

$$x = 0 \qquad \text{and} \qquad y = 0^2 = 0$$

or $\qquad\quad x = 2^{1/3} \qquad \text{and} \qquad y = (2^{1/3})^2 = 2^{2/3}$

Thus, the possible locations of local extrema are $(0,0)$ and $(2^{1/3}, 2^{2/3})$.

Applying the second derivative test, we see that at $(0,0)$,

$$F(0,0) = f_{xx}(0,0)\,f_{yy}(0,0) - [f_{xy}(0,0)]^2 = -36 < 0$$

and, therefore, f has *no* local extremum at $(0,0)$. Next, at $(2^{1/3}, 2^{2/3})$,

$$F(2^{1/3}, 2^{2/3}) = 12 \cdot 2^{1/3} \cdot 6 \cdot 2^{2/3} - 36$$
$$= 6 \cdot 12 \cdot 2 - 36 = 108 > 0$$

and $f_{xx}(2^{1/3}, 2^{2/3}) = 12 \cdot 2^{1/3} > 0$; hence, f has a local minimum at $(2^{1/3}, 2^{2/3})$.

Problems   In problems 1–11, find all the local extrema.

1. $f(x,y) = x^2 - 4x + y^2 - 8y + 5$
2. $f(x,y) = (x-2)^2 + (y+3)^2$
3. $f(x,y) = -x^3 + 9x - 4y^2$
4. $f(x,y) = -x^4 - 32x + y^3 - 12y + 7$
5. $f(x,y) = x^2 - 4xy + 2y^2$
6. $f(x,y) = (x-y)^2 + y^2$
7. $f(x,y) = 9xy - x^3 - y^3 + 10$
8. $f(x,y) = 2x^2 + y^2 - 2xy - 4x + 8$
9. $f(x,y) = 3xy + 4x^2 - 2y^2 - 5x - 7y + 5$
10. $f(x,y) = x^3 + y^3 - 3xy + 10$
11. $f(x,y) = xye^{-(x+y)}$, with $x > 0$ and $y > 0$
12. Find the shortest distance from the origin to the surface given by $y^2 - z^2 = 10$.
13. A rectangular box with no top is to have a fixed volume. What should its dimensions be if we want to use the least amount of material?
14. The total profit per acre on a wheat range is related to two cost factors: labor, and soil additives. Let $P(x,y)$ be the dollar profit per acre from spending x dollars per acre on labor and y dollars per acre on soil additives, and assume that

$$P(x,y) = 48x + 60y + 10xy - 10x^2 - 6y^2$$

Determine x and y for which $P(x,y)$ is maximum.
15. A store determines that its earnings $E(x,y)$ (in thousands of dollars), represented in terms of x, the investment in inventory (in thousands of dollars), and y, the space set aside for sales (in units of 10,000 square feet), is given by

$$E(x,y) = 4x + 5y + xy - x^2 - y^2$$

Find the maximum earnings, and the amount of inventory and space that afford this maximum.

SECTION 6
LAGRANGE MULTIPLIERS

In section 5, we considered the problem of minimizing the surface area of a rectangular box subject to the constraint that the box have fixed volume. This is an example of the problem of finding extreme values of a function subject to constraints on the domain of the function. Let us pursue this problem with another example, which leads us to a general procedure for finding extrema of functions subject to constraints on their domains.

EXAMPLE 1 Find the rectangular box of maximum volume if the surface area is fixed at a^2 square units.

SOLUTION Let x, y, and z be the dimensions of the box. Then for each (x,y,z), with $x > 0, y > 0, z > 0$,

$$f(x,y,z) = xyz$$

is the volume of the box. Our problem now is to maximize f, subject to the constraint

$$2(xy + yz + xz) = a^2$$

Finally, define g by

$$g(x,y,z) = 2(xy + yz + xz) - a^2$$

In terms of f and g, the problem now reads, Find the maximum of the numbers $f(x,y,z)$ subject to the constraint that the points (x,y,z) lie on a surface given by $g(x,y,z) = 0$. Because it is possible to put any constraint problem in this form, we shall pursue this problem in general.

We might explicitly solve for z in $g(x,y,z) = 0$, substitute this value for z in $f(x,y,z)$, and then maximize f; that is, solve for z in $g(x,y,z) = 0$ to obtain $z = u(x,y)$, substitute into f to obtain the numbers $f[x,y,u(x,y)]$ and then maximize this function of *two* variables. In Example 1, explicitly solving for z is possible but complicated; in some problems, it might be impossible.

A method devised by the 18th-century mathematician Joseph Louis Lagrange, which we shall call **Lagrange's method**, provides a way for solving many such problems. The method is as follows:

Let f and g be functions with continuous partial derivatives, and assume that (x,y,z) is a point for which $g(x,y,z) = 0$, and at least one of the derivatives $g_x(x,y,z)$, $g_y(x,y,z)$, $g_z(x,y,z)$ is not zero. Then, if the function f has a local extremum at a point (x,y,z) for which $g(x,y,z) = 0$ (that is, if f has a local extremum at (x,y,z)

subject to the constraining $g(x,y,z) = 0$), then there is a real number λ, called the **Lagrange multiplier**, *for which*

$$f_x(x,y,z) + \lambda g_x(x,y,z) = 0$$
$$f_y(x,y,z) + \lambda g_y(x,y,z) = 0$$
$$f_z(x,y,z) + \lambda g_z(x,y,z) = 0$$
$$g(x,y,z) = 0$$

In practice, the problem is to "solve" these equations to explicitly determine (x,y,z) and λ. Before illustrating the procedure, we shall sketch a proof that this method works.

In the spirit of implicit functions, let us assume that we have solved for z in the equation $g(x,y,z) = 0$, and have obtained $z = \mathbf{z}(x,y)$.* To maximize the numbers $f[x, y, \mathbf{z}(x,y)] = G(x,y)$, we must find (x,y) for which

$$G_x(x,y) = 0, \qquad G_y(x,y) = 0$$

By the chain rule, this becomes

(8.6)
$$G_x = f_x + f_z \cdot \mathbf{z}_x = 0$$
$$G_y = f_y + f_z \cdot \mathbf{z}_y = 0$$

However, if we substitute $\mathbf{z}(x,y)$ into the constraint equation $g(x,y,z) = 0$, we obtain $g[x,y, \mathbf{z}(x,y)] = 0$. Applying the chain rule, we obtain

$$g_x + g_z \cdot \mathbf{z}_x = 0 \qquad \text{and} \qquad g_y + g_z \cdot \mathbf{z}_y = 0$$

or, equivalently,

(8.7)
$$\mathbf{z}_x = -\frac{g_x}{g_z} \qquad \text{and} \qquad \mathbf{z}_y = -\frac{g_y}{g_z}$$

Substituting (8.7) into (8.6) yields

$$f_x + f_z \left(-\frac{g_x}{g_z} \right) = 0$$
$$f_y + f_z \left(-\frac{g_y}{g_z} \right) = 0$$

or, equivalently,

(8.8)
$$f_x - g_x \frac{f_z}{g_z} = 0$$
$$f_y - g_y \frac{f_z}{g_z} = 0$$

---

*The condition that the partial derivatives are not zero simultaneously is used to guarantee that there is a function $\mathbf{z}$.

Also, it is clear that

(8.9)
$$f_z - g_z \frac{f_z}{g_z} = 0$$

Let $\lambda = -f_z/g_z$; then (8.8) and (8.9) take the more symmetric form

$$f_x + \lambda g_x = 0$$
$$f_y + \lambda g_y = 0$$
$$f_z + \lambda g_z = 0$$
$$g(x,y,z) = 0$$

which is the set of equations we set out to derive.

We now have a system of four equations in the four unknowns x, y, z, and λ. In particular, the points (x,y,z) that solve this system are possible values at which f attains an extreme value. Although more sensitive tests exist that indicate whether or not such points are points at which f attains an extreme value, we shall be content here to rely on geometry and intuition.

We can now conclude our solution of Example 1, by maximizing

$$f(x,y,z) = xyz$$

subject to the constraint

$$g(x,y,z) = 2(xy + yz + xz) - a^2 = 0$$

Using Lagrange multipliers, we see that

$$f_x(x,y,z) + \lambda g_x(x,y,z) = yz + 2\lambda(y + z) = 0$$
$$f_y(x,y,z) + \lambda g_y(x,y,z) = xz + 2\lambda(x + z) = 0$$
$$f_z(x,y,z) + \lambda g_z(x,y,z) = xy + 2\lambda(x + y) = 0$$
$$g(x,y,z) = 2(xy + yz + xz) - a^2 = 0$$

From the first two equations we get

$$\frac{y}{x} = \frac{y + z}{x + z}$$

or, after some algebra,

$$z(y - x) = 0$$

Since z is positive, this implies that $y = x$. Applying the same reasoning to the second and third equations gives $z = y$. Hence, $x = y = z$, and we find that the desired box is a cube. Since $g(x,x,x) = 6x^2 - a^2 = 0$, we see that each edge of the cube is to have length $a/(6)^{1/2}$.

EXAMPLE 2   Find the shortest distance from the origin to the surface S given by $y^2 - z^2 = 9$.

SOLUTION   Let (x,y,z) be any point on the surface S. Then, $F(x,y,z)$, the square of the distance from $(0,0,0)$ to (x,y,z), is given by

$$F(x,y,z) = x^2 + y^2 + z^2$$

Note that if we minimize F, we shall at the same time minimize the distance itself. Let $g(x,y,z) = y^2 - z^2 - 9$, and note that (x,y,z) is on S if and only if $g(x,y,z) = 0$. Thus, we are to minimize $F(x,y,z)$ subject to the constraint that $g(x,y,z) = 0$. Using the method of Lagrange multipliers, we obtain the following set of equations:

$$F_x(x,y,z) + \lambda g_x(x,y,z) = 2x + \lambda \cdot 0 = 0$$
$$F_y(x,y,z) + \lambda g_y(x,y,z) = 2y + \lambda \cdot 2y = 0$$
$$F_z(x,y,z) + \lambda g_z(x,y,z) = 2z - 2\lambda z = 0$$
$$g(x,y,z) = y^2 - z^2 - 9 = 0$$

That is, $x = 0$, $y(1 + \lambda) = 0$, $z(1 - \lambda) = 0$,

and $y^2 - z^2 = 9$

Let us investigate some of the possibilities. First, x must be 0. Since $y(1 + \lambda) = 0$, either $y = 0$ or $1 + \lambda = 0$. Let us try $y = 0$. But if $y = 0$, then

$$9 = y^2 - z^2 = -z^2$$

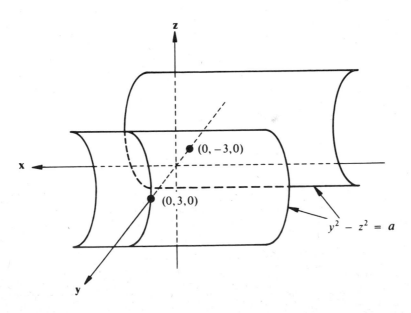

$(0, -3, 0)$

$(0, 3, 0)$

$y^2 - z^2 = a$

Figure 8-19

and has no real number solution. Hence, we must exclude the case $y = 0$. But then $1 + \lambda = 0$; hence, $\lambda = -1$. Thus,

$$z(1 - \lambda) = z(1 - (-1)) = 2z = 0$$

and $z = 0$. If $z = 0$, then

$$y^2 = 9 \qquad \text{and we have} \qquad y = \pm 3$$

Hence, $(0,3,0)$ and $(0,-3,0)$ are the points on S at which the distance from the origin is smallest (see Figure 8-19).

Recall that in the analysis of functions of one variable, we found that a continuous function has an *absolute extremum* on $[a,b]$ either at a point at which it has a *local* extremum in (a,b), or an *endpoint* at a or b. A similar result holds for functions of more than one variable; that is, a continuous function of, say, two variables defined on a "nice" region* attains its *absolute* extremum either at a local extremum (which can be found by using Lagrange multipliers, or the methods of the previous section), or at a point on the boundary of the region of interest. Very often, geometrical or algebraic considerations can settle the question, although sometimes the analysis is quite complicated. We shall not pursue this problem here.

Finally, it should be noted that Lagrange's method applies to functions other than those of three variables. For example, under the same types of assumptions as for functions of three variables, we have the following for a function f of two variables:

If f has a local extremum at a point (x,y) for which $g(x,y) = 0$, then there is a Lagrange multiplier, λ, for which

$$f_x(x,y) + \lambda g_x(x,y) = 0$$
$$f_y(x,y) + \lambda g_y(x,y) = 0$$

and
$$g(x,y) = 0$$

EXAMPLE 3   Let a, b, and c be real numbers. Find the shortest distance between the line given by $ax + by + c = 0$ and the origin.

SOLUTION   Let $f(x,y) = x^2 + y^2$ and $g(x,y) = ax + by - c$

The problem is equivalent to minimizing f subject to the constraint $g = 0$. By Lagrange's method, we obtain

$$f_x + \lambda g_x = 0: \quad 2x \qquad + a\lambda = 0$$
$$f_y + \lambda g_y = 0: \qquad \quad 2y + b\lambda = 0$$
$$g = 0: \quad ax + by \qquad = c$$

*For example, rectangular, triangular, circular, semicircular, and the like.

We must "solve three equations in three unknowns." Notice that by multiplying in the first equation by $-a/2$, the first and third equations are equivalent to

$$-\frac{a}{2} \cdot 2x \quad\quad -\frac{a}{2} \cdot a\lambda = 0$$

$$ax + by \quad\quad\quad = c$$

and thus we get, by adding 2 in the above equations

$$by - \frac{a^2}{2}\lambda = c$$

Hence, the system of equations is equivalent to

$$2x \quad\quad + a\lambda = 0$$
$$2y + \quad b\lambda = 0$$
$$by - \frac{a^2}{2}\lambda = c$$

Next, multiplying by $-b/2$ in the second equation, we obtain

$$2x \quad\quad + \quad\quad a\lambda = 0$$
$$-\frac{b}{2} \cdot 2y + \left(-\frac{b}{2}\right)b\lambda = 0$$
$$b \cdot y - \frac{a^2}{2}\lambda = c$$

or, adding the second and third equations,

$$2x \quad\quad\quad + a\lambda = 0$$
$$2y \quad\quad\quad + b\lambda = 0$$
$$-\tfrac{1}{2}(a^2 + b^2)\lambda = c$$

Hence, solving for λ, we obtain

$$\lambda = \frac{-2c}{a^2 + b^2}$$

from which we see that

$$y = \frac{-b\lambda}{2} = \frac{bc}{a^2 + b^2}$$

and

$$x = -\frac{a\lambda}{2} = \frac{ac}{a^2 + b^2}$$

Therefore, the minimum value of f is attained at $[bc/(a^2 + b^2), ac/(a^2 + b^2)]$; and thus, the minimum value of f is

$$f(x,y) = f\left(\frac{ac}{a^2 + b^2}, \frac{bc}{a^2 + b^2}\right)$$

$$= \left(\frac{ac}{a^2 + b^2}\right)^2 + \left(\frac{bc}{a^2 + b^2}\right)^2$$

$$= \frac{c^2}{a^2 + b^2}$$

Since $c^2/(a^2 + b^2)$ is the minimum value of the square of the distance, we see that

the shortest distance from the line given by $ax + by + c = 0$

and the origin is

$$\frac{|c|}{(a^2 + b^2)^{1/2}}$$

EXAMPLE 4 A producer buys "inputs" x and y to produce an "output," say $f(x, y)$. For example, x, the quantity of the first input, might be the quantity of labor; and y, the quantity of the second input, might be the quantity of some raw material. Assume that the price of the inputs x and y are a and b, respectively. Maximize the input f subject to a total budget of c (dollars).

SOLUTION The problem is to maximize f subject to the constraint $ax + by = c$. (Why?) Let $g(x, y) = c - ax - ay = 0$; then,

$$f_x + \lambda g_x = f_x + \lambda a = 0$$
$$f_y + \lambda g_y = f_y + \lambda b = 0$$
$$g(x, y) = c - ax - ay = 0$$

These equations reduce to

$$\frac{f_x}{a} = \frac{f_y}{b} = -\lambda$$

This equation is called the law of **equi-marginal productivity**, and its solution is called the **least-cost combination of inputs**.

Problems 1. Let $f(x, y, z) = 2x^2 + 4y^2 + z^2$. Find the extremal values of f subject to the constraint $4x - 8y + 2z = 10$.

2. Let $f(x, y, z) = x^2 + y^2 + z^2$ and let $g(x, y, z) = 4x^2 + 2y^2 + z^2 - 4$. Find the local extrema for f subject to the constraint $g = 0$.

3. Let $f(x, y, z) = (xyz)^2$. Show that the maximum value of f, subject to the constraint $x^2 + y^2 + z^2 = R^2$, is $(R^2/3)^3$.

4. Show that for all rectangles with fixed perimeter, the square encloses the largest area.

5. A rectangular box with no top is to be constructed from 96 square feet of material. What should be the dimensions of the box if it is to enclose a maximum volume?

6. Find the point on the plane $x + y + z = 1$ that is nearest the origin.

7. Find the shortest distance between the plane given by $x - 2y + 4z = 2$ and the origin.

8. Divide the number 12 into three parts in such a way that the product xy^2z^3 is a maximum.

9. Find the largest and smallest distances from the origin $(0,0,0)$ to the ellipsoid

$$\frac{x^2}{a^2} + \frac{y^2}{b^2} + \frac{z^2}{c^2} = 1$$

Assume $a > b > c > 0$.

10. Find the shortest distance from the origin to the plane $Ax + By + Cz + D = 0$.

11. Find the maximum and minimum distances from the origin $(0,0)$ to the ellipse $5x^2 - 6xy + 5y^2 - 4 = 0$.

12. Show that among all triangles having perimeter $2s$, the equilateral triangle has the greatest area.

13. A firm has \$100,000 to spend on labor and raw materials. Let x and y represent, respectively, the quantity of labor to be hired and the raw materials to be purchased. Assume that the *unit* price for the hiring of labor is \$2,000 and for the purchase of raw material is \$1,000. Thus, in thousands of dollars, the firm is operating on the constraint that the total expenditure on these two items is given by $2x + y = 100$ (Why?). If w, the output of the firm, is related to x and y by

$$w(x,y) = 5xy$$

use Lagrange multipliers to find the values of x and y that maximize w subject to the given budget constraint.

SECTION 7
DOUBLE INTEGRALS. BASIC PROPERTIES

The integral of a positive continuous function was used to define area. We seek an extension of the integral that will define volume. Let f be a function that is positive and continuous on a rectangular region $R: a \le x \le b, c \le y \le d$ (see Figure 8-20). We wish to define the volume bounded by G, the graph of f, the region R, and the (vertical) planes $x = a$, $x = b$, $y = c$, $y = d$ (Figure 8-20).

The plan is to approximate the volume by summing the volumes of appropriate parallelepipeds, and then to take a limit. We shall subdivide the rectangular region R into smaller rectangular regions and mount a parallelepiped on each of these smaller regions. Specifically, we divide $[a,b]$ into n subintervals by introducing the points $x_0, x_1, x_2, \ldots, x_n$, where $a = x_0 < x_1 < x_2 < \cdots < x_{n-1} < x_n = b$, and we divide $[c,d]$ into m subintervals by introducing the points $y_0, y_1, \ldots, y_m$, where $c = y_0 <

$y_1 < \cdots < y_{m-1} < y_m = d$; and then we form the grid pictured in Figure 8-21. Let R_{ij} be the subrectangle that is the set of (x,y), for which $x_{i-1} \leq x \leq x_i$, $y_{j-1} \leq y \leq y_j$, and let ΔR_{ij} be its area. If $\Delta x_i = x_i - x_{i-1}$ and $\Delta y_j = y_j - y_{j-1}$,

then $$\Delta R_{ij} = \Delta x_i \Delta y_j$$

Now, pick *any* point $(\bar{x}_i, \bar{y}_j)$ in R_{ij}. The volume of the parallel-epiped of height $f(\bar{x}_i, \bar{y}_j)$ and base R_{ij} is

$$f(\bar{x}_i, \bar{y}_j) \, \Delta R_{ij} = f(\bar{x}_i, \bar{y}_j) \, \Delta x_i \Delta y_j$$

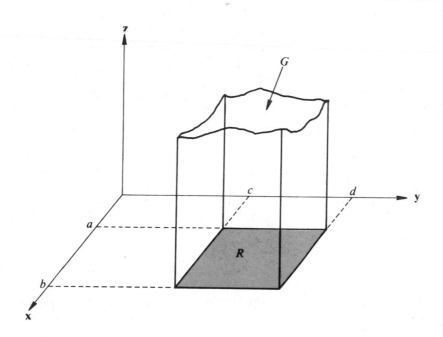

Figure 8-20

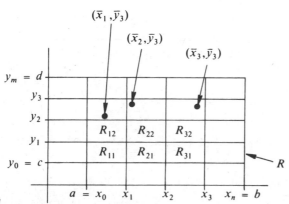

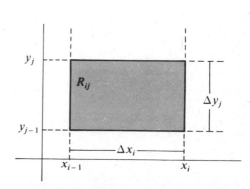

Figure 8-21

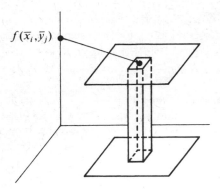

$f(\bar{x}_i, \bar{y}_j)$

Figure 8-22

(see Figure 8-22), and the sum of all the volumes is

$$\sum_{j=1}^{m} \sum_{i=1}^{n} f(\bar{x}_i, \bar{y}_j) \, \Delta x_i \Delta y_j$$

We now let the number of subrectangles increase indefinitely in such a way that the diagonal of the subrectangle with maximum area tends to zero. It can be shown that for *continuous* (not necessarily positive) f, this limit exists and is independent of the choice of grid or the choice of $(\bar{x}_i, \bar{y}_j)$. We denote the limit by

$$\int_c^d \int_a^b f(x,y)dx \, dy \qquad \text{or} \qquad \iint_R f$$

the **double-integral of f over R**. If $f \geq 0$ on R, then we define the volume described earlier by this double integral.

The double-integral shares some important properties with the ordinary definite integral. For instance,

Let f and g be continuous on a rectangular region R. Let c be a real number. Then:

(8.11) $$\iint_R (f + g) = \iint_R f + \iint_R g$$

(8.12) $$\iint_R [cf] = c\iint_R f$$

(8.13)   If $f(x, y) \geq g(x, y)$ for all (x, y) in R, then

$$\iint_R f \geq \iint_R g$$

Note that (8.11) asserts that the volume under the graph of $f + g$ is simply the volume under the graph of f added to the volume under the graph of g. A defense of (8.13) is to observe that if $f \geq 0$ on R, then $\iint_R f$ is a volume, and, accordingly, $\iint_R f \geq 0$. Since $f \geq g$ on R, it follows that $f - g \geq 0$ on R, so that $\iint_R (f - g) \geq 0$; from (8.11) and (8.12) we get $0 \leq \iint_R (f - g) = \iint_R f - \iint_R g$, which is the desired result.

EXAMPLE 1 Let f and g be functions with the property that for some region R,

$$\iint_R f = 5 \qquad \text{and} \qquad \iint_R g = -2$$

Then, $$\iint_R (6f) = 6\iint_R f = 30$$

and $$\iint_R (f + g) = 5 - 2 = 3$$

True or *false:* $f(x) \geq g(x)$ for x in R.

Recall two interesting facts about definite integrals:

1. If f is bounded on $[a,b]$, and continuous on $[a,b]$ except at a finite number of points, then $\int_a^b f$ exists.
2. $\int_a^a f = 0$; that is, the integral over a point is 0.

Roughly, since a point has no "thickness," there is no area under the graph of f over a point. A point or a finite set of points are each examples of sets on the line that have zero length. In fact, the previous two facts can be put in a somewhat stronger form as follows:

1. If f is bounded on $[a,b]$ and is continuous in $[a,b]$ except on a set of zero length, then $\int_a^b f$ exists.
2. Also, the integral of f over a set of zero length is equal to 0.

The analog for functions of two variables is this:

1. *If f is bounded on a rectangular region R of the plane, and is continuous in R except on a set G of zero area, then $\iint_R f$ exists.*
2. Also, $\iint_G f = 0$.

The sets in the plane with zero area that we will be concerned with are smooth curves (that is, curves with continuously turning tangent lines), or curves constructed by patching together a finite number of smooth curves by connecting the endpoint of one curve with the beginning point of the other curve. In Figure 8-23, each of the curves G_1, G_2, G_3, and G_4 have zero area.

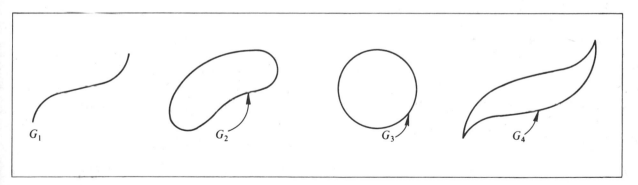

Figure 8-23

EXAMPLE 2   Let R be the rectangular region $\{(x,y): -2 \le x \le 2, -2 \le y \le 2\}$ and let G be the unit circle about the origin (see Figure 8-24).

Let $f(x,y) = 0$ for (x,y) in R and "outside" G, and let $f(x,y) = 1$ for all (x,y) in R and "inside" G. The graph of f is shaded in Figure 8-25. Thus, f is bounded on R, and continuous on R except on G, the circle, and G has zero area. Thus, $\iint_R f$ exists, and, in fact,

$$\iint_R f = \text{volume of the right circular cylinder of height}$$
$$\text{1, whose base is bounded by the circle } G^*$$
$$= \pi \text{ (cubic units)}$$

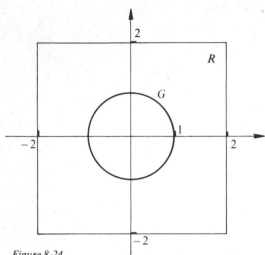

Figure 8-24

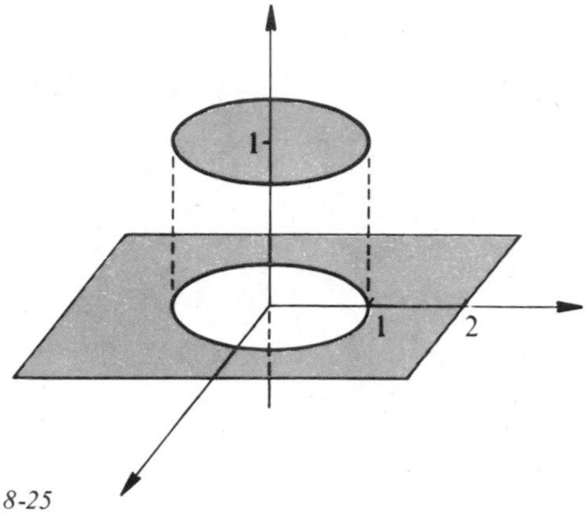

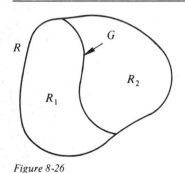

Figure 8-25

Finally, we observe that if a region R is divided into two regions R_1 and R_2 by a smooth curve G (see Figure 8-26), and if f is bounded on all of R, and continuous except possibly on G, then

$$\iint_R f = \iint_{R_1 \cup R_2} f = \iint_{R_1} f_1 + \iint_{R_2} f$$

Figure 8-26

Answers   Example 1: There are functions for which it is true and functions for which it is false.

Problems   1. Let R be a region in the plane, and assume that $\iint_R f = 7$ and $\iint_R g = 2$. Find
    a. $\iint_R (3f - 5g)$
    b. $\iint_R [2(f + g) - 3(f - g)]$
  2. Let R be the region given by

$$\{(x, y): -1 \le x \le 1, -1 \le y \le 1\}$$

*Recall the volume of a right circular cylinder whose base has radius r and whose height h is given by $\pi r^2 h$.

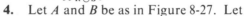

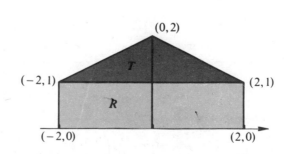

and let D be the disk given by

$$\{(x, y): x^2 + y^2 \leq 1\}$$

Let c be a positive real number, and let

$$f(x, y) = \begin{cases} c & \text{if } (x, y) \text{ is in } D \\ 0 & \text{if } (x, y) \text{ is in } R \text{ but not in } D \end{cases}$$

Find $\iint_R f$.

3. Let $R = \{(x, y): -2 \leq x \leq -1 \text{ or } 1 \leq x \leq 2, \text{ and } 0 \leq y \leq 2\}$

    and let $S = \{(x, y): -1 \leq x \leq 1, \text{ and } 0 \leq y \leq 1\}$.

    a. Sketch the sets R and S.

    b. Let $f(x, y) = \begin{cases} 1 & \text{for } (x, y) \in R \\ 2 & \text{for } (x, y) \in S \end{cases}$

    Find $\iint_{R \cup S} f$.

4. Let A and B be as in Figure 8-27. Let

$$f(x, y) = \begin{cases} 3 & \text{if } (x, y) \text{ is in } A \\ 5 & \text{if } (x, y) \text{ is in } B \end{cases}$$

Find $\iint_{A \cup B} f$.

5. Let R and T be as in Figure 8-28. (Assume that the boundary between R and T is in T.) Let

$$f(x, y) = \begin{cases} 10 & \text{if } (x, y) \text{ is in } R \\ 2 & \text{if } (x, y) \text{ is in } T \end{cases}$$

Find $\iint_{R \cup T} f$.

(0,1)

(1,1)

A

B

(0,0)

(1,0)

Figure 8-27

(0,2)

T

(−2,1)

(2,1)

R

(−2,0)

(2,0)

Figure 8-28

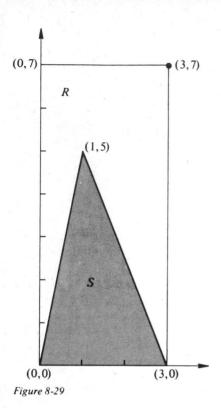

$(0,7)$ $(3,7)$

R

$(1,5)$

S

$(0,0)$ $(3,0)$

Figure 8-29

6. Let R and S be as in Figure 8-29. Let

$$f(x, y) = \begin{cases} 3 & \text{if } (x, y) \text{ is in } S \\ 0 & \text{if } (x, y) \text{ is in } R \text{ but not in } S \end{cases}$$

Find $\iint_R f$.

7. Let R be a region of the plane, and assume that f and g are functions for which $\iint_R fg = 3$. Find

$$\iint_R [(f + g)^2 - (f - g)^2]$$

SECTION 8

COMPUTATION OF DOUBLE INTEGRALS. ITERATED INTEGRATION

For the purpose of computation, the important fact about the double integral of continuous functions is that it is the iteration of two single integrals. Specifically,

If f is bounded on a rectangular region R, say given by $R = \{(x, y): a \le x \le b \text{ and } c \le y \le d\}$, and continuous on R except possibly on a set with zero area, then

$$\iint_R f = \int_c^d \left[\int_a^b f(x, y) \, dx \right] dy$$

$$= \int_a^b \left[\int_c^d f(x, y) \, dy \right] dx$$

EXAMPLE 1   Let $R = \{(x, y): 0 \le x \le 1 \text{ and } 0 \le y \le 2\}$, and let $f(x, y) = x + y$. Compute $\iint_R f$; that is, compute $\int_0^2 \int_0^1 (x + y) \, dx \, dy$.

SOLUTION   Since f is continuous on the entire plane, we see that

$$\iint_R f = \int_0^2 \left[\int_0^1 (x + y) \, dx \right] dy = \int_0^1 \left[\int_0^2 (x + y) \, dy \right] dx$$

To compute $\int_0^2 [\int_0^1 (x + y)\,dx]\,dy$, note that, treating y as fixed, since we are integrating with respect to x,

$$\int_0^1 (x + y)\,dx = \frac{x^2}{2} + xy \; \Big|_{x=0}^{x=1}$$

$$= \tfrac{1}{2} + y$$

Thus,

$$\int_0^2 \left[\int_0^1 (x + y)\,dx \right] dy = \int_0^2 (\tfrac{1}{2} + y)\,dy$$

$$= \frac{y + y^2}{2} \Big|_{y=0}^{y=2}$$

$$= 3$$

For completeness, compute $\int_0^1 [\int_0^2 (x + y)\,dy]\,dx$ as follows, treating x as fixed:

$$\int_0^2 (x + y)\,dy = \underline{\hspace{3cm}} \; \Big|_{y=0}^{y=2} = \underline{\hspace{3cm}}$$

Thus,

$$\int_0^1 \left[\int_0^2 (x + y)\,dy \right] dx = \int_0^1 \underline{\hspace{2cm}}\,dx$$

$$= \underline{\hspace{3cm}} \; \Big|_{x=0}^{x=1} = \underline{\hspace{3cm}}$$

EXAMPLE 2   Note that

$$\int_0^1 \int_1^2 (x^2 y^5)\,dx\,dy = \int_0^1 \left[\int_1^2 x^2 y^5\,dx \right] dy$$

Treating y as fixed,

$$\int_1^2 x^2 y^5\,dx = y^5 \int_1^2 x^2\,dx = \tfrac{7}{3} y^5$$

and, thus,

$$\int_0^1 \int_1^2 (x^2 y^5)\,dx\,dy = \frac{7}{3} \int_0^1 y^5\,dy$$

$$= \frac{7}{18}$$

In fact, note that

$$\int_0^1 \left[\int_1^2 (x^2 y^5)\,dx \right] dy = \left(\int_0^1 y^5\,dy \right) \left(\int_1^2 x^2\,dx \right)$$

$$= \tfrac{1}{6} \cdot \tfrac{7}{3} = \tfrac{7}{18}$$

Generalizing the last example, we see that if g is continuous on $[a,b]$, and h is continuous on $[c,d]$, then

$$\int_c^d \int_a^b g(x)h(y)\,dx\,dy = \left(\int_a^b g(x)\,dx\right)\left(\int_c^d h(y)\,dy\right)$$

EXAMPLE 3   Let $R = \{(x,y): 0 \le x \le 1 \quad \text{and} \quad 0 \le y \le 2\}$

Then, $\displaystyle\iint_R x^2 e^{xy}\,dx\,dy = \int_0^1 x^2\left(\int_0^2 e^{xy}\,dy\right)dx$

$$= \int_0^1 x^2\left(\frac{e^{xy}}{x}\Big|_{y=0}^{y=2}\right)dx$$

$$= \int_0^1 x(e^{2x} - 1)\,dx$$

The integral can be found by integrating by parts; we shall omit the details.

EXAMPLE 4   Let $R = \{(x,y): 0 \le x \le 1 \quad \text{and} \quad -1 \le y \le 0\}$

Complete the following:

$$\iint_R xy^2\,dx\,dy = \int_{\square}^{\square}\ \boxed{}\ dx \cdot \int_{\square}^{\square}\ \boxed{}\ dy$$

$$= \underline{}\ \Big|_{x=\square}^{x=\square} \cdot \underline{}\ \Big|_{y=\square}^{y=\square}$$

$$= \underline{}$$

Answers   Example 1:

$$\left(xy + \frac{y^2}{2}\right)\Big|_{y=0}^{y=2};\ 2x + 2;\ 2x + 2;\ x^2 + 2x\ \Big|_{x=0}^{x=1};\ 3$$

Example 4:

$$\int_0^1 x\,dx \cdot \int_{-1}^0 y^2\,dy = \frac{x^2}{2}\Big|_{x=0}^{x=1} \cdot \frac{y^3}{3}\Big|_{y=-1}^{y=0}$$

$$= \tfrac{1}{2}\cdot\tfrac{1}{3} = \tfrac{1}{6}$$

Problems   In problems 1–10, compute the given integral.

1. $\displaystyle\int_0^1 \int_0^1 x^5 y^{10} \, dx \, dy$

2. $\displaystyle\int_0^4 \int_0^1 (xy)^{1/2} \, dx \, dy$

3. $\displaystyle\int_{-1}^1 \int_{-1}^1 (x^2 + y^2) \, dx \, dy$

4. $\displaystyle\int_0^2 \int_1^4 (3x^2 + y^3) \, dy \, dx$

5. $\displaystyle\int_0^1 \int_0^1 (x + y)^3 \, dx \, dy$

6. $\displaystyle\int_0^1 \int_0^1 (x + y)^{1/2} \, dy \, dx$

7. $\displaystyle\int_0^a \int_0^b (ax + by) \, dx \, dy \qquad (a > 0, \, b > 0)$

8. $\displaystyle\int_0^1 \int_0^1 e^{x+y} \, dx \, dy$

9. $\displaystyle\int_0^2 \int_0^1 ye^{xy} \, dx \, dy$

10. $\displaystyle\int_0^1 \int_0^2 xe^{xy} \, dx \, dy$ (*Hint:* You can integrate by parts, or you can save some time by considering problem 9 and integrating first with respect to y.)

11. Let S be the rectangular region given by $0 \le x \le 1$, and $0 \le y \le 2$. Let

$$f(x, y) = x \sqrt{1 - x^2} \, e^{3y}$$

Compute $\iint_S f$.

SECTION 9

INTEGRATION OVER SOME NON-RECTANGULAR REGIONS. INTERCHANGING LIMITS

It is possible to apply the ideas involved in the construction of the integral over rectangular regions to obtain integrals over certain nonrectangular regions. The idea is as follows: Assume that u and v are functions with $u(x) \le v(x)$ for all $x \in [a,b]$, and let D be the region in the plane bounded by the graphs of u and v and the lines $x = a$, and $x = b$ (see Figure 8-30). Then, D is given by

$$\{(x,y) : u(x) \le y \le v(x), \qquad a \le x \le b\}$$

If f is continuous on D, then

$$\iint_D f = \int_a^b \left[\int_{u(x)}^{v(x)} f(x,y)\,dy \right] dx$$

and can be interpreted as the volume under the graph of f and over the plane region D (Figure 8-31).

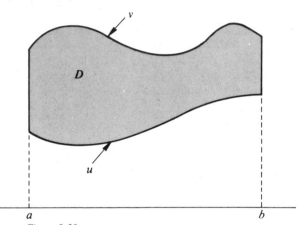

Figure 8-30

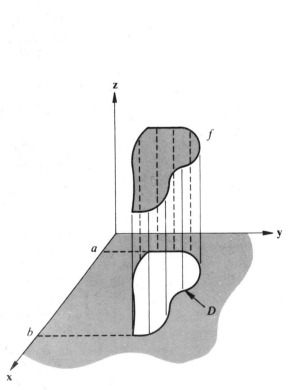

Figure 8-31

EXAMPLE 1   Let $f(x,y) = y$, all x,y. Let S be the region interior to and on the graph of $y = (1 - x^2)^{1/2}$ and the positive coordinate axes (Figure 8-32). Find $\int\int_S f$.

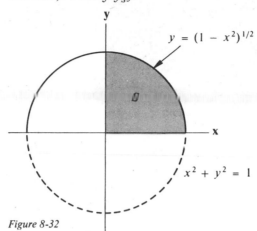

Figure 8-32

SOLUTION

$$\int\int_S f = \int_0^1 \left(\int_0^{\sqrt{1-x^2}} y \, dy \right) dx$$

$$= \int_0^1 \left\{ \frac{y^2}{2} \Big|_{y=0}^{y=\sqrt{1-x^2}} \right\} dx$$

$$= \frac{1}{2} \int_0^1 (1 - x^2) \, dx$$

$$= \frac{1}{3}$$

What we have found is the volume under the plane given by $z = y$ and on the region S.

EXAMPLE 2   Let S be the region given in Figure 8-33.   Let $f(x,y) = x^3 y^5$. Find $\int\int_S f$.

SOLUTION   One description of the region S is

$$x^2 \leq y \leq x \quad \text{and} \quad 0 \leq x \leq 1$$

Note that $y = x^2$ with $0 \leq x \leq 1$ if and only if $x = (y)^{1/2}$ with $0 \leq y \leq 1$; thus, viewing S from the y-axis, (see Figure 8-34), we see that another description of S is

$$y \leq x \leq (y)^{1/2} \quad \text{and} \quad 0 \leq y \leq 1$$

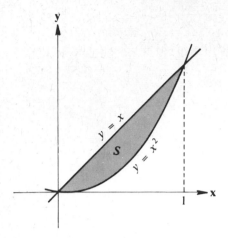

Figure 8-33

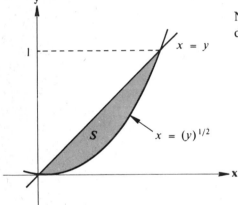

Figure 8-34

From the first description of S, we see that

$$\iint_S f = \int_0^1 \left(x^3 \int_{x^2}^x y^5 dy \right) dx$$

$$= \int_0^1 x^3 \cdot \left\{ \frac{y^6}{6} \Big|_{y=x^2}^{y=x} \right\} dx$$

$$= \int_0^1 x^3 \left(\frac{x^6}{6} - \frac{x^{12}}{6} \right) dx$$

$$= \frac{1}{6} \int_0^1 (x^9 - x^{15}) dx$$

$$= \frac{1}{6} \left(\frac{x^{10}}{10} - \frac{x^{16}}{16} \right) \Big|_0^1$$

$$= \frac{1}{6} \left(\frac{1}{10} - \frac{1}{16} \right)$$

$$= \frac{1}{160}$$

Notice that we might have computed the integral using the second description of S as follows:

$$\iint_S f = \int_0^1 \left(y^5 \int_y^{\sqrt{y}} x^3 dx \right) dy$$

$$= \int_0^1 y^5 \left(\frac{x^4}{4} \Big|_{x=y}^{x=\sqrt{y}} \right) dy$$

$$= \int_0^1 y^5 \left(\frac{y^2}{4} - \frac{y^4}{4} \right) dy$$

$$= \frac{1}{4} \int_0^1 (y^7 - y^9) dy$$

$$= \frac{1}{4} \left(\frac{y^8}{8} - \frac{y^{10}}{10} \right) \Big|_0^1$$

$$= \frac{1}{4} \left(\frac{1}{8} - \frac{1}{10} \right)$$

$$= \frac{1}{160}$$

In Example 2, two distinct descriptions were found for S, the region of integration, giving rise to two ways to compute the *same* integral. We relate the two integrals by saying that *the order of integration has been reversed*, or that *the limits have been interchanged.*

EXAMPLE 3   Write an integral which is the same as

$$\int_0^2 \left(\int_1^{e^x} dy \right) dx$$

but with the order of integration reversed.

SOLUTION   The region of interest S is sketched in Figure 8-35. Notice from Figure 8-35 that as x "goes from" 0 to 2, y "goes from" 1 to e^2; also, as y goes from 1 to e^x, we see that x goes from $\ln y$ to 2 (see Figure 8-36). Thus, viewing S from the y-axis, we can describe it by

$$\ln y \le x \le 2 \qquad \text{and} \qquad 1 \le y \le e^2$$

Thus,

$$\int_0^2 \left(\int_1^{e^x} dy \right) dx = \int_1^{e^2} \left(\int_{\ln y}^2 dx \right) dy$$

We leave it as an exercise to compute these integrals.

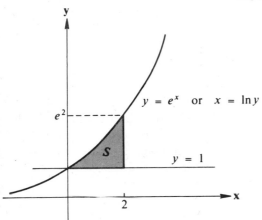

Figure 8-35

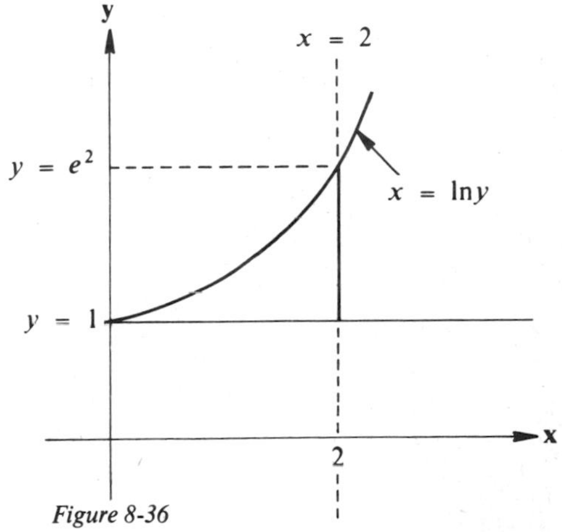

Figure 8-36

Problems   In problems 1–8, sketch the region of integration, and then compute the integral *two* ways: first as it is, and then with the limits of integration interchanged.

1. $\displaystyle\int_0^1 \int_{-x}^x (x^2 + 2y^2)\, dy\, dx$

2. $\displaystyle\int_0^4 \int_0^{4-x} (x^3 + 2y^3)\, dy\, dx$

3. $\displaystyle\int_0^3 \int_1^{4-x} (x^2 + y^3)\, dy\, dx$

4. $\displaystyle\int_{-1}^{1} \int_{-1}^{x} (3x + 2y)\,dy\,dx$

5. $\displaystyle\int_{0}^{1} \int_{x^3}^{1} (x + y^2)\,dy\,dx$

6. $\displaystyle\int_{0}^{4} \int_{0}^{\sqrt{x}} 3x\,dy\,dx$

7. $\displaystyle\int_{0}^{1} \int_{-x^2}^{x^2} (x - 3y)\,dy\,dx$

8. $\displaystyle\int_{1}^{e} \int_{0}^{\ln x} xy\,dy\,dx$

9. Sketch the region of integration, and write an equivalent integral with the limits interchanged.

$$\int_{0}^{2} \int_{0}^{\sqrt{4-x^2}} (x^2 + 4y^2)\,dy\,dx$$

(Do *not evaluate* these integrals.)

10. Let S be the region bounded by the graphs of the equations $y = 2x^2$ and $y = 2x$. Let $f(x,y) = 2xy + y^2$. Compute in *two* ways $\iint_S f$.

11. Let S be the region given, bounded by the graphs of the equations $y = x^2$ and $y = 2$. Let $f(x,y) = 2x + 3yx$. Compute *two* ways $\iint_S f$.

12. Let $f(x,y) = x + 2y + 4$. Let S be the region bounded by the lines given by $y = 2x$, $y = -x + 3$, and $y = 0$. Find the volume of the solid under the graph of f and above the region S. Perform the required integrations *two* ways (that is, with the order of integration reversed).

9 THE TRIGONOMETRIC FUNCTIONS

We shall study briefly an important class of functions for applications called the *trigonometric functions*. There are no new ideas in calculus developed here. Instead, the trigonometric functions are subjected to the same analysis as all the other functions already studied (for example, polynomial and logarithmic functions). The plan of action is as follows:

1. We define the "new" functions. Those of us who have studied trigonometry may recall that it is possible to define trigonometric functions in terms of ratios of sides of a right triangle. But this approach turns out to be inadequate for purposes of calculus, and it is replaced by viewing the trigonometric functions in terms of the coordinates of points on a circle of radius 1. This is studied in detail in section 1.

2. Once the trigonometric functions have been defined, they are graphed in a coordinate plane. After this is explained in section 2, we can begin the calculus. As in our earlier studies, calculus begins with a study of limits. Here, some special limits are established. Of course, the basic properties of limits are still valid (see pages 32 and 33 for a list of such properties). We next develop that special limit for each trigonometric function which is the derivative of that function; that is, we differentiate the trigonometric functions. Once the basic derivatives are known, we have at our disposal, as always, the sum rule, product rule, quotient rule, power rule, chain rule, and so on, for derivatives, as well as the notion of implicit functions. These calculations appear in sections 3 and 4. Now that we have graphs and derivatives, we can proceed to tangent lines, velocity, maximum and minimum points, and so on. However, these applications are postponed until section 6, after we develop the integral.

3. In section 5, we discuss the antiderivative, and then, equipped with the fundamental theorem of calculus, we investigate the integral. Again, the concepts are not in-

trinsically new, but only freshly applied to these "new" functions.

Trigonometric functions originated in the study of navigation, surveying, and other sciences that relied on the relationships between the angles and sides of triangles. Today, however, the major application of these functions is in the study of wave phenomena such as sound, heat, light, and electricity, and in nuclear physics and biology, or wherever "periodic" phenomena are studied. By periodic phenomena, we mean those situations in which a basic pattern is repeated over and over again.

SECTION 1

ANGLES. THE TRIGONOMETRIC FUNCTIONS

An angle in a plane can be given the following geometrical interpretation. Suppose two half-lines in a plane intersect at a point O. Then, the rotation about O that carries one line into the other is called an **angle.** If the rotation is counterclockwise, the angle is said to be a **positive angle,** whereas if the rotation is clockwise, the angle is said to be a **negative angle.** The point O is called the **vertex** of the angle. See Figure 9-1.

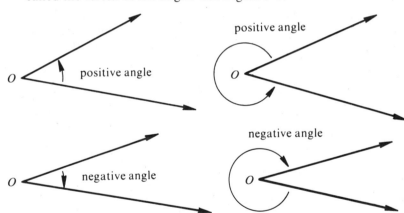

Figure 9-1

We shall now indicate some ways to measure angles. A common way, but for our purposes unsatisfactory, is to measure angles in terms of degrees. Specifically, one complete counterclockwise rotation of a half-line is assigned the degree measure of 360. We have

$$360° = \textbf{1 counterclockwise rotation}$$

Hence,

$$180° = \tfrac{1}{2} \text{ counterclockwise rotation}$$
$$90° = \tfrac{1}{4} \text{ counterclockwise rotation}$$
$$-90° = \tfrac{1}{4} \text{ clockwise rotation}$$

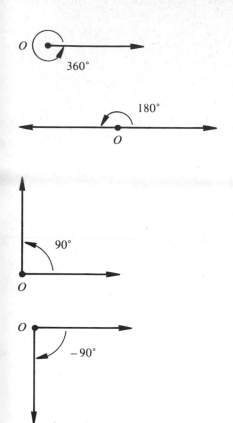

Figure 9-2

and so on. See Figure 9-2.

A more valuable measure of angle is the following one which is given in terms of signed lengths. Suppose θ is the angle obtained by rotating the half-line L_1 about the vertex O so that it coincides with the half-line L_2, as in Figure 9-3. Using O as a center, construct a circle of radius 1. *A circle whose radius is equal to 1 is called a* **unit circle**. The signed length of the arc of this unit circle that lies between L_1 and L_2 is called the **radian measure** of θ. By *signed* length we mean that the radian measure is positive if θ is counterclockwise and negative if it is clockwise. See Figure 9-3.

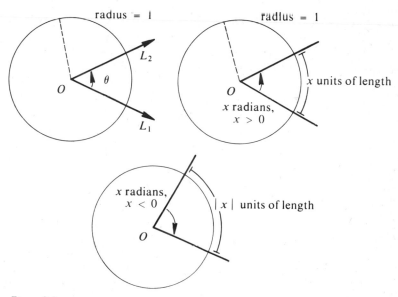

Figure 9-3

We know from plane geometry that the circumference of a unit circle is equal to 2π. Thus,

$$2\pi \text{ radians} = 360°$$

$$\pi \text{ radians} = 180°$$

$$\frac{\pi}{2} \text{ radians} = 90°$$

$$-\frac{\pi}{2} \text{ radians} = -90°$$

$$1 \text{ radian} = \left(\frac{360}{2\pi}\right)° = \left(\frac{180}{\pi}\right)°$$

and

$$5 \text{ radians} = \left(5 \cdot \frac{180}{\pi}\right)° = \left(\frac{900}{\pi}\right)°$$

Also, notice that

$$4\pi \text{ radians} = 720°$$

and

$$-\frac{5\pi}{2} \text{ radians} = \left(-\frac{5\pi}{2} \cdot \frac{180}{\pi}\right)° = -450°$$

See Figure 9-4 for some of these angles.

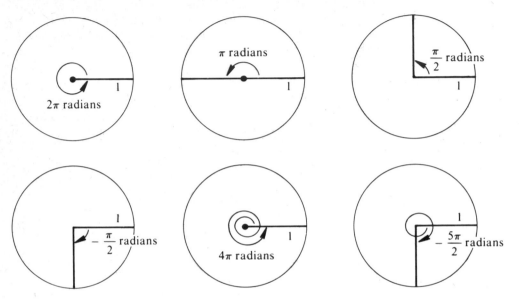

Figure 9-4

The important property of the radian measure of an angle is that it is defined in terms of signed *length,* and thus it is defined in terms of the real number line. For example, if we have an angle whose radian measure is 5, we can associate this angle measure with the point 5 on a number line, since in either context, we are talking about a segment whose length is 5 units.

We now turn to a class of functions whose domains are sets of real numbers which, as we shall see later, can be thought of as angles measured in radians. This class of functions is called the class of **trigonometric functions.** Because the domains of these functions are lengths of arcs on a unit circle (that is, radian measure of angles), these functions are sometimes referred to as **circular functions.** As before, when we studied functions, our setting is a coordinate plane.

Two fundamental trigonometric functions are **sine** and **cosine,** usually abbreviated **sin** and **cos.** They are defined as follows: Place a unit circle in a coordinate system so that the center of the circle is the origin (0,0). See Figure 9-5. Let P_x be the point on this unit circle whose arc distance from (1,0) is x (clockwise if $x < 0$, and counterclockwise if $x > 0$). Then, by definition, we

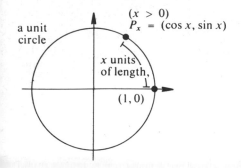

a unit circle

$(x > 0)$
$P_x = (\cos x, \sin x)$

x units of length

$(1, 0)$

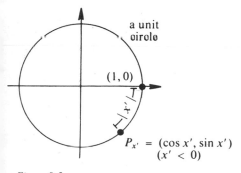

a unit circle

$(1, 0)$

$P_{x'} = (\cos x', \sin x')$
$(x' < 0)$

Figure 9-5

say that the point P_x has coordinates $(\cos x, \sin x)$. Equivalently, cos and sin are functions defined for every real number x by

$$\cos x \text{ is the first coordinate of } P_x$$
$$\sin x \text{ is the second coordinate of } P_x$$

Since each such arc distance on a unit circle corresponds to an angle measured in radians (see Figure 9-3), we see that we can think of $\sin x$ and $\cos x$ as functions of the angle x measured in radians.

Since the distance between $P_x = (\cos x, \sin x)$ and $(0, 0)$ is equal to 1, we see that

$$(\cos x - 0)^2 + (\sin x - 0)^2 = 1$$

It is customary to write $(\cos x)^2$ as $\cos^2 x$ and $(\sin x)^2$ as $\sin^2 x$. Thus, we have *for any real number* x,

$$\sin^2 x + \cos^2 x = 1$$

The sin and cos functions give rise to four other trigonometric functions, known as **tan** (**tangent**), **cot** (**cotangent**), **sec** (**secant**), and **csc** (**cosecant**). They are defined as follows:

$$\tan x = \frac{\sin x}{\cos x}$$

$$\cot x = \frac{1}{\tan x} = \frac{\cos x}{\sin x}$$

$$\sec x = \frac{1}{\cos x}$$

$$\csc x = \frac{1}{\sin x}$$

Of course, these definitions are meaningful only when the denominators are not equal to 0. We shall say more of this in later sections.

We shall now illustrate how we can compute $\cos x$ and $\sin x$ for at least some values x.

EXAMPLE 1   Compute

(a) $\cos 0$ (e) $\cos \pi$

(b) $\sin 0$ (f) $\sin \pi$

(c) $\cos \dfrac{\pi}{2}$ (g) $\cos \dfrac{3\pi}{2}$

(d) $\sin \dfrac{\pi}{2}$ (h) $\sin \dfrac{3\pi}{2}$

SOLUTION   Recall that for each x, $\cos x$ and $\sin x$ are, respectively, the first and the second coordinates of P_x. Refer to Figure 9-6. Thus,

(a) $\cos 0 = 1$, since $\cos 0$ is the first coordinate of $P_0 = (1, 0)$.
(b) $\sin 0 = 0$, since $\sin 0$ is the second coordinate of $P_0 = (1, 0)$.
Similarly,
(c) $\cos (\pi/2) = 0$ and (d) $\sin (\pi/2) = 1$, since $P_{\pi/2} = (0, 1)$.
(e) $\cos \pi = -1$ and (f) $\sin \pi = 0$, since $P_\pi = (-1, 0)$.
(g) $\cos(3\pi/2) = 0$ and (h) $\sin(3\pi/2) = -1$, since $P_{3\pi/2} = (0, -1)$.
Note also that, since $P_\pi = P_{-\pi}$, we have $\cos (-\pi) = -1$ and $\sin (-\pi) = 0$, and since $P_{3\pi/2} = P_{-\pi/2}$, we have

Figure 9-6

$$\cos\left(-\frac{\pi}{2}\right) = 0 \quad \text{and} \quad \sin\left(-\frac{\pi}{2}\right) = -1$$

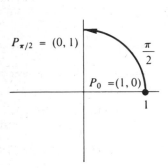

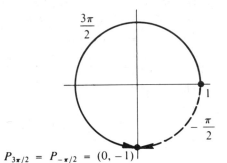

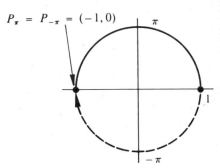

Figure 9-7

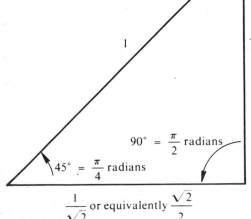

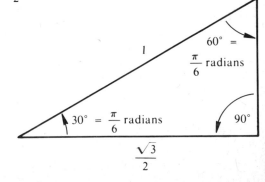

The following two examples use properties of 30°-60°-90° and 45°-45°-90° triangles. These properties are summarized in Figure 9-7 on page 294; in this figure the hypotenuse is equal to 1 unit, and the corresponding lengths of the other sides are indicated. The verification of these properties is a problem in plane geometry.

EXAMPLE 2   Compute $\sin \dfrac{\pi}{4}$ and $\cos \dfrac{\pi}{4}$.

SOLUTION   From Figure 9-8, we see that we want the coordinates of $P_{\pi/4}$, that is, the point whose arc length distance from $(1,0)$ on the unit circle is equal to $\pi/4$. From the 45°-45°-90° triangle in Figure 9-8, we see that

$$\cos \frac{\pi}{4} = \frac{\sqrt{2}}{2} = \sin \frac{\pi}{4}$$

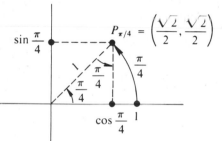

Figure 9-8

EXAMPLE 3   Compute $\sin \dfrac{\pi}{6}$ and $\cos \dfrac{\pi}{6}$.

SOLUTION   From Figure 9-9 we see that we can find the coordinate of $P_{\pi/6}$ by using the 30°-60°-90° (that is, $\dfrac{\pi}{6}$-$\dfrac{\pi}{3}$-$\dfrac{\pi}{2}$ radians) triangle POR. Thus, using the information in Figures 9-7 and 9-9,

$$\cos \frac{\pi}{6} = \frac{\sqrt{3}}{2} \qquad \text{and} \qquad \sin \frac{\pi}{6} = \frac{1}{2}$$

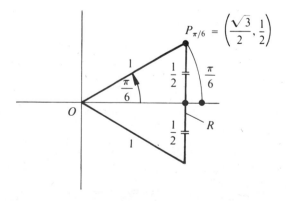

Figure 9-9

It is frequently convenient to look at trigonometric functions as functions whose domains consist of *angles* rather than numbers. Hence, *we define* **the cosine of an angle of x radians** *to be the cosine of the number x.* A similar definition may be made for the sine, and the other trigonometric functions.

EXAMPLE 4 $\sin 30° = \sin\left(\dfrac{\pi}{6} \text{ radians}\right) = \sin\dfrac{\pi}{6} = \dfrac{1}{2}$

Most of the computations listed in the following table have already been done. The angles considered here are the so-called standard angles. The computations for tan use the fact that $\tan x = \sin x / \cos x$.

Degrees	$-180°$	$-90°$	0	30°	45°	60°	90°	180°	270°	360°
Radians	$-\pi$	$-\dfrac{\pi}{2}$	0	$\dfrac{\pi}{6}$	$\dfrac{\pi}{4}$	$\dfrac{\pi}{3}$	$\dfrac{\pi}{2}$	π	$\dfrac{3}{2}\pi$	2π
sin	0	-1	0	$\dfrac{1}{2}$	$\dfrac{\sqrt{2}}{2}$	$\dfrac{\sqrt{3}}{2}$	1	0	-1	0
cos	-1	0	1	$\dfrac{\sqrt{3}}{2}$	$\dfrac{\sqrt{2}}{2}$	$\dfrac{1}{2}$	0	-1	0	1
tan	0	undef.	0	$\dfrac{1}{\sqrt{3}}$	$1\cdot$	$\sqrt{3}$	undef.	0	undef.	0

We shall list some of the basic properties of the sin, cos, and tan functions. These properties are verified in standard courses in trigonometry.

1. cos is an even function; that is, $\cos x = \cos(-x)$, for any number x
2. sin is an odd function; that is, $\sin(-x) = -\sin x$, for any number x
3. tan is an odd function; that is, $\tan(-x) = -\tan x$
4. $\cos\left(\dfrac{\pi}{2} - x\right) = \sin x$ and $\sin\left(\dfrac{\pi}{2} - x\right) = \cos x$

The **addition formulas:**

5. $\cos(x \pm y) = \cos x \cos y \mp \sin x \sin y$
6. $\sin(x \pm y) = \sin x \cos y \pm \cos x \cos y$
7. $\tan(x + y) = \dfrac{\tan x + \tan y}{1 - \tan x \tan y}$

EXAMPLE 5   (a) $\cos 75° = \cos(45° + 30°)$

$$= \cos 45° \cos 30° - \sin 45° \sin 30°$$

$$= \frac{\sqrt{2}}{2} \cdot \frac{\sqrt{3}}{2} - \frac{\sqrt{2}}{2} \cdot \frac{1}{2}$$

$$= \frac{\sqrt{6} - \sqrt{2}}{4}$$

(b) $\tan 2x = \tan(x + x) = \dfrac{\tan x + \tan x}{1 - (\tan x)(\tan x)} = \dfrac{2 \tan x}{1 - \tan^2 x}$

It is usually not so easy to find the values of the trigonometric functions by a direct use of the definitions in terms of the unit circle. It's just too difficult to determine explicitly the coordinates of points corresponding to any angle. One way around this is to determine the Taylor polynomials for these functions, and to use these polynomial approximations to evaluate the trigonometric functions. Taylor expansions for sin and cos are given in problem 7, page 234. To obtain these expansions, we must know how to differentiate the cos and sin functions; these computations appear in section 3. Two tables of trigonometric functions, one for radians (that is, arc length on the unit circle) and one for degrees, appear on pages 371–74 and 375, respectively.

EXAMPLE 6   Use tables to find (a) cos 1.2, (b) sin 24°, (c) sin 56°, (d) tan 35°, (e) cot 55°.

SOLUTION   (a) If we refer to the table on page 373, we see that cos 1.2 = 0.3624.

For (b)–(e), also refer to the table on page 375.

(b) To find sin 24°, look down the left column for θ until you reach 24°, and then move to the column whose top heading is sin. We find that sin 24° = 0.4067.

(c) We want sin 56°. But the left column for θ ends at 45°. So we move to the extreme right column of the table and notice that θ appears at the bottom. We read *up* that column until we get to 56°. Using the headings at the *bottom* of the table, read across to the sin column. We see that sin 56° = 0.8290.

(d) tan 35° = 0.7002.

(e) cot 55° = 0.7002.

Problems

1. $-180° = $ _____ radians

2. $135° = $ _____ radians

3. $1.2° = $ _____ radians

4. $(2\pi)° = $ _____ radians

5. 1.2 radians $= $ _____ °

6. 2 radians $= $ _____ °

7. $\tan \dfrac{\pi}{6} = $ _____

8. $\tan \dfrac{\pi}{4} = $ _____

9. $\sin 135° = $ _____

10. $\cos 135° = $ _____

11. $\sin 120° = $ _____

12. $\sin 210° = $ _____

13. $\cos 210° = $ _____

14. $\tan 210° = $ _____

15. $\tan 135° = $ _____

16. $\sec (-\pi) = $ _____

17. $\sec 0 = $ _____

18. $\sec \dfrac{\pi}{6} = $ _____

19. $\sec \dfrac{\pi}{4} = $ _____

20. $\csc \left(-\dfrac{\pi}{2}\right) = $ _____

21. $\csc \dfrac{\pi}{6} = $ _____

22. $\csc \dfrac{\pi}{4} = $ _____

23. $\csc \dfrac{\pi}{3} = $ _____

In problems 24–29, use the tables to evaluate the trigonometric function at the given value.

24. $\tan 1.03$

25. $\cot 1.35$

26. $\sin 39°$

27. $\cos 82°$

28. $\tan 89°$

29. $\cot 50°$

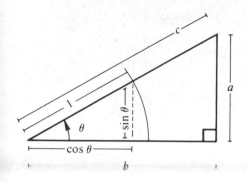

30. Suppose we have a right triangle whose hypotenuse has length c, and whose other sides have lengths a and b, as given in the accompanying diagram. Use the Figure to show that

$$\sin \theta = \frac{a}{c} = \frac{\text{length of side opposite } \theta}{\text{length of hypotenuse}}$$

$$\cos \theta = \frac{b}{c} = \frac{\text{length of side adjacent } \theta}{\text{length of hypotenuse}}$$

and

$$\tan \theta = \frac{a}{b} = \frac{\text{length of side opposite } \theta}{\text{length of side adjacent } \theta}$$

31. Use the addition formula to find
   a. $\sin 2x$
   b. $\cos 2x$

32. Use the addition formula for sin to verify that

$$\sin (x + 60°) = \frac{1}{2}\sin x + \frac{\sqrt{3}}{2}\cos x$$

33. Use the addition formula for cos to verify that

$$\cos x = -\cos (\pi - x)$$

SECTION 2
GRAPHS OF THE TRIGONOMETRIC FUNCTIONS

In this section we shall graph the trigonometric functions in a coordinate plane. Let us start with the functions sin and cos.

Note that since the unit circle is 2π units long, two angles whose radian measures differ by an integral multiple of 2π determine exactly the same point on the circle. In other words, if x is a number, then

$$P_x = P_{x+2n\pi}$$

for any integer n. See Figure 9-10. Now, since cos x and sin x are the coordinates of P_x, it follows that *for any integer n and real number x,*

$$\cos x = \cos (x + 2n\pi) \qquad \text{and} \qquad \sin x = \sin (x + 2n\pi)$$

that is, for any number x,

$$\cos x = \cos (x + 2\pi) = \cos (x + 4\pi) = \cos (x - 2\pi)$$
$$= \cos (x - 4\pi)$$

and so on, with similar results for sin. This fact is often described by saying that sin and cos are **periodic** functions with **period** 2π.

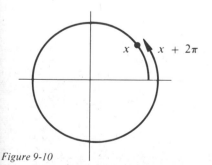

Figure 9-10

It can also be shown that tan $x = $ tan $(x + n\pi)$ for any integer n and any x in the domain of tan.

The essential consequence of this is that for any number x, the points $(x, \sin x)$ and $(x + 2n\pi, \sin (x + 2n\pi))$ on the graph of sin are at precisely the same height above (or below) the **x**-axis, with an analogous result for cos. For the function tan, the points $(x, \tan x)$ and $(x + n, \tan (x + n\pi))$ are the same height. We are now in a position to graph sin, cos, and tan.

EXAMPLE 1   Graph the cosine function.

SOLUTION   Since 2π is the period of this function, it is sufficient to graph it for x between $-\pi$ and π. Referring once more to the unit circle or to the table in section 1, we see that as increasing numbers x are chosen between 0 and $\pi/2$, cos x decreases from 1 to 0. For example,

$$\cos 0 = 1, \quad \cos \frac{\pi}{6} = \frac{\sqrt{3}}{2}, \quad \cos \frac{\pi}{4} = \frac{\sqrt{2}}{2}, \quad \cos \frac{\pi}{3} = \frac{1}{2}$$

$$\cos \frac{\pi}{2} = 0$$

See the heavy curve between 0 and $\frac{\pi}{2}$ in Figure 9-11.

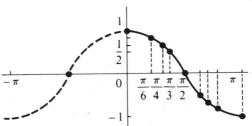

Figure 9-11

From this piece of the graph, we can obtain the graph for x between $\pi/2$ and π by using the fact that cos $x = -\cos (\pi - x)$ (see Figure 9-11). Lastly, since cos $(-x) = $ cox x we obtain the graph for x between $-\pi$ and 0. By periodicity, the graph can now be sketched over any interval. See Figure 9-12.

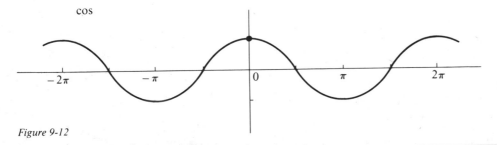

Figure 9-12

EXAMPLE 2   Graph the sine function.

SOLUTION   Note that for any real number x,

$$\sin x = \cos\left(\frac{\pi}{2} - x\right) = \cos\left(x - \frac{\pi}{2}\right).$$

Hence, for example, $\sin \frac{\pi}{2} = \cos\left(\frac{\pi}{2} - \frac{\pi}{2}\right) = \cos 0$, and $\sin \pi = \cos \frac{\pi}{2}$, and consequently, at $\frac{\pi}{2}$ and π, the values of the sin are, respectively, $\cos 0$ and $\cos \frac{\pi}{2}$. In general, a shift of the graph of cos by $\frac{\pi}{2}$ units to the right will coincide with the graph of sin. See Figure 9-13.

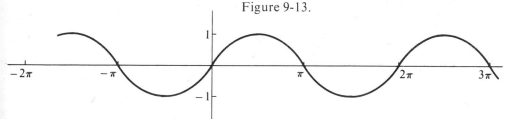

Figure 9-13

EXAMPLE 3   Graph the tangent function.

Figure 9-14

SOLUTION   Notice in Figure 9-14, $\triangle OPQ$ is similar to $\triangle ORS$, and thus,

$$\frac{\overline{RS}}{\overline{OS}} = \frac{\overline{PQ}}{\overline{OQ}}$$

that is,

$$\frac{\overline{RS}}{1} = \frac{\sin x}{\cos x}$$

so,

$$\overline{RS} = \tan x$$

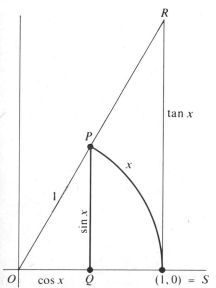

as indicated in Figure 9-14. From Figure 9-14, since $\overline{RS} = \tan x$, we can infer the following fact: As increasing numbers x are chosen between 0 and $\pi/2$, tan x is positive and increases indefinitely, and therefore the line given by $x = \pi/2$ is a vertical asymptote. Also, tan 0 = 0. See Figure 9-15.

We now use the fact that $\tan(-x) = -\tan x$ to obtain the graph of tan between $-\pi/2$ and $\pi/2$ (Figure 9-16) and then obtain the rest by periodicity (Figure 9-17).

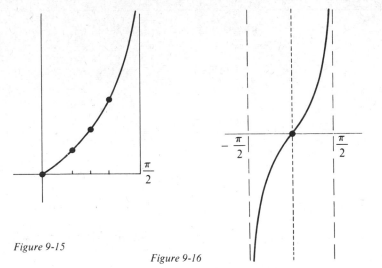

Figure 9-15

Figure 9-16

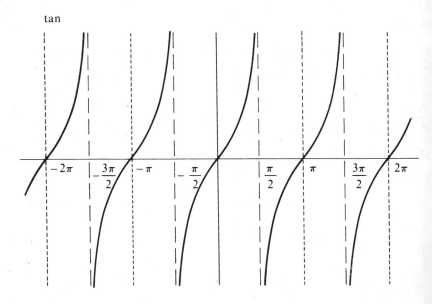

Figure 9-17

The functions cot, sec, and csc are graphed, respectively, in Figures 9-18, 9-19, and 9-20.

Figure 9-18   cot

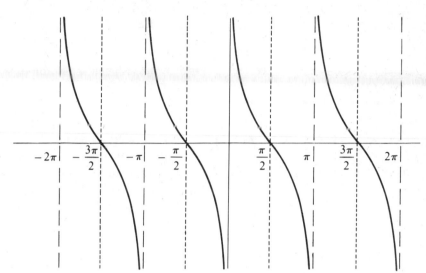

sec

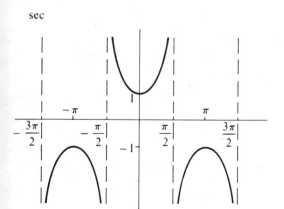

csc

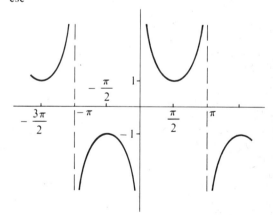

Figure 9-19

Figure 9-20

Problems   In problems 1–4, fill in the blanks:

1. $\sin 405° =$ _____   2. $\cos 750° =$ _____

3. $\cos \dfrac{9\pi}{4} =$ _____   4. $\cos \dfrac{13\pi}{3} =$ _____

5. Find all numbers x in $[0, 2\pi]$ for which $\sin 2x = 0$.

6. Find all numbers x in $[0, 6\pi]$ for which $\sin(x/3) = 0$.

In problems 7–13, sketch the graph of the given function.

7. $f(x) = -\sin x$ **8.** $f(x) = -\cos x$

9. $f(x) = 2 + \sin x$ **10.** $f(x) = 1 - \cos x$

11. $f(x) = 2 \sin x$ **12.** $f(x) = \sin 2x$

13. $f(x) = |\sin x|$

14. Let $f(x) = \sin 4x$.

    a. Determine the points in $[0, 2\pi]$ at which the graph of f crosses the x-axis.

    b. Sketch the graph of f in $[-2\pi, 2\pi]$.

SECTION 3
LIMITS AND CONTINUITY

The definition of the function *sin* led us to believe that the graph of *sin* is a smooth curve with no jumps or breaks. That is, our geometrical considerations led us to believe, and in fact it is the case, that sin is continuous everywhere. Similar reasoning leads us to believe that *cos* is continuous everywhere, and the other trigonometric functions are continuous everywhere in the domain. So, for example,

$$\lim_{x \to 0} \sin x = \sin 0 = 0$$

$$\lim_{x \to 0} \cos x = \cos 0 = 1$$

$$\lim_{x \to \pi/4} \cos x = \cos \frac{\pi}{4} = \frac{\sqrt{2}}{2}$$

and in general, for *any* real number a,

$$\lim_{x \to a} \sin x = \sin a \qquad \text{and} \qquad \lim_{x \to a} \cos x = \cos a$$

Notice that this is precisely what it means for sin and cos to be continuous at a. Another important limit involving sin is

$$\lim_{x \to 0} \frac{\sin x}{x} = 1$$

and thus,

$$\lim_{x \to 0} \frac{x}{\sin x} = 1 \qquad \text{(Why?)}$$

A geometrical consideration that makes this limit statement plausible rests on the following observations, based on Figure 9-21. Notive that Q is the point $(\cos x, \sin x)$, and since $\sin x$ is the second coordinate of that point, the segment from R to Q has length equal to $\sin x$. Also, $\triangle ORP$ is just the reflection in the

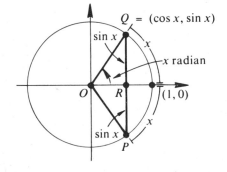

Figure 9-21

x-axis of $\triangle ORQ$. Thus, the arc from P to Q has length equal to $2x$, and the chord connecting P to Q has length equal to $2 \sin x$. Thus,

$$\frac{\text{length of chord from } P \text{ to } Q}{\text{length of arc from } P \text{ to } Q} = \frac{2 \sin x}{2x} = \frac{\sin x}{x}$$

Our geometrical intuition leads us to believe, and in fact it is the case, that as $x \to 0$, the lengths of the chord and arc tend to the same value, and thus the ratio of their lengths tends to 1. Hence, our result:

$$1 = \lim_{x \to 0} \frac{\text{length of chord}}{\text{length of arc}} = \lim_{x \to 0} \frac{\sin x}{x}$$

From these facts, it is possible to evaluate various limits involving the trigonometric functions. The following examples illustrate some of these limits.

EXAMPLE 1   Find $\lim\limits_{x \to 0} \tan x$.

SOLUTION $$\lim_{x \to 0} \tan x = \lim_{x \to 0} \frac{\sin x}{\cos x} = \frac{\sin 0}{\cos 0} = \frac{0}{1} = 0$$

EXAMPLE 2   Find $\lim\limits_{x \to 0} \dfrac{\tan x}{x}$.

SOLUTION
$$\lim_{x \to 0} \frac{\tan x}{x} = \lim_{x \to 0} \frac{\sin x}{x \cos x}$$
$$= \lim_{x \to 0} \frac{\sin x}{x} \cdot \frac{1}{\cos x}$$
$$= 1 \cdot 1 = 1$$

EXAMPLE 3   Compute $\lim\limits_{x \to 0} \dfrac{\sin 2x}{x}$

SOLUTION
$$\lim_{x \to 0} \frac{\sin 2x}{x} = \lim_{x \to 0} 2 \cdot \frac{\sin 2x}{2x}$$
$$= 2 \lim_{x \to 0} \frac{\sin 2x}{2x}$$

Since $x \to 0$ is equivalent to $2x \to 0$, we see that

$$2 \lim_{x \to 0} \frac{\sin 2x}{2x} = 2 \lim_{2x \to 0} \frac{\sin 2x}{2x} = 2 \cdot 1 = 2$$

Thus,

$$\lim_{x \to 0} \frac{\sin 2x}{x} = 2$$

EXAMPLE 4   Find $\lim\limits_{x \to 0} \dfrac{\sin 2x}{\sin 3x}$

SOLUTION

$$\lim_{x \to 0} \frac{\sin 2x}{\sin 3x} = \lim_{x \to 0} \frac{\sin 2x}{2x} \cdot 2x \cdot \frac{3x}{\sin 3x} \cdot \frac{1}{3x}$$

$$= \lim_{x \to 0} \frac{2x}{3x} \cdot \frac{\sin 2x}{2x} \cdot \frac{3x}{\sin 3x}$$

$$= \frac{2}{3} \cdot 1 \cdot 1 = \frac{2}{3}$$

EXAMPLE 5   Find $\lim\limits_{x \to 0} \dfrac{1 - \cos x}{x}$

SOLUTION

$$\lim_{x \to 0} \frac{1 - \cos x}{x} = \lim_{x \to 0} \frac{1 - \cos x}{x} \cdot \frac{1 + \cos x}{1 + \cos x}$$

$$= \lim_{x \to 0} \frac{1 - \cos^2 x}{x(1 + \cos x)}$$

$$= \lim_{x \to 0} \frac{\sin^2 x}{x(1 + \cos x)}$$

$$= \lim_{x \to 0} \sin x \cdot \frac{\sin x}{x} \cdot \frac{1}{1 + \cos x}$$

$$= 0 \cdot 1 \cdot \frac{1}{2}$$

$$= 0$$

We have already observed that sin and cos are continuous everywhere on the line. For completeness, it should be noted that *all the trigonometric functions are continuous wherever they are defined (that is, throughout their domains).*

For example,

$$\lim_{x \to a} \tan x = \tan a \qquad \text{for any } a \text{ in the domain of tan}$$

(that is, $a \neq n\pi/2$ where n is any integer).

Problems   Compute:

1. $\lim\limits_{x \to 0} \dfrac{\sin 3x}{\sin 4x}$

2. $\displaystyle\lim_{x\to 0} \frac{\sin 7x}{x}$

3. $\displaystyle\lim_{x\to 0} \frac{2 + \cos x}{1 + 3\cos x}$

4. $\displaystyle\lim_{x\to 0} x \csc x$

5. $\displaystyle\lim_{x\to 0} x^2 \csc x$

6. $\displaystyle\lim_{x\to 0} \left(\frac{\sin 10x}{2x} + 3\cos^2 x \right)$

7. $\displaystyle\lim_{x\to 0} \frac{x}{\tan 3x}$

8. $\displaystyle\lim_{x\to 0} x \cot \frac{x}{2}$

9. $\displaystyle\lim_{x\to 0} \frac{\sin^2 x}{x}$

10. $\displaystyle\lim_{x\to 0} \frac{\sin^2 4x}{x^2}$

11. $\displaystyle\lim_{x\to\infty} x \sin \frac{1}{x}$

     (*Hint:*   Let $u = 1/x$.)

12. $\displaystyle\lim_{x\to 0} \frac{1 - \cos x}{x^2}$

13. $\displaystyle\lim_{x\to 0} \ln \frac{\sin x}{x}$

14. $\displaystyle\lim_{x\to 0} \exp\left(\frac{\sin^2 x}{x} \right)$

SECTION 4

DERIVATIVES OF TRIGONO- METRIC FUNCTIONS

We now have enough information to find the derivatives of the trigonometric functions. We shall first show that

$$D \sin x = \cos x$$
$$D \cos x = -\sin x$$
$$D \sin f(x) = (\cos f(x)) \cdot Df(x)$$
$$D \cos f(x) = (-\sin f(x)) Df(x)$$

Remark   The last two formulas are valid at all points x for which $Df(x)$ exists.

We shall first verify that $D \sin x = \cos x$. To do this, we must start from scratch and view the derivative as a limit of a difference quotient. Going all the way back to first principles, we see that

$$D \sin x = \lim_{h \to 0} \frac{\sin(x + h) - \sin x}{h}$$

The computation of this limit uses the addition formula for sin and example 5, of section 4, on limits. It goes as follows:

$$D \sin x = \lim_{h \to 0} \frac{\sin(x + h) - \sin x}{h}$$

$$= \lim_{h \to 0} \frac{\sin x \cdot \cos h + \cos x \cdot \sin h - \sin x}{h}$$

(by the addition formula for sin)

$$= \lim_{h \to 0} \frac{(-\sin x + \sin x \cos h) + \cos x \sin h}{h}$$

$$= \lim_{h \to 0} \left[-\sin x \left(\frac{1 - \cos h}{h} \right) + \cos x \cdot \frac{\sin h}{h} \right]$$

Now, we know that

$$\lim_{h \to 0} \frac{1 - \cos h}{h} = 0 \qquad \text{(see example 5, section 3)}$$

and

$$\lim_{h \to 0} \frac{\sin h}{h} = 1$$

Thus,

$$\lim_{h \to 0} \left[-\sin x \left(\frac{1 - \cos h}{h} \right) + \cos x \cdot \frac{\sin h}{h} \right]$$

$$= (-\sin x) \cdot 0 + (\cos x) \cdot 1$$

$$= \cos x$$

That is, $D \sin x = \cos x$, which is what we wanted to prove.

The fact that $D \sin f(x) = (\cos f(x)) \cdot f'(x)$ is a consequence of the chain rule; we shall omit the details.

We can show that $D \cos x = -\sin x$ by using the identities: for all real x,

$$\cos x = \sin \left(\frac{\pi}{2} - x \right) \qquad \text{and} \qquad \sin x = \cos \left(\frac{\pi}{2} - x \right)$$

Thus,

$$D \cos x = D \sin \left(\frac{\pi}{2} - x \right)$$

We see that $\sin(\frac{1}{2}\pi - x)$ is of the form $\sin f(x)$, where $f(x) = \frac{1}{2}\pi - x$. Thus,

$$D \cos x = D \sin\left(\frac{\pi}{2} - x\right)$$
$$= \cos\left(\frac{\pi}{2} - x\right) \cdot D\left(\frac{\pi}{2} - x\right)$$
$$= (\sin x)(-1)$$
$$= -\sin x$$

which is the desired result.

The fact that $D \cos f(x) = (-\sin f(x))f'(x)$ is again an application of the chain rule.

With the basic differentiation formulas for sin and cos, and using such facts as the power rule, the product rule, the quotient rule (which are all valid for any differentiable function, and so in particular for sin and cos), we are in a position to differentiate all the trigonometric functions. The next examples illustrate this.

EXAMPLE 1   Compute $D \sin(3x + 2)$

SOLUTION   Notice that $\sin(3x + 2)$ is of the form $\sin f(x)$, where $f(x) = 3x + 2$. Thus,

$$D \sin(3x + 2) = (\cos(3x + 2)) \cdot D(3x + 2)$$
$$= 3 \cos(3x + 2)$$

EXAMPLE 2   Compute $D \sin(x^5)$

SOLUTION   Notice that $\sin(x^5)$ is of the form $\sin f(x)$, where $f(x) = x^5$. Hence,

$$D \sin(x^5) = (\cos(x^5)) Dx^5$$
$$= 5x^4 \cos(x^5)$$

EXAMPLE 3   Compute $D \sin^5 x$.

SOLUTION   We see that $\sin^5 x$ is a power of $\sin x$. Thus, we use the power rule and obtain

$$D \sin^5 x = (5 \sin^4 x)D \sin x$$
$$= 5(\sin^4 x)(\cos x)$$

EXAMPLE 4   Show that

$$D \tan x = \sec^2 x$$

SOLUTION   Using the quotient rule, we obtain

$$D \tan x = D \frac{\sin x}{\cos x}$$

$$= \frac{(\cos x)D \sin x - (\sin x)D \cos x}{\cos^2 x}$$

$$= \frac{\cos^2 x + \sin^2 x}{\cos^2 x}$$

$$= \frac{1}{\cos^2 x}$$

$$= \sec^2 x$$

EXAMPLE 5   Show that

$$D \sec x = \tan x \sec x$$

SOLUTION $$D \sec x = D \frac{1}{\cos x}$$

$$= \frac{\sin x}{\cos^2 x} \qquad \text{(quotient rule)}$$

$$= \frac{\sin x}{\cos x} \cdot \frac{1}{\cos x}$$

$$= (\tan x)(\sec x)$$

EXAMPLE 6   Compute $Dx^2 \cos x$

SOLUTION $$Dx^2 \cos x = x^2 D \cos x + (Dx^2) \cos x \qquad \text{(product rule)}$$

$$= -x^2 \sin x + 2x \cos x$$

EXAMPLE 7   Compute $Dx^{\sin x}$ for numbers x for which $x > 0$.

SOLUTION   We use logarithmic differentiation (see page 163–4). Let $y = x^{\sin x}$. The problem is to find Dy. We see that

$$\ln y = \ln (x^{\sin x}) = \sin x \cdot \ln x$$

Thus,

$$D \ln y = D (\sin x \cdot \ln x)$$

$$\frac{y'}{y} = \sin x \cdot \frac{1}{x} + (\ln x)(\cos x)$$

$$y' = y\left[\frac{\sin x}{x} + (\ln x)(\cos x)\right]$$

so

$$Dx^{\sin x} = x^{\sin x}\left[\frac{\sin x}{x} + (\ln x)(\cos x)\right]$$

We now summarize the differentiation facts about the trigonometric functions. Some of them have been verified, and some will be verified in the problem sets.

$D \sin x = \cos x$	$D \cos f(x) = (-\sin f(x))Df(x)$
$D \cos x = -\sin x$	$D \tan f(x) = (\sec^2 f(x))Df(x)$
$D \tan x = \sec^2 x$	$D \cot f(x) = (-\csc^2 f(x))Df(x)$
$D \cot x = -\csc^2 x$	$D \sec f(x) = (\tan f(x))$
$D \sec x = (\tan x)(\sec x)$	$\qquad (\sec f(x)) \cdot Df(x)$
$D \csc x = -(\cot x)(\csc x)$	$D \csc f(x) = -(\cot f(x))$
$D \sin f(x) = (\cos f(x))Df(x)$	$\qquad (\csc f(x)) \cdot Df(x)$

Remark These formulas are valid whenever x and $f(x)$ are in the domain of the given trigonometric function, and $f'(x)$ exists.

EXAMPLE 8 Find y' if $y = \cos (x + y)$.

SOLUTION We assume y is a function of x; say, for convenience, $y = y(x)$. Thus,

$$y(x) = \cos (x + y(x))$$

and hence

$$Dy(x) = D \cos (x + y(x))$$
$$Dy(x) = (-\sin (x + y(x))) \cdot D(x + y(x))$$
$$Dy(x) = (-\sin (x + y)) \cdot (1 + Dy(x))$$
$$Dy(x) = -\sin (x + y) - \sin (x + y)Dy(x)$$
$$Dy(x) + \sin (x + y)Dy(x) = -\sin (x + y)$$
$$Dy(x) (1 + \sin (x + y)) = -\sin (x + y)$$

and so, finally,

$$Dy(x) = -\frac{\sin(x + y)}{1 + \sin(x + y)}$$

Notice that we have used implicit differentiation to solve this problem.

Problems   In problems 1–20, find $f'(x)$.

1. $f(x) = 3 \sin 7x$

2. $f(x) = 5 \sin 2x$

3. $f(x) = \cos(5x + 2)$

4. $f(x) = (\sin x)(\cos x)$

5. $f(x) = \cos(4x^2 - 1)$

6. $f(x) = \sin^2(3x)$

7. $f(x) = x^2 \sin 5x$

8. $f(x) = \sin^5(x^2 + 1)$

9. $f(x) = \sec(x^2)$

10. $f(x) = \sqrt{\cos x}$

11. $f(x) = \ln|\sin 5x|$

12. $f(x) = \ln|x + \tan x|$

13. $f(x) = \ln|\sec 2x|$

14. $f(x) = \ln|(\sin x)(\cos x)|$

15. $f(x) = \tan^3\left(\dfrac{x}{2}\right)$

16. $f(x) = \dfrac{\sin 2x}{1 + \cos 2x}$

17. $f(x) = \dfrac{\sin x}{1 - 2 \cos x}$

18. $f(x) = e^{\sin x}$

19. $f(x) = e^{-\sin x \cos x}$

20. $f(x) = \sin(\cos x)$

21. Use implicit differentiation to find dy/dx if $\sin x = \cos y$.

22. Use implicit differentiation to find y', given that $x \cos y + y \cos x = 1$.

23. Use implicit differentiation to find y' if $x \sin 2y = y \cos 2x$.

In problems 24–27, find $f'(x)$.

24. $f(x) = x^{\sin x}$

25. $f(x) = (\cos x)^{x^2}$

26. $f(x) = \sin x^{\sin x}$

27. $f(x) = \sin x^{\tan x}$

28. Show that $D \cot x = -\csc^2 x$

29. Show that $D \csc x = (-\csc x)(\cot x)$

30. Let $f(x,y) = (\cos x)(\sin y)$. Find

    a. $\dfrac{\partial f(x,y)}{\partial x}$
    b. $\dfrac{\partial f(x,y)}{\partial y}$

    c. $\dfrac{\partial^2 f(x,y)}{\partial x^2}$
    d. $\dfrac{\partial^2 f(x,y)}{\partial y^2}$

    e. $\dfrac{\partial^2 f(x,y)}{\partial x\, \partial y}$

31. Let $f(x,y) = \sin(2x - 3y)$. Find:

    a. $\dfrac{\partial f(x,y)}{\partial x}$
    b. $\dfrac{\partial f(x,y)}{\partial y}$

    c. $\dfrac{\partial^2 f(x,y)}{\partial x^2}$
    d. $\dfrac{\partial^2 f(x,y)}{\partial y^2}$

    e. $\dfrac{\partial^2 f(x,y)}{\partial x\, \partial y}$

SECTION 5
INTEGRATION

Now that we have found the derivatives of the trigonometric functions, we are in a position to hunt for antiderivatives. As in our earlier work with antiderivatives, we shall find that after we have written down some of the basic antiderivatives, we shall have to rely on such techniques as substitution and integration by parts to find more complicated antiderivatives. There is an ingredient here that was absent from our earlier work; namely, there are times when a seemingly complicated trigonometric form can be simplified by using a trigonometric identity. Of course, this means that one should have available a fund of such identities. They can be found in books on trigonometry, but in general we shall try to avoid computations in which such identities are used.

Since $D \sin x = \cos x$, we see that

$$\int \cos x \, dx = \sin x + c$$

Since $D(-\cos x) = -(-\sin x) = \sin x$, we see that

$$\int \sin x \, dx = -\cos x + c$$

In a similar way, using facts about derivatives, we get the following antiderivatives:

$$\int \cos x \, dx = \sin x + c$$
$$\int \sin x \, dx = -\cos x + c$$
$$\int \sec^2 x \, dx = \tan x + c$$
$$\int \csc^2 x \, dx = -\cot x + c$$
$$\int (\tan x)(\sec x) \, dx = \sec x + c$$
$$\int (\cot x)(\csc x) \, dx = -\csc x + c$$

Of course, since our trigonometric functions are continuous throughout their domains, we can use the fundamental theorem of calculus to compute integrals.

EXAMPLE 1   Find the area under the graph of sin from 0 to π.

SOLUTION   This area is equal to

$$\int_0^\pi \sin x \, dx$$

Using the fundamental theorem,

$$\int_0^{\pi} \sin x \, dx = \int \sin x \, dx \Big|_0^{\pi}$$

$$= -\cos x \Big|_0^{\pi}$$

$$= -\cos \pi + \cos 0$$

$$= -(-1) + 1$$

(since $\cos \pi = -1$ and $\cos 0 = 1$)

$$= 2$$

The next examples illustrate the use of substitution and integration by parts to evaluate antiderivatives.

EXAMPLE 2   Compute $\int \cos 3x \, dx$.

SOLUTION   We use substitution as follows:

let $u(x) = 3x$, then $du = 3 \, dx$, so

$$\int \cos 3x \, du = \tfrac{1}{3} \int \cos u \, du$$

$$= \tfrac{1}{3} \sin u + c$$

$$= \tfrac{1}{3} \sin 3x + c$$

EXAMPLE 3   Compute $\int \tan x \, dx$.

SOLUTION   Since $\tan x = \sin x / \cos x$, our problem reduces to finding

$$\int \frac{\sin x}{\cos x} \, dx$$

We use substitution. Let $u(x) = \cos x$. Then, $du/dx = -\sin x$, so $-du = \sin x \, dx$. Thus,

$$\int \frac{\sin x}{\cos x} \, dx = - \int \frac{1}{u} \, du$$

$$= -\ln |u| + c$$

$$= -\ln |\cos x| + c$$

Hence,

$$\int \tan x \, dx = -\ln |\cos x| + c$$

EXAMPLE 4   Compute $\int \cos x \sqrt{1 + \sin x} \, dx$.

SOLUTION   Let $u = 1 + \sin x$. Then, $du = \cos x\,dx$. Thus, using substitution,

$$\int \cos x \sqrt{1 + \sin x}\,dx = \int u^{1/2}\,du$$

$$= \frac{2}{3}\,u^{3/2} + c$$

$$= \frac{2}{3}\,(1 + \sin x)^{3/2} + c$$

EXAMPLE 5   Compute $\int x \sin x\,dx$.

SOLUTION   The reader is invited to try the more obvious substitutions; after a little work, it should be clear that they won't work. So, we try integration by parts. Let

$$f(x) = x \quad \text{and} \quad g'(x) = \sin x$$

then

$$f'(x) = 1 \quad \text{and} \quad g(x) = -\cos x$$

Thus,

$$\int x \sin x\,dx = -x \cos x - \int 1 \cdot (-\cos x)\,dx$$

$$= -x \cos x + \int \cos x\,dx$$

$$= -x \cos x + \sin x + c$$

EXAMPLE 6   Compute $\int_0^{\pi/2} x \sin x\,dx$.

SOLUTION   From example 5, we see that

$$\int_0^{\pi/2} x \sin x\,dx = (-x \cos x + \sin x)\Big|_0^{\pi/2}$$

$$= \left(-\frac{\pi}{2} \cos \frac{\pi}{2} + \sin \frac{\pi}{2}\right) - (-0 \cdot \cos 0 + \sin 0)$$

$$= \left(-\frac{\pi}{2} \cdot 0 + 1\right) - (0 + 0)$$

$$= 1$$

In practice, many integrals are computed by using a table of integrals. The table of integrals at the end of the book includes trigonometric forms (formulas 91 to 112), trigonometric reduction formulas (formulas 113 to 124), and some exponential formulas (formulas 132 and 133). Some problems using the tables appear in the problem set.

Problems   In problems 1–12, evaluate the integral.

1. $\int (2 \sin x - 3 \cos x)\, dx$ 2. $\int (\sin 2x + \cos 3x)\, dx$
3. $\int \tan 3x\, dx$ 4. $\int \cos x \tan x\, dx$
5. $\int \cos x \, \sin^8 x\, dx$ 6. $\int \cos x \, \sqrt[3]{\sin x}\, dx$
7. $- \int \sin x \, \sqrt[3]{(\cos x) + 1}\, dx$

8. $\int \dfrac{\cos x}{\sin^2 x}\, dx$

9. $\int \dfrac{\cos x}{\sqrt{\sin x}}\, dx$

10. $\int \dfrac{\sin x}{1 - \cos x}\, dx$

11. $\int \dfrac{\cos (\ln x)}{x}\, dx$

12. $\int (\sin x)e^{\cos x}\, dx$

In problems 13–17, use integration by parts to evaluate the given integral.

13. $\int x \sin x\, dx$ 14. $\int x \cos x\, dx$
15. $\int x^2 \sin x\, dx$
    (*Hint:* Let $f(x) = x, g'(x) = \sin x$, and use problem 13.)
16. $\int e^x \sin x\, dx$
17. $\int \sin^2 x\, dx$
    (*Hint:* At the appropriate time, use the fact that $\cos^2 x = 1 - \sin^2 x$.)
18. Compute

$$\int (\sin x) \sin (\cos x)\, dx$$

19. Compute

$$\int \sin^5 x\, dx$$

    (*Hint:* $\sin^5 x = (\sin^4 x) \sin x$ and $\sin^2 x = 1 - \cos^2 x$.)

20. Compute

$$\int \sin^2 x \cos^3 x\, dx$$

    (*Hint:* Notice that $\cos^3 x = (\cos^2 x)(\cos x)$ and $\cos^2 x = 1 - \sin^2 x$.)

Use the table of integrals to compute
21. $\int \cos^2 x\, dx$
22. $\int \cos^3 x\, dx$
23. $\int \cos^4 x\, dx$
24. $\int \cos^5 x\, dx$

25. Use the table of integrals to compute

$$\int \frac{dx}{\sqrt{5} + 3 \sin x}$$

26. Use the table of integrals to show that:

    a. for *any* integers h and k,

$$\int_0^{2\pi} (\sin hx)(\cos kx)\, dx = 0$$

    b. for integers h and k with $h \neq k$

$$\int_0^{2\pi} (\sin hx)(\sin kx)\, dx = 0$$

27. Let $f(x,y) = \sin x \cos y$. Find

$$\int_0^{\pi/2} \int_0^{\pi/2} f(x,y)\, dx\, dy$$

28. Compute

$$\int_0^{\pi} \int_0^{\pi} \cos(x + y)\, dy\, dx$$

SECTION 6
APPLICATIONS

Where do we stand at this point? We know how to differentiate and integrate the trigonometric functions (if the integrals aren't too complicated). We can now discuss a variety of applications that depend on our ability to perform these computations. We can investigate the trigonometric functions as they relate to problems in tangent lines, velocity, acceleration, extrema, area, volume of revolution, and so on. In fact, in principle we could go back to each application of the derivative and see how the trigonometric functions fit into these applications. We shall pick some of these areas in the following examples.

EXAMPLE 1 Find the equation of the line tangent to the graph of sin at

$$\left(\frac{\pi}{6}, \sin \frac{\pi}{6}\right) = \left(\frac{\pi}{6}, \frac{1}{2}\right)$$

SOLUTION Recall that an equation for the line tangent to the graph of a function f at the point $(a, f(a))$ is

$$y - f(a) = f'(a)(x - a)$$

Here, $f(x) = \sin x$ and, therefore, $f'(x) = \cos x$; thus,

$$f'\left(\frac{\pi}{6}\right) = \cos \frac{\pi}{6} = \frac{\sqrt{3}}{2}$$

Thus, the tangent line we want has equation

$$y - \frac{1}{2} = \frac{\sqrt{3}}{2}\left(x - \frac{\pi}{6}\right)$$

or

$$y = \frac{\sqrt{3}}{2}x + \frac{6 - \pi\sqrt{3}}{12}$$

EXAMPLE 2   Let

$$f(x) = x \sin x, \qquad -\frac{\pi}{2} \le x \le \frac{\pi}{2}$$

Determine the extrema of f.

SOLUTION   Recall that since f is continuous on $[-\pi/2,\pi/2]$, it attains both a maximum and a minimum on $[-\pi/2,\pi/2]$, and these must be either at an endpoint or a critical value of f. To find the critical values, we set $f'(x) = 0$. We see that $f'(x) = x \cos x + \sin x$ and thus $f'(x) = 0$ becomes

$$x \cos x + \sin x = 0$$
$$x \cos x = -\sin x$$
$$-x = \tan x$$

If we draw the graph of $y = -x$ and $y = \tan x$, $-\pi/2 \le x \le \pi/2$, we see that the only solution of this is $x = 0$. See Figure 9-22.

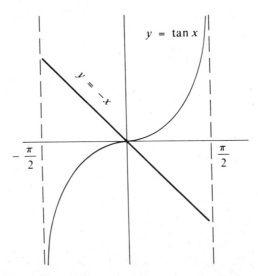

Figure 9-22

Thus, 0 is the only critical point. Checking f at the endpoints and critical points, we see that

$$f\left(-\frac{\pi}{2}\right) = -\frac{\pi}{2}\sin\left(-\frac{\pi}{2}\right)$$

$$= \frac{\pi}{2}\sin\left(\frac{\pi}{2}\right) \qquad \left(\text{since } \sin\left(-\frac{\pi}{2}\right) = -\sin\frac{\pi}{2}\right)$$

$$= \frac{\pi}{2} \qquad \left(\text{since } \sin\left(\frac{\pi}{2}\right) = 1\right)$$

$$f\left(\frac{\pi}{2}\right) = \frac{\pi}{2}$$

$$f(0) = 0$$

Thus, the maximum of f on $[-\pi/2, \pi/2]$ is $\pi/2$, and the minimum is 0.

EXAMPLE 3   A light source hanging from the end of a pendulum casts a light on a horizontal floor. Suppose the distance $f(t)$ the light beam has moved along the floor at time t (in seconds) is given by

$$f(t) = 2\sin\left(\frac{\pi}{2}t\right)$$

Discuss the motion of the light.

SOLUTION   Before getting into the problem of velocity and acceleration, let us directly investigate f, the distance function. At time 0, we see that $f(0) = 0$, so the light is at the starting "equilibrium" position, and the pendulum is hanging straight down. Since sin is an increasing function from 0 to $\pi/2$, we see that f is increasing from $t = 0$, at which time the distance is $f(0)$, which is equal to 0, to $t = 1$, at which time the distance is $f(1) = 2\sin(\pi/2) = 2$. That is, the light moves to the right a distance of 2 units as time goes from 0 to 1. But when we just pass 1 second, we see that f is *decreasing*. Thus, the light goes 2 units to the right, and then starts *back again*. At time 2 seconds, the distance is

$$f(2) = 2\sin\frac{\pi}{2}\cdot 2 = 2\sin\pi = 0$$

so the light is back to its initial position, and again the pendulum is hanging straight down.

What happens when we just pass 2 seconds? When t is slightly past 2, we see that $2\sin(\pi/2)t$ is slightly bigger than $2\sin\pi$, and hence is negative, and, is decreasing through negative values. Thus, the light is now moving to the *left*. At 3 seconds, the distance $f(3)$ is equal to $2\sin(3\pi/2)$, which is equal to -2. That is, at the end of 3 seconds, the light is 2 units to the *left*. The reader should convince himself that from time 3 to 4 seconds, the light is returning to equilibrium position. Thus, a distance function involving sin (or, in fact, cos) displays an **oscillatory motion,** sometimes referred to as **simple harmonic motion.** See Figure 9-23.

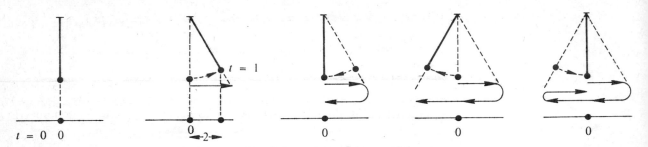

Figure 9-23

We shall next investigate $f'(t)$, the velocity of the light along the floor. We see that

$$f'(t) = \pi \cos \frac{\pi}{2} t$$

Keep Figure 9-23 in mind as we carry out the computations. Notice that as t goes from 0 to 1, the velocity $f'(t)$ goes from π to 0. Thus, as we expect, the light is "at rest" when $t = 1$. As t goes from 1 to 2, $f'(t)$ goes from 0 to -1. Thus, from $t = 1$ to $t = 2$, the velocity is negative, so the motion of the light over $1 \leq t \leq 2$ is opposite its motion over the time interval $0 \leq t \leq 1$. From time 2 to time 3, the velocity goes from -1 to 0, and thus is slowing down. At $t = 3$, the light is again "at rest". From time 3 to time 4, the velocity is again positive, so again the motion is reversed. At time 4, the light is in the same position as it was at time 0, and then the entire process starts again.

EXAMPLE 4   Let $f(x) = \sec x$, $0 \leq x \leq \pi/4$. Find the volume obtained by revolving the graph of f about the **x**-axis.

SOLUTION   Recall that the volume of revolution is

$$\pi \int_0^{\pi/4} [f(x)]^2 \, dx = \pi \int_0^{\pi/4} \sec^2 x \, dx$$
$$= \pi \tan x \big|_0^{\pi/4}$$
$$= \pi \qquad \text{(square units)}$$

Problems

1. Let $f(x) = x \cos x$. Find an equation for the line tangent to the graph of f at $(\pi/2, 0)$.

2. Let $f(x) = \sec x$. Find an equation for the line tangent to the graph of f at $(\pi/4, \sqrt{2})$.

3. Show that the line given by $y = x$ is tangent to the curve given by $y = x \sin x$ whenever $\sin x = 1$.

4. Let $f(x) = \sin^2 x + \cos x$, $0 \le x \le 2\pi$. Find all local extrema for f.

5. Let $f(x) = \sin x + \cos x$, $0 \le x \le 2\pi$. Find all local extrema for f.

6. Let $f(x) = \cos^3 x - 4 \cos^2 x$, $-0.1 \le x \le \pi + 0.1$. Find all local maximum and local minimum points.

7. Let

$$f(x) = \frac{8}{\sin x} + \frac{27}{\cos x}, \qquad 0 < x < \frac{\pi}{2}$$

Explain why f must attain a minimum in this open interval. Show that the point x at which the minimum is attained satisfies the condition $\tan x = \frac{3}{2}$.

8. Find the volume obtained by revolving the graph of sin from 0 to π about the **x**-axis.

9. Find the volume obtained by revolving about the **y**-axis the region bounded by the curves given by $y = \sin x$, the **x**-axis, the **y**-axis, and $x = \pi$.

10. Let $f(x) = \cos^2 x$. Find the volume obtained by revolving the graph of f between $x = 0$ and $x = \pi$ about the **x**-axis. (*Hint:* See problem 23, section 5.)

11. Suppose an object moves in a straight line path in such a way that at time t, its position $f(t)$ is given by

$$f(t) = 3 \cos 2t \qquad 0 \le t \le 2\pi$$

a. What is the velocity of the object?
b. What is the acceleration of the object?
c. When is the object at rest?
d. Describe the motion of the object.

12. A particle moves on a curve in a plane in such a way that the x- and y-coordinates of the particle are given at time t by

$$x(t) = r \cos bt \qquad y(t) = r \sin bt$$

when $r > 0$ and b are "fixed constants." The functions $x''(t)$ and $y''(t)$ are called, respectively, the components of the acceleration in the x- and y-directions.

a. Show that the particle is moving on the circumference of a circle.

b. Show that the x- and y-components of acceleration are given, respectively, by $-b^2x$ and $-b^2y$.

13. Use Taylor's theorem with remainder to find the Taylor polynomial about 0 for sin. Show that

$$\lim_{n \to \infty} |R_n(x)| = 0$$

for any x. Derive the Taylor series about 0 (that is, Maclaurin series) for f.

Appendix

This is a two-part appendix that serves to provide reference material for topics used in calculus, and a quick review of such topics.

Part 1 is a presentation of an elementary set theory. It is concerned with the basic properties of sets. To provide some applications, elementary counting problems are introduced.

Part 2 is a short treatment of those areas of algebra used in calculus. Not meant to be a theoretical, or even thorough, treatment of algebra, the material is deliberately brief, and centered around solved examples.

SETS   Part 1

This chapter is intended to establish some basic ideas and notation that will be useful in the study of probability and calculus. For example, a basic probability space is a set, and the elementary events are the elements of this set. Of central importance to this entire book is the *set* of real numbers. Thus, the main interest of this chapter is to establish a basic vocabulary.

SECTION 1
SETS AND SUBSETS

Any collection of objects may form a **set.** For example, the first three letters of the alphabet form a set which is denoted by $\{a,b,c\}$. If we call this set S then we write $S = \{a,b,c\}$ and say that a, b, and c are **elements** of S. This is denoted by writing, for example, $a \in S$, read as "a belongs to S" or "a is an element of S". Similarly, in this example, $b \in S$ and $c \in S$.

One way to describe a set is by the use of the following notation:

$$\{ x : \qquad \}$$

which is read "the set of all x for which" with the description of the set following the colon. For example,

$$\{x : x \text{ is (or was) a president of the United States}\}$$

is the set of all objects x for which x is or was a president of the

United States; that is, it is the set of all United States presidents. If M is the set of male Americans, then

$$\{x: \ x \in M \text{ and (the age of } x) = 21\}$$

is the set of all 21-year-old male Americans.

A set B is a **subset** of a set A if and only if* every element of A is an element of B. A is indicated as a subset of B, by the notation $A \subset B$.

EXAMPLE 1   If $S = \{1,2,3\}$ the subsets of S are $\{1,2,3\}$, $\{1,2\}$, $\{1,3\}$, $\{2,3\}$, $\{1\}$, $\{2\}$, $\{3\}$, $\{\ \}$. This last set $\{\ \}$ is called the **empty set** (the set which has no elements) and it is agreed that the empty set is a subset of every set. Note that $\{1,2,3\} \subset S$, $\{1,2\} \subset S$, $\{\ \} \subset S$, etc. This example shows that a set of three elements has exactly eight subsets.

Note that $\{a\}$ is a set having exactly one element and that $\{a\} \neq a$. This is a matter of definition; we simply agree that $\{a\}$ is a different kind of object from a, in the same way that a physics class with exactly one student is different from the student himself, for while the class may be abolished (for insufficient enrollment) the student will continue to exist.

Also, it is agreed that two sets are *equal* if and only if they have the same elements. It can then be argued that there is only one empty set since any two empty sets have the same elements; hence, are equal. Finally the empty set is denoted by the symbol ϕ.

EXAMPLE 2   The set $\{1,2\}$ has four subsets:
$$\{1,2\}, \{1\}, \{2\}, \phi$$

EXAMPLE 3   The set $\{1\}$ has two subsets:
$$\{1\}, \phi$$

EXAMPLE 4   The set ϕ has one subset: ϕ.

It is of some interest to know how many subsets of various sizes a set has. We illustrate by example.

---

*The use of the phrase "if and only if" to connect two statements means that the statements are logically equivalent (that is, they are simultaneously true or false).

EXAMPLE 5   How many two-element subsets does a set of four elements have?

SOLUTION   The set $\{1,2,3,4\}$ has exactly 6 two-element subsets: $\{1,2\}$, $\{1,3\}$, $\{1,4\}$, $\{2,3\}$, $\{2,4\}$, $\{3,4\}$; and, similarly, any set of 4 elements has 6 two-element subsets.

EXAMPLE 6   How many subsets of each size does a four-element set have?

SOLUTION   The set $\{1,2,3,4\}$ has

One subset with no elements:	$\emptyset$
Four subsets with one element:	$\{1\}$, $\{2\}$, $\{3\}$, $\{4\}$
Six subsets with two elements:	$\{1,2\}$, $\{1,3\}$, $\{1,4\}$, $\{2,3\}$, $\{2,4\}$, $\{3,4\}$
Four subsets with three elements:	$\{2,3,4\}$, $\{1,3,4\}$, $\{1,2,4\}$, $\{1,2,3\}$
One subset with four elements:	$\{1,2,3,4\}$

It is not mere coincidence that the numbers 1, 4, 6, 4, 1 appearing in this last example are the fourth row* of the array known as *Pascal's triangle*.

$$
\begin{array}{ccccccccccccc}
 & & & & & & 1 & & & & & & \\
 & & & & & 1 & & 1 & & & & & \\
 & & & & 1 & & 2 & & 1 & & & & \\
 & & & 1 & & 3 & & 3 & & 1 & & & \\
 & & 1 & & 4 & & 6 & & 4 & & 1 & & \\
 & 1 & & 5 & & 10 & & 10 & & 5 & & 1 & \\
1 & & 6 & & 15 & & 20 & & 15 & & 6 & & 1
\end{array}
$$

It is easy to form new rows to this triangle when we notice that each number in the array is the sum of the two numbers above it. For example, $3 + 3 = 6$ and in the triangle we have $\begin{smallmatrix} 3 & & 3 \\ & 6 & \end{smallmatrix}$, similarly $4 + 6 = 10$ and in the triangle we have $\begin{smallmatrix} 4 & & 6 \\ & 10 & \end{smallmatrix}$. In fact, Pascal's triangle is defined by this relationship, for then starting with $\begin{smallmatrix} & 1 & \\ 1 & & 1 \end{smallmatrix}$ the rest of the triangle can be generated.

The following theorem is convenient for a large variety of computations involving numbers which are not too large.

---

*The top row consisting of just the number *one* is called *zero*th row.

Theorem*   *Each number in the nth row of Pascal's triangle is the number of subsets of a fixed size of a set of n elements. (For example, the numbers* 1, 6, 15, 20, 15, 6, 1 *in the sixth row tell us that the set* {1,2,3,4,5,6} *has* 1 *empty subset,* 6 *one-element subsets,* 15 *two-element subsets,* 20 *three-element subsets,* 15 *four-element subsets,* 6 *five-element subsets, and* 1 *six-element subset.)*

EXAMPLE 7   How many committees of three can be chosen from a group of six people.

SOLUTION   The number of committees is the number of three-element subsets of a set of six elements. The sixth row of Pascal's triangle is 1, 6, 15, 20, 15, 6, 1; hence, the number of three-element subsets is 20, so that the number of committees is 20.

EXAMPLE 8   There are nine judges on the Supreme Court. How many different lists of judges could appear on a majority report?

SOLUTION   Any subset of 5 through 9 judges constitutes a majority. The numbers in the ninth row of Pascal's triangle are 1, 9, 36, 84, 126, 126, 84, 36, 9, 1. The number of five-element subsets is 126, of six-element subsets is 84, etc. Therefore, there are $126 + 84 + 36 + 9 + 1 = 256$ different majorities. A different method of solution is to remember that every subset is either a majority or a minority and that there are equal numbers of each; so that one-half the number of all possible subsets, 2^9, are majorities, or $\frac{1}{2}(2^9) = 256$.

EXAMPLE 9   How many diagonals does a polygon of seven sides have?

SOLUTION   If we consider the vertices as points, then any two distinct points determine a line which is either a side of the polygon or a diagonal. Since the number of two-point subsets of the seven points is 21, there are 21 possible lines (diagonals plus sides). Hence, there are $21 - 7 = 14$ diagonals.

---

*Here is a sketch of the proof of this theorem: Let $A = \{1, 2, \ldots, m\}$ and $B = \{1, 2, \ldots, n, n+1\}$. We want to see that the number of r-element subsets of B is equal to the number of $r - 1$ element subset of A plus the number of r-element subsets of A (for this is the relation which determines Pascal's triangle). However, each r-element subset of B is either an r-element subset of A or is obtained from an $r - 1$ element subset of A by adjoining the element $n + 1$. For example, the two-element subsets of $\{1,2,3,4,5\}$ are $\{1,2\}$, $\{1,3\}$, $\{1,4\}$, $\{2,3\}$, $\{2,4\}$, $\{3,4\}$, and $\{1,5\}$, $\{2,5\}$, $\{3,5\}$, $\{4,5\}$, and $6 + 4 = 10$.

Problems

1. List the three-element subsets of $\{1,2,3,4,5\}$.
2. List the two-element subsets of $\{1,2,3,4,5,6\}$.
3. How many three-element subsets does a set of seven elements have?
4. How many subsets does a set of 100 elements have?
5. How many different ways can a 10-question *true-false* examination be answered if exactly one-half the answers are *false*.
6. Five people meet at a party and each shakes hands exactly once with each of the other four. How many handshakes are there?
7. How many subsets does a set of n elements have?
8. How many $(n - 1)$-element subsets does a set of n elements have?
9. Ten boys choose up sides to play basketball. How many different teams are possible?
10. In poker, a flush is a hand of 5 cards all of the same suit. An ordinary deck has 52 cards and four suits. How many hands are there? How many flushes?

SECTION 2

SET ALGEBRA

Operations of arithmetic such as addition and multiplication have their analog in set theory. If A and B are sets, there are several operations which are both natural and useful to define.

1. **Intersection:** $A \cap B$ *(read "A intersection B") is the set of all elements which belong to A and B.*

2. **Union:** $A \cup B$ *(read "A union B") is the set of all elements which belong to A or B (or both).*

3. **Cartesian Product:** $A \times B$ *(read "A cross B") is the set of all ordered pairs (two-tuples) of the form (a,b) where $a \in A$, $b \in B$; that is, all two-tuples with first component an element of A and second component an element of B.*

EXAMPLE 1   If $A = \{1,2,a,b,c\}$ and $B = \{2,a,c,7,d\}$, then

$$A \cap B = \{2,a,c\}$$
$$A \cup B = \{1,2,7,a,b,c,d\}$$
$$A \times B = \{(1,2), (1,a), (1,c), (1,7), (1,d), (2,2),$$
$$(2,a),(2,c)(a,2)(a,a),(a,c) \text{ etc.}\}$$

Notice that $A \times B$ has $7 \cdot 3 = 21$ elements.

The union and intersection may be indicated schematically by Venn diagrams as in Figure A-1.

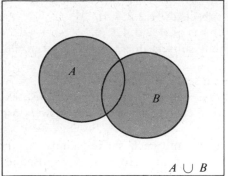

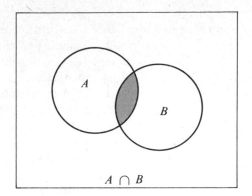

Figure A-1

If subsets of a fixed set S are being studied, then a third operation, complement, may be defined as:

A' *(read "the* **complement** *of A") is the set of all elements not in A but in S.*

EXAMPLE 2   If $S = \{1,2,3,4,5\}$ and $A = \{3,4\}$, then $A' = \{1,2,5\}$.

The Venn diagram indicates the complement relationship in Figure A-2.

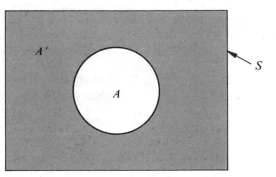

Figure A-2

The complement is not well-defined unless the fixed set (sometimes called the **universal** set) is specified. Unfortunately, there is no one set large enough to serve as a fixed universal set for all sets.

The three operations of union, intersection, and complement satisfy the following laws:

1. Commutative $A \cup B = B \cup A$,
                       $A \cap B = B \cap A$

2. Associative $(A \cup B) \cup C = A \cup (B \cup C),$
$A \cap (B \cap C) = (A \cap B) \cap C$

3. Distributive $A \cap (B \cup C)$
$= (A \cap B) \cup (A \cap C)$

4. Idempotent $A \cap A = A, A \cup A = A$

5. Identity $A \cap \emptyset = \emptyset, A \cup \emptyset = A$

6. De Morgan's Laws $(A \cup B)' = A' \cap B',$
$(A \cap B)' = A' \cup B'$

7. Complements $(A')' = A$

Problems

1. Draw Venn diagrams to illustrate the distributive law.
2. Draw Venn diagrams to illustrate De Morgan's Laws.
3. If A and B are sets and $A \cap B = A$, can it be inferred that A is a subset of B? (*Answer*: Yes. For if A is not a subset of B, then $A \cap B$ is a subset of A but not all of A.)
4. If A and B are sets and $A \cup B = A$, what can be inferred?
5. *True* or *false*. If A, B, and C are sets, then

$$A \cup (B \cap C) = (A \cup B) \cap (A \cup C)$$

6. *True* or *false*. If A, B, and C are sets, then

$$A \times (B \cup C) = (A \times B) \cup (A \times C)$$

A REVIEW OF ALGEBRA

Part 2

This chapter is a review of those areas of algebra that will be used in the study of probability and calculus. No effort has been made to approach the subject axiomatically, although when simple explanations of *theorems* are available, they will be included. For other verifications of theorems, or for a more in-depth discussion of a topic, any standard college algebra text can be consulted.* The emphasis here is on the use of algebraic results in the solution of algebraic problems; that is, the emphasis is on drill. One use of this chapter is that of a reference as the book is read.

SECTION 1
REAL NUMBERS. THE REAL LINE. ORDER

The numbers we shall be interested in throughout this book are the **real numbers**. We shall outline this system. The set of **natural numbers** (or **counting numbers**) **N** is the set

$$\{1,2,3,4,5,\ldots\}$$

and the set of integers **I** is

$$\{\ldots,-5,-4,-3,-2,-1,0,1,2,3,4,5,\ldots\}$$

Note that the natural numbers are precisely the positive integers; thus, $N \subset I$. A number is said to be a **rational number** if it can be expressed as the ratio of integers (of course, with nonzero denominator). We shall denote the set of rational numbers by **Q**. Thus, $x \in Q$ means that x can be expressed as p/q with $p \in I$, $q \in I$, $q \neq 0$. For example, $\frac{1}{2} \in Q$, $\frac{-10}{3} \in Q$, $0.15 = \frac{15}{100} \in Q$ and $17 = \frac{17}{1} \in Q$. Note that every integer n can be expressed as $n/1$. Therefore, every integer is a rational number; that is, $I \subset \mathbf{Q}$.

The *real numbers* are those that can be written as the sum of an integer and a decimal; that is, can be written as

$$n\,.abcd\ldots$$

where $n \in I$, and a, b, c, d, etc., are any of the digits from 0 to 9, and the dots "$\ldots$" signify that the "decimal expansion" may continue indefinitely. Denote the set of real numbers by **R**. We see

---

*For example, Steven Bryant, Jack Karush, Leon Nower, and Daniel Saltz, *College Algebra and Trigonometry* (Pacific Palisades, California: Goodyear Publishing Company, Inc., 1971).

that $\frac{1}{2} = 0.5 \in \mathbf{R}$, and $\frac{1}{3} = 0.333\ldots \in \mathbf{R}$. The numbers π, $(2)^{1/2}$, $(3)^{1/2}$ are in $\mathbf{R}$ (we shall not give their decimal expansions here, but π, for example, has been tabulated to a several thousand decimal place accuracy); these numbers, unlike 0.5 and 0.333..., *cannot* be written as the ratio of integers, and are called **irrational numbers**. Thus, a number is irrational if it is a real number that is not rational. If we denote the set of irrational numbers by $\mathbf{Z}$, we see that

$$\mathbf{R} = \mathbf{Z} \cup \mathbf{Q} \quad \text{and} \quad \mathbf{Z} \cap \mathbf{Q} = \phi$$

Henceforth, *we shall refer to real numbers simply as numbers.* Note that "$x \in \mathbf{R}$" and "x is a number" mean the same thing.

We next construct a geometrical analog to the real numbers which we shall call the **real line**. Draw a horizontal straight line (obviously, all that we can actually draw on paper is a segment of that line). Pick a point on this line, and label it zero; pick a point distinct from zero and to its right, and label it one. Then the length of the interval 0–1 will be a unit of length. Reproduce this length indefinitely to either side of zero and label the points 2, 3, 4,... to the right of 1, and −1, −2, −3,... to the left of 0 (see Figure A-3). Observe that the distinction here between 1 and −1 is the direction from 0 from which they are measured. The fundamental principle is that *to each real number there corresponds a unique point on the line, and that each point on the real line corresponds to exactly one real number.*

The line may be pictured as in Figure A-4 (the arrow points in the direction of increasing numbers). Henceforth, we shall make no distinction between a real number and the point that corresponds to that number on the (real) line.

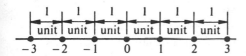

Figure A-3

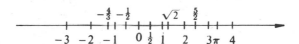

Figure A-4

Some of the basic properties of addition and multiplication of real numbers are listed below. For all $a \in \mathbf{R}$, $b \in \mathbf{R}$, and $c \in \mathbf{R}$,

1. $a + b = b + a$ and $ab = ba$ (referred to by saying addition and multiplication are **commutative**)
2. $(a + b) + c = a + (b + c)$ and $a(bc) = (ab)c$ (addition and multiplication are **associative**)
3. $a(b + c) = ab + ac$ (multiplication is **distributive** over addition)
4. $a + 0 = 0 + a = a$, $a \cdot 1 = 1 \cdot a = a$
5. $a + (-a) = (-a) + a = 0$

6. If $a \neq 0$, then $a \cdot (1/a) = (1/a) \cdot a = 1$
7. $a = b$ if and only if* $a + c = b + c$ and $a - c = b - c$
8. $a = b$ if and only if $a \cdot c = b \cdot c$ (where $c \neq 0$)

If a and b are real numbers, we shall write $a < b$ (read "*a is less than b*") or $b > a$ (read "*b is greater than a*") *to mean that a is to the left of b on the number line.* Similarly,

$$\mathbf{a} \leq \mathbf{b} \qquad \text{means} \qquad a < b \qquad \text{or} \qquad a = b$$

(clearly not both). Also,

$$a < x < b \qquad \text{means} \qquad a < x \qquad \text{and} \qquad x < b$$

For example, $1 < 5$ and $5 > 1$; $3 < 7$ and $7 < 10$, and thus $3 < 7 < 10$; $4 \leq 8$ and $4 \leq 4$. We say that the number x is **positive** if $x > 0$, and it is **negative** if $x < 0$. The following properties of inequalities are basic to their study.

Let a, b, c, u, and w be (real) numbers.

1. *If $a < b$, then $a + c < b + c$, and $a - c < b - c$, and conversely.*
2. *Assume that $u > 0$. If $a < b$, then $au < bu$, and conversely.*
3. *Assume that $w < 0$. If $a < b$, then $aw > bw$, and conversely.*

EXAMPLE 1   We shall illustrate our results on inequalities.

1. $10 > 2$, and of course $10 + 37 > 2 + 37$, and $10 - 18 > 2 - 18$
2. $2 < 10$, and of course $2 + 37 < 10 + 37$, and $2 - 18 < 10 - 18$
3. $-1 < 2$, and of course $-1 + 5 < 2 + 5$
4. $-2 < 4$, and of course $-2 - 1 < 4 - 1$
5. $6 > 2$, and of course $6 \cdot 3 > 2 \cdot 3$
6. $6 > -4$, and of course $6 \cdot 5 > -4 \cdot 5$
7. $6 > 2$, and $-3 < 0$, and we see that $6 \cdot (-3) < 2 \cdot (-3)$
8. $-1 > -3$ and $-4 < 0$, and we see that $-1 \cdot -4 < -3 \cdot -4$

EXAMPLE 2   Let x be a number. Then

$$3 - 2x < -5$$

if and only if

$$(3 - 2x) - 3 < -5 - 3$$

$$-2x < -8$$

and, finally, multiplying by $-\frac{1}{2}$

$$x > 4$$

---

*Recall that, when the phrase "if and only if" connects two statements, we mean that these statements are logically equivalent.

EXAMPLE 3   Find all numbers x (or, equivalently, find all $x \in \mathbf{R}$) for which $3x - 5 = x + 3$ (that is, solve the equation $3x - 5 = x + 3$).

SOLUTION   Observe that for any number x, both $3x - 5$ and $x + 3$ are simply numbers. In the following computation we shall not specify which of the eight laws (pp. 331–32) is being used. We now see that x is a real number for which

$$3x - 5 = x + 3$$

if and only if

$$3x - 5 + 5 = x + 3 + 5$$

$$3x = x + 8$$

$$3x - x = x + 8 - x$$

$$2x = 8$$

$$\tfrac{1}{2} \cdot 2x = \tfrac{1}{2} \cdot 8$$

$$x = 4$$

Problems   In problems 1–12, answer true or false, and explain.

1. $\dfrac{\pi}{3} \in \mathbf{Q}$
2. $-8 \in \mathbf{Q}$
3. $2\sqrt{2} \in \mathbf{Q}$
4. $\mathbf{Q} \subset \mathbf{Z}$
5. If $x \in \mathbf{Z}$ and $y \in \mathbf{Z}$, then $xy \in \mathbf{Z}$ (that is, the product of irrational numbers is also an irrational number)
6. If $x \in \mathbf{I}$ and $y \in \mathbf{I}$, then $x/y \in \mathbf{I}$
7. $-5 > -3$
8. $6 \leq 6$
9. $0 < -8$
10. $-x < 3$ if and only if $x > -3$
11. $2x < -4$ if and only if $x > -2$
12. $-x > 2$ if and only if $3x < -6$
13. Note that $\tfrac{1}{2} \in \mathbf{Q}$ and $\tfrac{3}{4} \in \mathbf{Q}$. Is $\tfrac{1}{2} + \tfrac{3}{4} \in \mathbf{Q}$? Is $\tfrac{1}{2} \cdot \tfrac{3}{4} \in \mathbf{Q}$? Is $\tfrac{1}{2} \div \tfrac{3}{4} \in \mathbf{Q}$? Guess some generalizations of these observations.

In problems 14–19, find the number x for which

14. $2x + 1 = x + 5$
15. $5x - 4 = -x + 8$
16. $10x + 20 = 9x - 10$
17. $1 - 3x = 4x + 1$
18. $x^2 + 8x - 2 = x^2 - 2x + 8$
19. $3x^2 + 4x + 1 = 3x^2 + 2x + 15$

In problems 20–27, find all real numbers x for which

20. $3x - 4 < 5$
21. $5x + 1 \leq 16$
22. $1 - x \leq -3$
23. $-6 - 2x < -4$
24. $-3 + 2x < -7$
25. $8 + 5x \leq -12$
26. $3 < 1 + 2x < 5$
27. $-5 < 3 - x < 8$

SECTION 2
ABSOLUTE VALUE

For any real number x, define the *absolute value* of x, denoted as $|x|$, by the distance from x to 0; that is,

$$|x| = \begin{cases} x, \text{ if } x \text{ is positive or zero (that is, if } x \geq 0) \\ -x, \text{ if } x \text{ is negative (that is, if } x < 0) \end{cases}$$

For example $|-5| = -(-5) = 5, |5| = 5$ (see Figure A-5), $|0| = 0$, $|-7| = |7| = 7$, $|(-2)^{1/2}| = |(2)^{1/2}| = (2)^{1/2}$; in fact, for any a, $|a| = |-a| \geq 0$

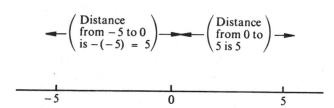

Figure A-5

EXAMPLE 1   The distance from -3 to 0 is $|-3| = 3$. The distance from 3 to 0 is $|3| = 3$. Thus $|-3| = |3| = 3$.

The distance between the points (that is, numbers) a and b is defined as $|a - b|$.

Note that the distance between a and b is the same as the distance between b and a, which is reflected in the fact that $|a - b| = |b - a|$.

EXAMPLE 2   The distance between 5 and 3 is $|5 - 3| = |3 - 5| = 2$. The distance between -5 and -3 is $|-5 - (-3)| = |-3 - (-5)| = 2$. See Figure A-6.

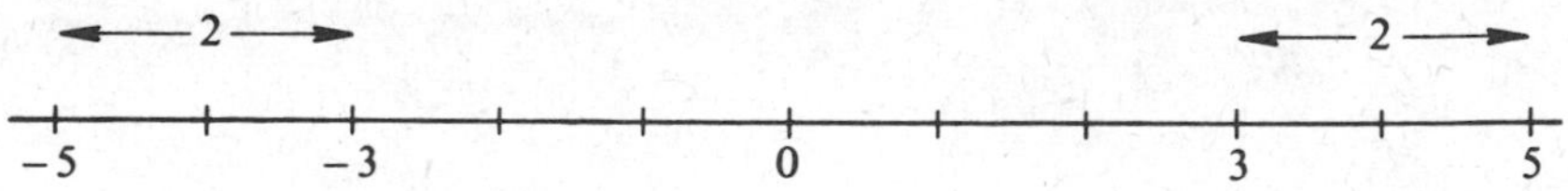

Figure A-6

EXAMPLE 3   Find all numbers x for which $|x - 2| = 5$.

SOLUTION   There are precisely two possibilities for the number $x - 2$ in order that $|x - 2| = 5$: either $x - 2 = 5$, in which case $|x - 2| = |5| = 5$, or $x - 2 = -5$, in which case $|x - 2| = |-5| = 5$. Now, $x - 2 = 5$ if and only if $(x - 2) + 2 = 5 + 2$ if and only if $x = 7$, and $x - 2 = -5$ if and only if $x = -3$. Thus, there are two numbers x for which $|x - 2| = 5$, namely, 7 and -3. Note that, as expected, $|7 - 2| = |5| = 5$ and $|-3 - 2| = |-5| = 5$.

Problems   In problems 1–4, find the distance between the given points.
   1. -5 and 17   2. -3 and 4
   3. -2 and -8 4. -7 and -28
In problems 5–10, find all real numbers x for which
   5. $|x - 5| = 10$ 6. $|x - 2| = 1$
   7. $|2 - x| = 4$ 8. $|5 - x| = 5$
   9. $|1 - 2x| = 5$ 10. $|2 - 5x| = 3$
  11.   When does $|5 - 2x|$ equal $5 - 2x$?   When does $|5 - 2x|$ equal $2x - 5$?

SECTION 3

PRODUCTS. FACTORING

One of the basic properties of real numbers that we shall need is the *distributive property: if a, b, and c are numbers, then $a(b + c) = ab + ac$.*

   Thus, for example, if x, y, x', and y' are numbers, then, by the distributive property,

(A.1) $(x + y)(x' + y') = xx' + xy' + yx' + yy'$

since $(x + y)(x' + y') = (x + y) \cdot x' + (x + y) \cdot y'$
                              $= x'(x + y) + y'(x + y)$
                              $= x'x + x'y + xy' + yy'$

From (A.1) we can derive each of the following:

(A.2) $(x + y)^2 = x^2 + 2xy + y^2$

$\qquad\qquad\qquad$ (in (A.1), let $x = x'$ and $y = y'$)

(A.3) $(x + y)(x - y) = x^2 - y^2$

$\qquad\qquad\qquad$ (in (A.1), let $x' = x$ and $y' = -y$)

(A.4) $x^2 + (a + b)x + ab = (x + a)(x + b)$

$\qquad\qquad\qquad$ (in (A.1), let $y = a$, $x' = x$, and $y' = b$)

The equation given in (A.3) is called the **difference of squares** equation.

EXAMPLE 1   For any real number x,

(A.5) $$(x - 3)^2 = x^2 + 2(-3)x + (-3)^2$$
$$= x^2 - 6x + 9 \qquad \text{(from (A.2))}$$

(A.6) $$(2x + 5)^2 = (2x)^2 + 2 \cdot 5(2x) + 5^2$$
$$= 4x^2 + 20x + 25 \qquad \text{(from (A.2))}$$

(A.7) $(x - 5)(x + 5) = x^2 - 5^2 = x^2 - 25$ (from (A.3))

(A.8) $(x + \sqrt{2})(x - \sqrt{2}) = x^2 - (\sqrt{2})^2 = x - 2$

$\qquad\qquad\qquad$ (from (A.3))

(A.9) $$(x + 3)(x + 2) = x^2 + (3 + 2)x + 3 \cdot 2$$
$$= x^5 + 5x + 6 \qquad \text{(from (A.4))}$$

(A.10) $$(x - 3)(x + 2) = x^2 + (-3 + 2)x + (-3) \cdot 2$$
$$= x^2 - x - 6 \qquad \text{(from (A.4))}$$

(A.11) $$(2x + 3)(x - 4) = (2x) \cdot x + (2x)(-4)$$
$$+ 3x + 3(-4) = 2x^2 - 5x - 12$$
$$\text{(from (A.1))}$$

(A.12) $$(3x - 1)(2x + 5) = (3x)(2x) + (3x) \cdot 5 + (-1)(2x)$$
$$+ (-1)(5) = 6x^2 + 13x - 5$$
$$\text{(from (A.1))}$$

In the following examples, note that if a and b are real numbers, then:

$$ab = 0 \text{ if and only if } a = 0 \text{ or } b = 0$$

EXAMPLE 2   Find all real numbers for which $x^2 + 5x + 6 = 0$.

SOLUTION   From equation (A.9) we see that for *any* real number x, we have $x^2 + 5x + 6 = (x + 3)(x + 2)$. Thus,

$$x^2 + 5x + 6 = 0$$

$$\text{if and only if} \qquad (x + 3)(x + 2) = 0$$
$$\text{if and only if} \qquad x + 3 = 0 \quad \text{or} \quad x + 2 = 0$$
$$\text{or, equivalently,} \qquad x = -3 \quad \text{or} \quad x = -2$$

EXAMPLE 3   Find all $x \in \mathbf{R}$ for which $x^2 + 2x - 15 = 0$; that is, solve the quadratic equation $x^2 + 2x - 15 = 0$.

SOLUTION   As in equation (A,4), we want to find numbers a and b for which, for any x in $\mathbf{R}$,

$$x^2 + 2x - 15 = (x + a)(x + b) = x^2 + (a + b)x + ab.$$

Thus, ab must be -15, and $a + b$ must be 2. Hopefully, there are integers a and b for which this is true. In order for $ab = -15$, we see that the possibilities are these: $a = 1$, $b = -15$; or, $a = 3$, $b = -5$; or, $a = 5$, $b = -3$, etc. Only $a = 5$, $b = -3$ works, since $a + b = 5 - 3 = 2$. Thus, for x in $\mathbf{R}$,

$$(x^2 + 2x - 15) = (x + 5)(x - 3)$$

and hence the desired numbers are -5 and 3.

EXAMPLE 4   Find all numbers x for which $x^2 - 4x + 3 = 0$.

SOLUTION   After some trial and error, we see that for any x in $\mathbf{R}$, we can write $x^2 - 4x + 3 = (x - 1)(x - 3)$. Thus, the desired numbers are 1 and 3.

EXAMPLE 5   Find all real numbers x for which $4x^2 - x - 3 = 0$.

SOLUTION   Let x be in $\mathbf{R}$. We want to find real numbers a, b, c, and d, for which

$$4x^2 - x - 3 = (ax + b)(cx + d)$$

Thus, ac must be 4 and bd must be -3. Some possibilities using natural numbers are

$$(2x + 1)(2x - 3), (4x - 1)(x + 3), \text{ and } (4x + 3)(x - 1)$$

We see that $(4x + 3)(x - 1)$ works; that is, for any x in $\mathbf{R}$,

$$4x^2 - x - 3 = (4x + 3)(x - 1)$$

Thus, the required solutions are $-\frac{3}{4}$ and 1.

Remarks   To observe that 6 can be written as $3 \cdot 2$ (more precisely, that $6 = 3 \cdot 2$) is to understand what is sometimes referred to as *factoring* 6. Similarly, for example, to observe that for any x in **R**, we have $x^2 + 5x + 6 = (x + 3)(x + 2)$, or $4x^2 - x - 3 = (4x + 3)(x - 1)$, is called **factoring**.

Trial and error is of course an uncertain procedure. We shall later find a more systematic procedure for solving quadratic equations using the quadratic formula.

For completeness, we also include the following "special products" (each can be verified by multiplication). In each case, x and y are in **R**.

(A.13) $$x^3 - y^3 = (x - y)(x^2 + xy + y^2)$$

(A.14) $$x^3 + y^3 = (x + y)(x^2 - xy + y^2)$$

(A.15) $$(x + y)^3 = x^3 + 3x^2y + 3xy^2 + y^3$$

In section 9, we shall investigate the number $(x + y)^n$ where n is any positive integer.

EXAMPLE 6   For x in **R**,

(A.16) $$x^3 - 8 = x^3 - 2^3 = (x - 2)(x^2 + 2x + 4)$$

(A.17) $$x^3 + 8 = x^3 + 2^3 = (x + 2)(x^2 - 2x + 4)$$

(A.18) $$(x + 1)^3 = x^3 + 3x^2 + 3x + 1$$

Problems   In problems 1–16, find the indicated product. In each case, x is a real number.

1. $2x(x - 7)$
2. $-3x(4 - 5x)$
3. $(x + 2)(x + 4)$
4. $(x - 1)(x + 3)$
5. $(x - 5)(x - 4)$
6. $(4x + 3)^2$
7. $(2x - 7)^2$
8. $(2x + 3)(3x + 2)$
9. $(-3x + 4)(2x - 3)$
10. $(-x - 4)(-x + 4)$
11. $(x + \sqrt{3})(x - \sqrt{3})$
12. $(x - 2\sqrt{2})(x + 2\sqrt{2})$
13. $(2x - 6)(2x + 6)$
14. $(2 - 3x)(2 + 3x)$
15. $(x + 1)(x^2 - x + 1)$
16. $(x + 3)(x^2 - 3x + 9)$

In problems 17–25, find the indicated product. In each case, x and y are real numbers.

17. $(2x + 3y)^2$
18. $(3x - y)^2$

19. $(4x - 5y)^2$ 20. $(3x + 2y)^2$
21. $(3x + 5y)(3x - 5y)$
22. $(\sqrt{3}x - \sqrt{2}y)(\sqrt{3}x + \sqrt{2}y)$
23. $(5x + y)(3x + 2y)$ 24. $(-x + 3y)(x - 3y)$
25. $(2x - 3y)(3x - 2y)$

In problems 26–40, find all real numbers x for which

26. $x^2 + 2x = 0$ 27. $-6x^2 - 21x = 0$
28. $3x^3 - x^2 = 0$ 29. $7x^3 - 5x^2 = 0$
30. $x^3 - 16x = 0$ 31. $x^4 - 9x^2 = 0$
32. $x^2 - 3x + 2 = 0$ 33. $x^2 - x - 12 = 0$
34. $x^2 - 7x - 8 = 0$ 35. $x^2 + 4x + 3 = 0$
36. $x^2 - 7x + 12 = 0$ 37. $2x^2 + x - 3 = 0$
38. $2x^2 + x - 6 = 0$ 39. $4x^2 + 8x + 3 = 0$
40. $4x^2 + 3x - 1 = 0$

SECTION 4
FRACTIONS

Recall that for any nonzero real number a, the number $1/a$, called the **reciprocal*** of a, is *the*** number with the property that

(A.19)
$$a \cdot \frac{1}{a} = \frac{1}{a} \cdot a = 1$$

But by the symmetry in this equation, if $1/a$ is the reciprocal of a, it also follows that a is the reciprocal of $1/a$. Thus, for example, $\frac{1}{2}$ is the reciprocal of 2, since $2 \cdot \frac{1}{2} = \frac{1}{2} \cdot 2 = 1$; and for the same reason, 2 is the reciprocal of $\frac{1}{2}$, so $2 = \frac{1}{\frac{1}{2}}$. In general, then, if $a \neq 0$, then

$$a = \frac{1}{1/a}$$

We can now show that the product of reciprocals is the reciprocal of the product; that is, if $a \neq 0$ and $b \neq 0$, then

(A.20)
$$\frac{1}{a} \cdot \frac{1}{b} = \frac{1}{ab}$$

To see this, note first that by definition $1/ab$ is the reciprocal of ab, and thus all we have to show is that $1/a \cdot 1/b$ is also the reciprocal of ab. Since ab has precisely one reciprocal, it then follows that $1/ab = 1/a \cdot 1/b$. Now,

$$ab \left(\frac{1}{a} \cdot \frac{1}{b} \right) = a \cdot \frac{1}{a} \cdot b \cdot \frac{1}{b} = 1 \cdot 1 = 1$$

thus, $1/a \cdot 1/b$ is the reciprocal of ab.

---

*The number $1/a$ is also called the *multiplicative inverse* of a.
**It is possible to show that a nonzero real number has precisely one reciprocal, which can therefore be designated by $1/a$.

Division* can now be defined as follows: If a and b are real numbers, and $b \neq 0$, then the fraction a/b is defined by

(A.21)
$$\frac{a}{b} = a \cdot \frac{1}{b}$$

that is, "a divided by b" means "a times the reciprocal of b". We can now investigate the addition and multiplication of fractions. In what follows, $a, b, c,$ and d are real numbers.

(A.22)
$$\frac{a}{c} + \frac{b}{c} = \frac{a + b}{c}$$

since, using the distributive property as well as equations (A.19) and (A.21),

$$\frac{a}{c} + \frac{b}{c} = a \cdot \frac{1}{c} + b \cdot \frac{1}{c} = (a + b) \cdot \frac{1}{c} = \frac{a + b}{c}$$

(A.23)
$$\frac{a}{b} + \frac{c}{d} = \frac{ad + bc}{bd}$$

Since,
$$\frac{a}{b} + \frac{c}{d} = a \cdot \frac{1}{b} + c \cdot \frac{1}{d} = a \cdot \frac{1}{b} \cdot \frac{d}{d} + c \cdot \frac{1}{d} \cdot \frac{b}{b}$$

$$= ad \cdot \frac{1}{bd} + cb \cdot \frac{1}{bd}$$

$$= (ad + cb) \cdot \frac{1}{bd}$$

$$= \frac{ad + cb}{bd}$$

(A.24)
$$\frac{a}{b} \cdot \frac{c}{d} = \frac{ac}{bd}$$

Since, using (A.20)

$$\frac{a}{b} \cdot \frac{c}{d} = a \cdot \frac{1}{b} \cdot c \cdot \frac{1}{d} = ac \frac{1}{b} \cdot \frac{1}{d} = ac \cdot \frac{1}{bd} = \frac{ac}{bd}$$

---

*We have already used this definition informally. It is stated here only for completeness.

(A.25)
$$\frac{a}{b} - \frac{c}{d} = \frac{ad - bc}{bd}$$

It is also possible to show that

(A.26)
$$\frac{a/b}{c/d} = \frac{a}{b} \cdot \frac{d}{c} = \frac{ad}{bc}$$

(A.27)
$$\frac{-a}{b} = \frac{a}{b} = \frac{a}{-b}$$

(A.28)
$$\frac{-a}{b} = \frac{a}{b}$$

EXAMPLE 1 Find all real numbers x for which

$$\frac{3 - 2x}{4} = x - \frac{7}{4}$$

SOLUTION Note that for any x in **R**, both $3 - 2x$ and $4x - 7$ are simply *real numbers*. Thus, x is a number for which

$$\frac{3 - 2x}{4} = x - \frac{7}{4}$$

if and only if

$$\frac{4(3 - 2x)}{4} = 4(x - \tfrac{7}{4})$$

$$3 - 2x = 4x - 7$$

$$3 - 2x - 3 = 4x - 7 - 3$$

$$-2x = 4x - 10$$

$$-2x - 4x = 4x - 10 - 4x$$

$$-6x = -10$$

$$(-\tfrac{1}{6})(-6x) = (-\tfrac{1}{6}) \cdot (-10)$$

and, finally,
$$x = \frac{-10}{-6} = \frac{5}{3}$$

EXAMPLE 2   If $x \neq 1$, then

$$\frac{2x}{x-1} + \frac{x}{1-x} = \frac{2x}{x-1} + \frac{x}{-(x-1)}$$

$$= \frac{2x}{x-1} - \frac{x}{x-1}$$

$$= \frac{2x-x}{x-1}$$

$$= \frac{x}{x-1}$$

EXAMPLE 3   If $x \neq 1$ or -1, then

$$\frac{-2}{x^2-1} + \frac{x}{x-1} = \frac{-2}{x^2-1} + \frac{x}{x-1} \cdot \frac{x+1}{x+1}$$

$$= \frac{-2}{x^2-1} + \frac{x^2+x}{x^2-1}$$

$$= \frac{x^2+x-2}{x^2-1}$$

$$= \frac{(x+2)(x-1)}{(x+1)(x-1)}$$

$$= \frac{x+2}{x+1}$$

EXAMPLE 4   If $x \neq -1$ or -2, then

$$\frac{x^3-4x}{x^2+2x+1} \cdot \frac{x+1}{x+2} = \frac{x(x+2)(x-2)}{(x+1)^2} \cdot \frac{x+1}{x+2}$$

$$= \frac{x(x-2)}{x+1} \cdot \frac{x+2}{x+2} \cdot \frac{x+1}{x+1}$$

$$= \frac{x(x-2)}{x+2}$$

EXAMPLE 5   If $x \neq 0$ and if $x \neq -h$, then

$$\frac{1}{x+h} - \frac{1}{x} = \frac{x-(x+h)}{x(x+h)} = \frac{-h}{x(x+h)}$$

Problems   Compute.

1. $\frac{1}{2} + \frac{1}{3}$ 2. $\frac{3}{5} - \frac{1}{7}$

3. $\frac{4}{3} - \frac{2}{9} + \frac{2}{27}$ 4. $\frac{5}{8} + \frac{3}{2} - \frac{7}{4}$

5. $\dfrac{\frac{3}{5}}{\frac{3}{8}}$ 6. $\dfrac{\frac{3}{7}}{\frac{9}{14}}$

In problems 7–22, simplify as in 7. In each problem, a, b, and c are real numbers for which the appropriate denominators are not 0.

7. $\dfrac{2a}{b} + \dfrac{a}{2b} = \dfrac{2}{2} \cdot \dfrac{2a}{b} + \dfrac{a}{2b} = \dfrac{5a}{2b}$

8. $\dfrac{5a}{12b} - \dfrac{2a}{b}$

9. $\dfrac{7c}{ab} + \dfrac{2a}{bc}$

10. $\dfrac{2}{a + c} - \dfrac{1}{2a + 2c}$

11. $\dfrac{a}{a - c} + \dfrac{c}{c - a}$

12. $\dfrac{3a}{a + c} + \dfrac{5}{a - c}$

13. $\dfrac{a + c}{a - c} + \dfrac{a}{(a + c)}$

14. $\dfrac{a}{a^2 - 1} + \dfrac{3a^2}{a + 1}$

15. $\dfrac{2a}{a^2 - 16} - \dfrac{3a^2 + 1}{a - 4}$

16. $\dfrac{4}{a^2 + 5a + 6} + \dfrac{a - 2}{a + 3}$

17. $\dfrac{-5a - 6}{a^2 + 2a - 3} + \dfrac{a + 2}{a - 1}$

18. $\dfrac{a - 1}{a + 2} - \dfrac{a + 3}{a - 1}$

19. $\dfrac{2a + 1}{a - 1} + \dfrac{a - 3}{a + 4}$

20. $\dfrac{a - c}{a + c} - \dfrac{a - 2c}{2a + c}$

21. $\dfrac{a - 2b}{2a + b} + \dfrac{-a + 2b}{a - 2b}$

22. $\dfrac{1}{(a + b)^2} - \dfrac{1}{a^2}$

In problems 23–27, compute:

23. $\dfrac{x^2 + x - 6}{x(x + 3)} \cdot \dfrac{x^3}{(x + 2)}$ $(x \neq 0, -2, -3)$

24. $\dfrac{x^3 - 1}{x} \cdot \dfrac{x^2 - 8x}{x - 1}$ $(x \neq 0, x \neq 1)$

25. $\dfrac{x + a}{x + 3a} \cdot \dfrac{x^2 - 9a^2}{a^2 - x^2}$ x in $\mathbf{R}$, a in $\mathbf{R}$
$(x \neq a, -a, -3a)$

26. $\dfrac{\dfrac{1}{x} - \dfrac{1}{a}}{\dfrac{1}{x} + \dfrac{1}{a}}$ $(a \neq 0, x \neq 0)$

27. $\dfrac{1}{1 + \dfrac{1}{x}} \cdot \dfrac{1}{x}$ $(x \neq 0, x \neq -1)$

In problems 28–37, find all real numbers x for which

28. $\dfrac{2 - 3x}{4} = \dfrac{x - 1}{3}$

29. $\dfrac{2x + 1}{2} = -\dfrac{x + 1}{5}$

30. $\dfrac{x}{2} + \dfrac{x + 3}{10} = -\dfrac{3x + 1}{5}$

31. $\dfrac{2x}{3} - \dfrac{x - 1}{9} = \dfrac{3x + 1}{18}$

32. $\dfrac{2}{x} + \dfrac{3}{x + 3} = \dfrac{-x + 4}{x^2 + 3x}$

33. $\dfrac{5}{2x} + \dfrac{2}{x - 1} = \dfrac{3x + 2}{2x^2 - 2x}$

34. $\dfrac{2}{x-1} + \dfrac{3}{x+1} = \dfrac{x+5}{x^2-1}$

35. $\dfrac{-2}{x-4} + \dfrac{1}{x+4} = \dfrac{2x-1}{x^2-16}$

36. $\dfrac{4}{x+2} - \dfrac{3}{x-1} = \dfrac{2x}{x^2+x-2}$

37. $\dfrac{5}{x+3} - \dfrac{2}{x+4} = \dfrac{2x+3}{x^2+7x+12}$

Answer true or false (in each, a, b, and c are real numbers).

38. $\dfrac{a+b}{a} = \dfrac{\cancel{a}+b}{\cancel{a}} = b$ $\qquad$ $(a \neq 0)$

39. $\dfrac{ac+b}{c} = \dfrac{a\cancel{c}+b}{\cancel{c}} = a+b$ $\qquad$ $(c \neq 0)$

40. $\dfrac{a+b}{a} = 1 + \dfrac{b}{a}$ $\qquad$ $(a \neq 0)$

41. $\dfrac{ac+b}{c} = a + \dfrac{b}{c}$ $\qquad$ $(c \neq 0)$

SECTION 5
EXPONENTS AND ROOTS

Recall that if a is any real number, then $a^1 = a$, $a^2 = a \cdot a$, $a^3 = a \cdot a \cdot a$, and, in general, if n is any positive integer, then

$$a^n = \underbrace{a \cdot a \cdots a}_{n \text{ times}}$$

If $a \neq 0$, then it turns out to be useful to define

$$a^0 = 1$$

Next, if $a \neq 0$, then we define

$$a^{-1} = \frac{1}{a}, \qquad a^{-2} = \frac{1}{a^2}, \qquad a^{-3} = \frac{1}{a^3}$$

and, in general, if n is any integer, then (by definition)

$$a^{-n} = \frac{1}{a^n}$$

EXAMPLE 1 $2^{-3} = \dfrac{1}{2^3} = \dfrac{1}{8}$; $3^{-1} = \dfrac{1}{3}$; $3^{-2} = \dfrac{1}{3^2} = \dfrac{1}{9}$; $5^3 = 5^{-(-3)} = \dfrac{1}{5^{-3}}$; $6^2 = \dfrac{1}{6^{-2}}$; and $17^0 = 1$.

With these definitions, it is possible to prove the important "rules of exponents": if $x \neq 0$ and $y \neq 0$, and if m and n are integers, then

(A.29) $x^n \cdot x^m = x^{n+m}$

(A.30) $(x^m)^n = x^{m \cdot n}$

(A.31) $(xy)^n = x^n y^n$

(A.32) $\dfrac{x^m}{x^n} = x^m \cdot x^{-n} = x^{m-n}$

(A.33) $\left(\dfrac{x}{y}\right)^n = \dfrac{x^n}{y^n}$

The proofs can be found in most algebra books.

EXAMPLE 2   We shall illustrate (A.29)–(A.33).

(A.34) $3^5 \cdot 3^7 = 3^{5+7} = 3^{12}$

(A.35) $7^{-6} \cdot 7^4 = 7^{-6+4} = 7^{-2} = \dfrac{1}{7^2} = \dfrac{1}{49}$

or, $7^{-6} \cdot 7^4 = \dfrac{1}{7^6} \cdot 7^4 = \dfrac{7^4}{7^6} = \dfrac{1}{7^2} = \dfrac{1}{49}$

(A.36) $\dfrac{7^{10}}{7^{11}} = 7^{10-11} = 7^{-1} = \dfrac{1}{7}$

(A.37) $(5^3)^6 = 5^{18}$

(A.38) $\dfrac{5^{-3}}{5^{-5}} = 5^{-3-(-5)} = 5^2 = 25$

(A.39) $(5^{-3})^6 = 5^{-18}$ or $(5^{-3})^6 = \left(\dfrac{1}{5^3}\right)^6 = \dfrac{1^6}{5^{18}} = \dfrac{1}{5^{18}}$

EXAMPLE 3   If a, b, and c are positive numbers, then

(A.40) $(a^2 b^3 c^{-3})(a^7 b^{-3} c^{-5}) = a^{7+2} b^{3-3} c^{-3-5}$
                $= a^9 c^{-8}$ (or, a^9/c^8)

$$(A.41) \qquad \left(\frac{a^2b^3}{c^5}\right)^4 = \frac{(a^2b^3)^4}{(c^5)^4} = \frac{a^8b^{12}}{c^{20}}$$

or,

$$(a^2b^3c^{-5})^4 = a^8b^{12}c^{-20}$$

$$(A.42) \qquad \left(\frac{a^{-2}b^3c^4}{a^5b^{-5}c^2}\right)^{-2} = (a^{-2-5}\, b^{3-(-5)}\, c^{4-2})^{-2}$$

$$= (a^{-7}\, b^8\, c^2)^{-2}$$

$$= a^{14}b^{-16}c^{-4} \qquad \left(\text{or,} \; \frac{a^{14}}{b^{16}c^4}\right)$$

$$(A.43) \qquad \frac{a^2b^3c + ab^2c^2 - abc}{ab^2c^3} = \frac{a^2b^3c}{ab^2c^3} + \frac{ab^2c^2}{ab^2c^3} - \frac{abc}{ab^2c^3}$$

$$= \frac{ab}{c^2} + \frac{1}{c} - \frac{1}{bc^2}$$

We now extend the definition of exponent to include rational number exponents. The definitions are constructed in such a way that the laws of exponents remain valid! First, we define $(x)^{1/n}$ as the **nth root** of x (where, of course, n is a positive integer); that is, *if n is an even positive integer, then*

$$y = (x)^{1/n} \textbf{ if and only if } y > 0 \textbf{ and } y^n = x$$

whereas, if n is an odd positive integer, then

$$y = (x)^{1/n} \text{ if and only if } y^n = x$$

Note that $(x)^{2/2} = |x|$.

EXAMPLE 4

$$
\begin{aligned}
(16)^{1/2} &= 4 & &\text{since } 4 > 0 \text{ and } 4^2 = 16 \\
(16)^{1/4} &= 2 & &\text{since } 2 > 0 \text{ and } 2^4 = 16 \\
(8)^{1/3} &= 2 & &\text{since } 2^3 = 8 \\
(-8)^{1/3} &= -2 & &\text{since } (-2)^3 = -8 \\
(-27)^{1/3} &= -3 & &\text{since } (-3)^3 = -27 \\
(3)^{10/10} &= 3 & &\text{(Why?)} \\
(5)^{16/16} &= 5
\end{aligned}
$$

Thus, we have defined rational number exponents. If $(a)^{1/n}$ is meaningful (that is, if $(a)^{1/n}$ is a real number), then

$$a^{1/n} = \sqrt[n]{a} = (\text{the } n\text{th root of } a) = a^{1/n}$$

So, for example, $4^{1/2} = \sqrt[2]{4} = 2$, whereas $(-4)^{1/2}$ is meaningless, since $\sqrt{-4}$ is not a real number (that is, there is no real number whose square is -4).

Since the rules of exponents are to hold, we define

$$a^{m/n} = (a^{1/n})^m = (\sqrt[n]{a})^m$$

whenever $a^{m/n}$ and $(a^{1/n})^m$ are meaningful.

With these definitions, it can be shown that the "laws of exponents" are still valid; that is, if $x \neq 0$ and $y \neq 0$ and r and s are *rational* numbers, then

(A.44) $$x^{-r} = 1/x^r$$

(A,45) $$x^r \cdot x^s = x^{r+s}$$

(A.46) $$(x^r)^s = x^{rs}$$

(A.47) $$(xy)^r = x^r y^r$$

(A.48) $$x^r / x^s = x^{r-s}$$

EXAMPLE 5

(A.49) $$8^{2/3} = (8^{1/3})^2 = (\sqrt[3]{8})^2 = 2^2 = 4$$

(A.50) $$8^{-2/3} = \frac{1}{8^{2/3}} = \frac{1}{4}$$

(A.51) $$16^{-1/2} = \frac{1}{16^{1/2}} = \frac{1}{\sqrt{16}} = \frac{1}{4}$$

(A.52) $$(-8)^{2/3} = [(-8)^{1/3}]^2 = (-2)^2 = 4$$

(A.53) $$(32)^{0.8} = (2^5)^{8/10} = (2^5)^{4/5} = 2^{5 \cdot 4/5} = 2^4 = 16$$

EXAMPLE 6   If a, b, and c are positive real numbers, then

(A.54) $$(a^4 b^6 c^{2/3})^{1/2} = (a^4)^{1/2}(b^6)^{1/2}(c^{2/3})^{1/2} = a^2 b^3 c^{1/3}$$

(A.55)
$$(a^{1/2} b^{1/4} c^{-1/8})^4 = (a^{1/2})^4 (b^{1/4})^4 (c^{-1/8})^4$$
$$= a^2 b c^{-1/2}$$
$$= \frac{a^2 b}{\sqrt{c}}$$

(A.56)
$$(a^{1/2} b^{1/3})(a^{3/4} b^{1/2}) = a^{1/2+3/4} b^{1/3+1/2}$$
$$= a^{5/4} b^{5/6}$$
$$= a^{4/4} \cdot a^{1/4} \cdot b^{5/6}$$
$$= b^{5/6} a \sqrt[4]{a}$$

(A.57)
$$a^{5/2} b^{17/3} c^{20/7} = a^{4/2} \cdot a^{1/2} \cdot b^{15/3} b^{2/3} c^{14/7} c^{6/7}$$
$$= a^2 b^5 c^2 \sqrt[2]{a} \sqrt[3]{b^2} \sqrt[7]{c^6}$$

$$(A.58) \qquad \sqrt[4]{16ab^{10}c^{23}} = \sqrt[4]{2^4 a \cdot b^2 b^8 c^{20} c^3}$$
$$= \sqrt[4]{2^4 (b^2)^4 (c^5)^4 \cdot ab^2 c^3}$$
$$= 2b^2 c^5 \cdot \sqrt[4]{ab^2 c^3}$$

$$(A.59) \qquad \sqrt[4]{16a^5 b^{10}c^{23}} = (16a^5 b^{10}c^{23})^{1/4}$$
$$= 16^{1/4} a^{5/4} b^{10/4} c^{23/4}$$
$$= 2 a^{4/4} \cdot a^{1/4} b^{8/4} b^{2/4} c^{20/4} c^{3/4}$$
$$= 2 a b^2 c^5 (ab^2 c^3)^{1/4}$$

We conclude this section with two examples that furnish a useful observation about factoring.

EXAMPLE 7   Let $x > 0$, and consider the number $x^{-1/2} + x^{1/2}$. Note that it is possible to "factor out" either $x^{1/2}$ or $x^{-1/2}$; in fact,

$$x^{-1/2} + x^{1/2} = x^{1/2}(x^{-1} + 1)$$

and $\qquad\qquad\qquad x^{-1/2} + x^{1/2} = x^{-1/2}(1 + x)$

Of the two factorizations, the latter turns out to be the more useful.

EXAMPLE 8   Let $a > 0$ and $b > 0$. Then,

$$a^{1/2} x^{-1/2} + x^{1/2} a^{-1/2} = a^{-1/2} x^{-1/2} (a + x)$$

Problems   Compute problems 1–9.

1. $3^3 \cdot 3^8$ 2. $4^7 \cdot 4^5$

3. $(2^3)^5$ 4. $(8^2)^3$

5. $\dfrac{5^8}{5^3}$ 6. $\dfrac{3^{-5}}{3^8}$

7. $\dfrac{3^{-2}}{3^{-3}}$ 8. $8^{2/3} \cdot 4^{1/2}$

9. $16^{3/4} \cdot 8^{-1/3}$

Compute problems 10–28 (assume that a, b, and c are positive real numbers).

10. $(a^5 b^2 c^8)(ab^3 c^{10})$ 11. $(a^{-3} b^{-2} c^5)(a^4 b^{-5} c^{-5})$

12. $(a^{-5} b^{10} c^{-1})(a^4 b^{-9} c)$ 13. $(a^2 b^3 c)^5$

14. $(a^3 bc^5)^3$ 15. $\left(\dfrac{ab^2}{c^3} \right)^4$

16. $\left(\dfrac{a^2 c^5}{b^3} \right)^5$ 17. $\left(\dfrac{a^{-3} c}{b^{-2}} \right)^4$

18. $\left(\dfrac{a^3 b^{-2} c^3}{ab^2 c} \right)^{-3}$ 19. $\left(\dfrac{a^{-2} b^{-3} c^{-4}}{a^2 b^{-3} c^5} \right)^{-1}$

20. $\dfrac{a^2b^3c - abc^5 + abc}{ab^2c^2}$ 21. $\dfrac{a^3bc^5 + a^2(bc)^3 - abc^2}{(abc)^2}$

22. $\dfrac{a^{-2}b^{-3}c^2 + ab^{-3} + a^2b^3}{a^2b^2c}$ 23. $(a^2b^4c)^{1/2}$

24. $\left(\dfrac{a^6b^9}{c^{15}}\right)^{1/3}$ 25. $(a^{1/4}\, b^{1/2}\, c^{-1/2})^4$

26. $(a^{1/8}\, b^{-1/2}\, c^{1/16})^{16}$ 27. $(a^{1/3}\, b^{2/5}\, c^{1/2})(a^{1/4}\, b^{3/5}\, c)$

28. $(a^{1/5}\, b^{-1/5}\, c^{3/4})(a^{2/5}\, bc^{1/8})$

Simplify (as in Example 6, formulas (A.57) and (A.58)). (Assume that a, b, and c are positive real numbers.)

29. $a^{7/2}\, b^{10/3}\, c^{61/5}$ 30. $a^{-10/3}\, b^{9/4}\, c^{11/2}$

31. $\sqrt{a^3b^8c^5}$ 32. $\sqrt[3]{a^{11}b^5c^{28}}$

Factor problems 33–36.

33. $2a^{1/2}\, b^{-1/2} + 3a^{-1/2}\, b^{1/2}$ 34. $-3a^{1/2}\, b^{-1/2} + 7a^{-1/2}\, b^{1/2}$

35. $a^{1/2}\, b + a^{-1/2}\, b^2$ 36. $a^{1/2}\, b^2c + a^{-1/2}\, bc^2$

Answer *true* or *false*.

37. $\sqrt{x + y} = \sqrt{x} + \sqrt{y}$ $(x \geq 0, y > 0)$

38. $(x + h)^2 - x^2 = x^2 + h^2 - x^2 = h^2$ $(x, h \text{ in } \mathbf{R})$

39. $\dfrac{a^2 + b^2}{a^2} = b^2$ $(a, b \text{ in } \mathbf{R}, a \neq 0)$

40. $\dfrac{a^2}{a^2 + b^6} = \dfrac{1}{b^6}$ $(a \neq 0, b \neq 0)$

41. $\sqrt{(a + b)^2} = \sqrt{a^2} + \sqrt{b^2}$ $(a, b \text{ in } \mathbf{R})$

42. $\sqrt{16} = \pm 4$

SECTION 6
RATIONALIZING SQUARE ROOTS

The difference-of-squares equation has an interesting application to square roots. Notice that if a and b are positive real numbers, then

$$(\sqrt{a} + \sqrt{b})(\sqrt{a} - \sqrt{b}) = (\sqrt{a})^2 - (\sqrt{b})^2 = a - b$$

Or, $$(\sqrt{a} + \sqrt{b})(\sqrt{a} - \sqrt{b}) = a - b$$

Thus, we can rewrite the number $\sqrt{a} - \sqrt{b}$ as follows:

$$\sqrt{a} - \sqrt{b} = (\sqrt{a} - \sqrt{b}) \cdot \dfrac{(\sqrt{a} + \sqrt{b})}{\sqrt{a} + \sqrt{b}}$$

$$= \dfrac{a - b}{\sqrt{a} + \sqrt{b}}$$

The foregoing computation is often referred to as **rationalizing** the number $\sqrt{a} - \sqrt{b}$. Of course, a similar computation is possible for the number $\sqrt{a} + \sqrt{b}$.

EXAMPLE 1

$$\sqrt{5} - \sqrt{3} = \frac{(\sqrt{5} - \sqrt{3})(\sqrt{5} + \sqrt{3})}{\sqrt{5} + \sqrt{3}}$$

$$= \frac{5 - 3}{\sqrt{5} + \sqrt{3}}$$

$$= \frac{2}{\sqrt{5} + \sqrt{3}}$$

We shall concentrate on rationalization techniques that prove to be useful in calculus. Specifically, we shall see that performing such rationalizations are invaluable when finding certain types of limits.

EXAMPLE 2   If x and h are real numbers, with $x > 0$ and $x + h > 0$, then

$$\sqrt{x + h} - \sqrt{x} = \frac{(\sqrt{x + h} - \sqrt{x})(\sqrt{x + h} + \sqrt{x})}{\sqrt{x + h} + \sqrt{x}}$$

$$= \frac{(x + h) - x}{\sqrt{x + h} + \sqrt{x}}$$

$$= \frac{h}{\sqrt{x + h} + \sqrt{x}}$$

EXAMPLE 3   If x and h are real numbers with $x > 0$ and $x + h > 0$, then

$$\frac{1}{\sqrt{x + h}} - \frac{1}{\sqrt{x}} = \frac{\sqrt{x} - \sqrt{x + h}}{\sqrt{x + h} \cdot \sqrt{x}}$$

$$= \frac{\sqrt{x} - \sqrt{x + h}}{\sqrt{x + h}\sqrt{x}} \cdot \frac{\sqrt{x} + \sqrt{x + h}}{\sqrt{x} + \sqrt{x + h}}$$

$$= \frac{x - (x + h)}{\sqrt{x}\sqrt{x + h}(\sqrt{x} + \sqrt{x + h})}$$

$$= \frac{-h}{\sqrt{x}\sqrt{x + h}(\sqrt{x} + \sqrt{x + h})}$$

Equivalently,

$$\frac{1}{\sqrt{x + h}} - \frac{1}{\sqrt{x}} = \frac{\left(\dfrac{1}{\sqrt{x + h}} - \dfrac{1}{\sqrt{x}}\right)\left(\dfrac{1}{\sqrt{x + h}} + \dfrac{1}{\sqrt{x}}\right)}{\dfrac{1}{\sqrt{x + h}} + \dfrac{1}{\sqrt{x}}} \qquad \text{(etc.)}$$

Computations similar to those in Examples 2 and 3 appear in limit problems in calculus.

Problems   Rationalize.

1. $\sqrt{5} - \sqrt{3}$ 2. $\sqrt{5} + 1$

3. $\sqrt{5} + \sqrt{3}$ 4. $3 - \sqrt{3}$

5. $\dfrac{1}{\sqrt{7} + \sqrt{2}}$ 6. $\dfrac{2}{\sqrt{3} - \sqrt{5}}$

7. $\dfrac{3}{\sqrt{5} - 2}$ 8. $\dfrac{\sqrt{7} + \sqrt{2}}{\sqrt{7} - \sqrt{2}}$

In each of the following, suppose x, h, and $x + h$ are positive real numbers. Rationalize as in Example 2.

9. $\sqrt{x + h + 1} - \sqrt{x + 1}$
10. $\sqrt{x + h + 5} - \sqrt{x + 5}$
11. $\sqrt{2(x + h)} - \sqrt{2x}$
12. $\sqrt{3x + 3h + 1} - \sqrt{3x + 1}$

13. $\dfrac{1}{\sqrt{x + h + 1}} - \dfrac{1}{\sqrt{x + h}}$

In problems 14–17, n is a positive integer. Rationalize.

14. $\sqrt{n + 1} - \sqrt{n}$
15. $\sqrt{n^2 + 1} - n$ $\left(Hint:\ n = \sqrt{n^2}\right)$
16. $\sqrt{n + 2\sqrt{n}} - \sqrt{n}$
17. $\sqrt{n^4 + n^2} - n^2$ $\left(Hint:\ n^2 = \sqrt{n^4}\right)$
18. If x is any real number and if h is any nonzero real number, show that

$$|x + h| - |x| = \frac{h(2x + h)}{|x + h| + |x|}$$

$$\left(Hint:\ \text{Multiply by } \frac{|x + h| + |x|}{|x + h| + |x|}.\right)$$

SECTION 7

COMPLETING THE SQUARE. THE QUADRATIC FORMULA

The only technique we have developed for factoring so far is that of trial and error. We shall see that completing the square provides a more systematic procedure for factoring the *quadratic* $ax^2 + bx + c$. The observation needed is that for any $x \in \mathbf{R}$,

$$x^2 + kx + \left(\frac{k}{2}\right)^2 = \left(x + \frac{k}{2}\right)^2$$

Hence $x^2 + kx + (k/2)^2$ is a perfect square. For example,

$$x^2 + 2x + 1 = (x + 1)^2 \quad \text{(with } k = 2\text{)}$$

$$x^2 - 2x + 1 = (x - 1)^2 \quad \text{(with } k = -2\text{)}$$

$$x^2 + 3x + \left(\tfrac{3}{2}\right)^2 = \left(x + \tfrac{3}{2}\right)^2 \quad \text{(with } k = 3\text{)}$$

That is, $x^2 + 3x + \tfrac{9}{4} = \left(x + \tfrac{3}{2}\right)^2$

We shall now show how this observation about perfect squares can be applied to factoring quadratics. The procedure is called **completing the square**.

EXAMPLE 1   Factor $x^2 + 3x + 2$.

SOLUTION   Note that for any $x \in \mathbf{R}$

$$x^2 + 3x + 2 = x^2 + 3x + \tfrac{9}{4} - \tfrac{9}{4} + 2$$
$$= (x + \tfrac{3}{2})^2 - \tfrac{1}{4}$$

But $(x + \tfrac{3}{2})^2 - \tfrac{1}{4} = (x + \tfrac{3}{2})^2 - (\sqrt{\tfrac{1}{4}})^2 = (x + \tfrac{3}{2})^2 - (\tfrac{1}{2})^2$

is a difference of squares, and thus

$$x^2 + 3x + 2 = [(x + \tfrac{3}{2}) + \tfrac{1}{2}][(x + \tfrac{3}{2}) - \tfrac{1}{2}]$$
$$= (x + 2)(x + 1)$$

EXAMPLE 2   Solve the equation $2x^2 - 6x - 5 = 0$.

SOLUTION   For any $x \in \mathbf{R}$,

$$2x^2 - 6x - 5 = 2(x^2 - 3x - \tfrac{5}{2})$$

Now, with $k = -3$, we see that

$$x^2 - 3x + (\tfrac{3}{2})^2 - (\tfrac{3}{2})^2 = (x - \tfrac{3}{2})^2 - \tfrac{9}{4}$$

Hence, $2(x^2 - 3x - \tfrac{5}{2}) = 2([x - \tfrac{3}{2}]^2 - \tfrac{9}{4} - \tfrac{5}{2})$
$$= 2([x - \tfrac{3}{2}]^2 - \tfrac{19}{4})$$

Thus, $2x^2 - 6x - 5 = 0$, if and only if

$$2([x - \tfrac{3}{2}]^2 - \tfrac{19}{4}) = 0$$
$$(x - \tfrac{3}{2})^2 - \tfrac{19}{4} = 0$$
$$(x - \tfrac{3}{2})^2 = \tfrac{19}{4}$$
$$x - \frac{3}{2} = \pm\left(\frac{19}{4}\right)^{1/2}$$
$$x = \frac{3}{2} \pm \left(\frac{19}{4}\right)^{1/2} = \frac{3}{2} \pm \frac{(19)^{1/2}}{2}$$

That is, $x = \dfrac{3 + (19)^{1/2}}{2}, \qquad x = \dfrac{3 - (19)^{1/2}}{2}$

If the technique of completing the square is applied to the general quadratic $ax^2 + bx + c$, we get the following important result which gives a *formula* for finding the solutions to $ax^2 + bx + c = 0$. The formula is called the **quadratic formula**. Specifically,

If a, b, and c are real numbers with a $\neq$ 0 and $b^2 - 4ac > 0$, then there are two numbers x for which $ax^2 + bx + c = 0$, and they are

$$x = \frac{-b + (b^2 - 4ac)^{1/2}}{2a} \quad \text{and} \quad x = \frac{-b - (b^2 - 4ac)^{1/2}}{2a}$$

If $b^2 - 4ac = 0$, then the equation $ax^2 + bx + c = 0$ has precisely one solution, and it is $x = -b/2a$.

If $b^2 - 4ac < 0$, then there are no real numbers x for which $ax^2 + bx + c = 0$.

EXAMPLE 3 Find all $x \in \mathbf{R}$ for which $x^2 + 3x + 1 = 0$.

SOLUTION Here, $a = 1$, $b = 3$, and $c = 1$. Thus, $b^2 - 4ac = 5$, and so we obtain the two solutions

$$x = \frac{-3 + (5)^{1/2}}{2} \quad \text{and} \quad x = \frac{-3 - (5)^{1/2}}{2}.$$

EXAMPLE 4 Solve the equation $2x^2 + 3x - 2 = 0$.

SOLUTION Here $a = 2$, $b = 3$, and $c = -2$, and thus $b^2 - 4ac = 9 - 4(-2)(2) = 25$. The two solutions are

$$\frac{-3 + (25)^{1/2}}{4} = \frac{1}{2} \quad \text{and} \quad \frac{-3 - (25)^{1/2}}{4} = -2$$

EXAMPLE 5 Solve the equation $x^2 - 2\pi x + \pi^2 = 0$.

SOLUTION With $a = 1$, $b = -2\pi$, and $c = \pi^2$, we see that $b^2 - 4ac = 4\pi^2 - 4\pi^2 = 0$. There is precisely one solution, and it is $-[(-2\pi)/2] = \pi$.

EXAMPLE 6 Solve the equation $x^2 + (2)^{1/2}x + \pi = 0$.

SOLUTION With $a = 1$, $b = (2)^{1/2}$, and $c = \pi$, we see that $b^2 - 4ac = 2 - 4\pi < 0$. Thus, there are no (real) solutions.

By completing the square in problems 1–6, find all $x \in \mathbf{R}$ for which

1. $x^2 - 4x - 2 = 0$ **2.** $x^2 - 4x + 2 = 0$

3. $x^2 + 3x - 2 = 0$ **4.** $x^2 - 2 = -3x$

5. $2x^2 = 2x - 1$ **6.** $-4x^2 + 3x + 2 = 0$

In problems 7–16, find all real numbers x for which

7. $x^2 + x - 1 = 0$ **8.** $x^2 - 3x + 1 = 0$

9. $x^2 - 4x + 2 = 0$ **10.** $x^2 - 7x + 8 = 0$

11. $4x^2 = 6x + 7$ **12.** $x^3 + (2)^{1/2}x^2 = \frac{1}{4}x$

13. $\dfrac{x}{x+2} + \dfrac{1}{x-4} = 0$ **14.** $\dfrac{x}{x+1} - \dfrac{2x}{x^2-1} = 0$

15. $\dfrac{x}{x+2} + \dfrac{x-2}{x+3} = 0$ **16.** $\dfrac{1}{x^2+x-12} + \dfrac{x}{x-3} = 0$

17. Let x and y be real numbers. By completing the square, find real numbers a and b for which

$$x^2 + y^2 - 4x + 6y + 4 = (x-a)^2 + (y-b)^2 - 9$$

SECTION 8

INTERVALS AND HALF-LINES. INEQUALITIES

In our subsequent work, we shall need not only the entire real line, but also certain subsets of the line, called intervals and half-lines. The following will describe these subsets, as well as establish some notation for them. Recall that

$$a < b \text{ means that } a \textbf{ is less than } b$$

and $\quad a \le b$ means that a **is less than or equal to** b

Let a and b be real numbers with $a < b$. Then, $[a,b]$ *is the set of all numbers x for which $a \le x \le b$; it is called the* **closed interval** *from a to b; (a,b) is the set of all x for which $a < x < b$; it is called the* **open interval** *from a to b; $[a,b)$ is the set of all x for which $a \le x < b$; $(a,b]$ is the set of all x for which $a < x \le b$.*

Equivalently,

$$[a,b] = \{x: a \le x \le b\} = \text{(the closed interval from } a \text{ to } b\text{)}$$

$$(a,b) = \{x: a < x < b\} = \text{(the open interval from } a \text{ to } b\text{)}$$

and $\quad [a,b) = \{x: a \le x < b\}, \qquad (a,b] = \{x: a < x \le b\}$

The **half-lines** are given by

$$[a,\infty) = \{x: x \ge a\}, \text{ a } \textbf{closed half-line}$$

$$(a,\infty) = \{x: x > a\}, \text{ an } \textbf{open half-line}$$

The closed half-line $(-\infty, b]$ and the open half-line $(-\infty, b)$ are defined similarly. In each of the preceding sets, a and b are

Include $[0,1]$

(a)

Omit $(0,1)$

(b)

$[1, \infty)$

(c)

Figure A-7

called the left and right endpoints of the interval, respectively (or half-line, although, of course, a half-line has only *one* endpoint); thus, an endpoint of an interval (or half-line) need *not* be in the interval (or half-line). For example, in $(0,1]$, the left endpoint is not in the interval, but the right endpoint is. Figure A-7 sketches some intervals and half-lines.

EXAMPLE 1   The closed interval $[-1,1]$ is the set of all x for which $-1 \le x \le 1$. Both endpoints, -1 and 1, are in $[-1,1]$. The open interval $(-1,1)$ is the set of all x such that $-1 < x < 1$. Therefore, 0 is in $(-1,1)$, but 1 is *not* in $(-1,1)$. The half-line $[3,\infty)$ is the set of all x such that $x \ge 3$; it is a closed half-line.

EXAMPLE 2   Find all x such that $x^2 - 2x - 3 > 0$.

SOLUTION   Since $x^2 - 2x - 3 = (x - 3)(x + 1)$, all x, the problem reduces to an investigation of the factors $x - 3$ and $x + 1$. Let x be a real number. We see that $x^2 - 2x - 3 > 0$ provided the numbers $x - 3$ and $x + 1$ have the same sign. Now,

$$x - 3 \ge 0 \text{ if and only if } x \ge 3$$

and $\qquad\qquad x - 3 < 0 \text{ if and only if } x < 3$

In Figure A-8, we write beside $x - 3$ the appropriate sign for each x. Similarly,

$$x + 1 \ge 0 \text{ if and only if } x \ge -1$$

$$x + 1 < 0 \text{ if and only if } x < -1$$

and again in Figure A-8, we have written beside $x + 1$ the appropriate sign. By inspection, we see from Figure A-8 that $x - 3$ and $x + 1$ have the same sign if and only if $x < -1$ or $x > 3$. Hence, $x^2 - 2x - 3 > 0$ if and only if $x < -1$ or $x > 3$; that is, $x^2 - 2x - 3 > 0$ if and only if x is on either of the open half-lines $(3,\infty)$ or $(-\infty,-1)$.

We can also observe from Figure A-8 that x is a real number for which $x^2 - 2x - 3 < 0$ if and only if x is in the open interval $(-1,3)$.

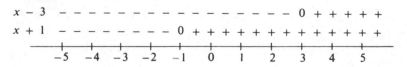

Figure A-8

Problems   Complete, as in 1.

1. $\{x: x \le -1\} = (-\infty, -1]$
2. $\{x: x \ge 3\} = $ _____
3. $\{x: x > -10\} = $ _____
4. $\{x: x < 35\} = $ _____
5. $\{x: -100 \le x < 0\} = $ _____
6. $\{x: -10 < x < 2\pi\} = $ _____
7. $\{x: \sqrt{2} \le x \le \pi + 1\} = $ _____
8. $\{x: -\pi < x \le \pi\} = $ _____
9. Which of the sets in problems 1–8 are open intervals? Closed intervals? Open half-lines? Closed half-lines?

Answer *true* or *false*.

10. $3 \in (0,5)$
11. $-1 \in [-1,1)$
12. $1 \in [-1,1)$
13. $\mathbf{R} = (-\infty,3) \cup (3,\infty)$
14. $\mathbf{R} = (-\infty,3] \cup (3,\infty)$
15. $(-10,5) \cap [5,10) = \emptyset$
16. $[-5,3) \cup (0,5] = [-5,5]$
17. $[-5,3) \cap [0,5] = [0,3]$

In problems 18–29, find all real numbers x which satisfy the given condition. Express your answer as intervals or half-lines.

18. $x^2 - 16 < 0$
19. $4x^2 - 9 \ge 0$
20. $x^3 - 25x < 0$
21. $x^4 - x^2 \le 0$
22. $x^2 + x - 6 > 0$
23. $x^2 + 4x - 5 \le 0$
24. $x^3 + 7x^2 + 12x < 0$
25. $x^4 - 3x^3 + 2x^2 > 0$
26. $(x - 1)(x - 2)(x - 3) < 0$
27. $(x + 4)(x - 3)(x + 2) > 0$

28. $\dfrac{(x + 5)(x - 1)}{x^2 - 4} \ge 0$
29. $\dfrac{x^2 + x - 6}{x^2 + 4x - 5} > 0$

SECTION 9

THE BINOMIAL THEOREM

It is relatively simple to carry out the details to show that for any numbers x and y

$$(x + y)^2 = x^2 + 2xy + y^2$$

a little more complicated to show that

$$(x + y)^3 = x^3 + 3x^2y + 3xy^2 + y^3$$

and unpleasant to expand $(x + y)^{21}$ by repeated multiplication. The binomial theorem gives a single expression for $(x + y)^n$ where n is any positive integer. The proof of this theorem is algebraic, and can be found in any standard college algebra text. We shall need the following definitions.

For any positive integer n,

(A.59) $n! = 1 \cdot 2 \cdot 3 \cdot \ldots \cdot (n - 1) \cdot n$

and is called **n-factorial**.

(A.60) $0! = 1$ (this is a definition)

For nonnegative integers n and k with $n \geq k$,

(A.61) $$\binom{n}{k} = \frac{n!}{k!(n-k)!}$$

and is called a **binomial coefficient**.

(A.62) $$x^0 = 1$$

for any real number $x \neq 0$.

EXAMPLE 1   We shall illustrate the notation in (A.59) and (A.61).

(A.63) $$4! = 1 \cdot 2 \cdot 3 \cdot 4 = 24$$

(A.64) $$\binom{n}{1} = \frac{n!}{1!(n-1)!} = \frac{1 \cdot 2 \cdot 3 \cdot \ldots \cdot (n-1) \cdot n}{1 \cdot 2 \cdot 3 \cdot \ldots \cdot (n-1)} = n$$

The main result is **the binomial theorem**. *Let n be any positive integer and let x and y be any two nonzero real numbers. Then*

(A.65) $$(x+y)^n = x^n + \binom{n}{1} x^{n-1} y + \binom{n}{2} x^{n-2} y^2$$
$$+ \cdots + \binom{n}{n-1} xy^{n-1} + y^n$$

To emphasize the symmetry of the equation, we could write it in the equivalent form

(A.66) $$(x+y)^n = \binom{n}{0} x^n y^0 + \binom{n}{1} x^{n-1} y^1 + \binom{n}{2} x^{n-2} y^2$$
$$+ \cdots + \binom{n}{n-1} x^1 y^{n-1} + \binom{n}{n} x^0 y^n$$

You should verify that (A.65) and (A.66) are identical. From (A.66) we observe the following characteristics of the equation: The powers of x begin at n and descend to 0, the powers of y begin at 0 and ascend to n, the sum of the exponents appearing in any summand is n, the lower entry in each binomial coefficient corresponds to the exponent of y in that summand, the top entry of each binomial coefficient minus the lower entry in the same coefficient is the exponent of x in that summand, and the lower entries in the binomial coefficients go from 0 to n as we go from one summand to the next.

EXAMPLE 2 We shall use (A.65) to expand $(x + y)^2$ (where, of course, $x \in \mathbf{R}$ and $y \in \mathbf{R}$). Here $n = 2$. Thus,

$$(x + y)^2 = x^2 + \binom{2}{1} xy + y^2 = x^2 + 2xy + y^2$$

since

$$\binom{2}{1} = \frac{2!}{1!(2 - 1)!} = 2$$

EXAMPLE 3 For $x \in \mathbf{R}$ and $y \in \mathbf{R}$, find $(x + y)^3$.

SOLUTION

$$(x + y)^3 = x^3 + \binom{3}{1} x^2y + \binom{3}{2} xy^2 + y^3$$

$$= x^3 + \frac{3!}{1!2!} x^2y + \frac{3!}{1!2!} xy^2 + y^3$$

$$= x^3 + 3x^2y + 3xy^2 + y^3$$

Problems 1. Let $a \in \mathbf{R}, b \in \mathbf{R}$. Find $(a + b)^4, (a - b)^4, (a^2 + 2b)^3$.
2. Write an expression for $(x + h)^n - x^n$ where n is any positive integer, and $x \in \mathbf{R}$.
3. Find $(2a - 3b)^5, [(2/3)a^2 - (b)^{1/2}]^3, [(2/a) + (1/b)]^4$ (in each case, a and b are numbers for which the problems make sense).

Table of Integrals

For each integration $\int f(x)\,dx$, note that $\int f(x)\,dx + c$, c any real number, is also correct; that is, the tables omit the so-called "constant of integration."

Whenever the signs $\pm$ or $+$ appear more than once in any formula, it is understood that the *upper* (or, respectively, *lower*) signs apply simultaneously throughout. For example, $a \pm b = c \mp d$ means the *two* statements. $u + b - c - d$ and $a \quad b - c + d$.

Although this book has restricted its attention to powers and roots of rational, exponential, and log functions, the tables include the trigonometric, inverse trigonometric, hyperbolic, and inverse hyperbolic functions for completeness, and for providing a more substantial table for those whose future labors might involve the use of such tables. For those who are familiar with some (or all) of these functions, the tables indicate how they arise in integration (25, 29, and 31 show how these functions arise when integrating certain *rational* functions).

In any formulas, arctan u, arcsin u, arcsec u, and $\cosh^{-1} u$ represent function values on the intervals indicated below:

$$-\tfrac{1}{2}\pi \le \arcsin u \le \tfrac{1}{2}\pi; \qquad (\arcsin u < 0 \text{ if } u < 0)$$

$$-\tfrac{1}{2}\pi < \arctan u < \tfrac{1}{2}\pi; \qquad (\arctan u < 0 \text{ if } u < 0)$$

$$0 \le \text{arcsec } u < \tfrac{1}{2}\pi \text{ if } u > 0; \qquad -\pi \le \text{arcsec } u < -\tfrac{1}{2}\pi \text{ if } u < 0;$$

$$0 \le \cosh^{-1} u. \qquad (u \ge 1 \text{ } always)$$

**Integrals
of Rational Functions
of u and $(a + bu)$**

1. $\displaystyle \int u^n\,du = \frac{u^{n+1}}{n+1}.$

2. $\displaystyle \int \frac{du}{u} = \ln|u|.$

3. $\displaystyle \int (a + bu)^n\,du = \frac{(a + bu)^{n+1}}{b(n+1)}.$

4. $\displaystyle \int \frac{du}{a + bu} = \frac{1}{b}\ln|a + bu|.$

5. $\displaystyle \int \frac{du}{u(a + bu)} = \frac{1}{a}\ln\left|\frac{u}{a + bu}\right|.$

6. $\displaystyle\int \frac{du}{u^2(a + bu)} = -\frac{1}{au} + \frac{b}{a^2} \ln \left| \frac{a + bu}{u} \right|.$

7. $\displaystyle\int \frac{du}{u(a + bu)^2} = \frac{1}{a(a + bu)} + \frac{1}{a^2} \ln \left| \frac{u}{a + bu} \right|.$

8. $\displaystyle\int \frac{du}{u^2(a + bu)^2} = -\frac{a + 2bu}{a^2 u(a + bu)} + \frac{2b}{a^3} \ln \left| \frac{a + bu}{u} \right|.$

9. $\displaystyle\int \frac{u\, du}{(a + bu)^2} = \frac{1}{b^2} \left(\ln | a + bu | + \frac{a}{a + bu} \right).$

10. $\displaystyle\int \frac{u\, du}{(a + bu)^3} = \frac{1}{b^2} \left[-\frac{1}{a + bu} + \frac{a}{2(a + bu)^2} \right].$

11. $\displaystyle\int \frac{u^2\, du}{(a + bu)^3} = \frac{1}{b^3} \left[\ln | a + bu | + \frac{2a}{a + bu} - \frac{a^2}{2(a + bu)^2} \right].$

12. $\displaystyle\int \frac{du}{u^m(a + bu)^n} = \frac{-1}{a(m - 1)\, u^{m-1}(a + bu)^{n-1}}$

$$- \frac{b(m + n - 2)}{a(m - 1)} \int \frac{du}{u^{m-1}(a + bu)^n}.$$

13. $\displaystyle\int \frac{du}{u^m(a + bu)^n} = \frac{1}{a(n - 1)\, u^{m-1}(a + bu)^{n-1}}$

$$+ \frac{(m + n - 2)}{a(n - 1)} \int \frac{du}{u^m(a + bu)^{n-1}}.$$

14. $\displaystyle\int \frac{1}{u^2 - a^2}\, du = \frac{1}{2a} \ln \left| \frac{u - a}{u + a} \right| + C.$

15. $\displaystyle\int \frac{1}{(au + b)(cu + d)}\, du = \frac{1}{bc - ad} \ln \left| \frac{cu + d}{au + b} \right| + C,$

$$(bc - ad \neq 0).$$

16. $\displaystyle\int \frac{u}{(au + b)(cu + d)}\, du$

$$= \frac{1}{bc - ad} \left\{ \frac{b}{a} \ln | au + b | - \frac{d}{c} \ln | cu + d | \right\} + C$$

$$(bc - ad \neq 0).$$

17. $\displaystyle\int \frac{1}{(au + b)^2(cu + d)}\, du$

$$= \frac{1}{bc - ad} \left\{ \frac{1}{au + b} + \frac{c}{bc - ad} \ln \left| \frac{cu + d}{au + b} \right| \right\} + C$$

$$(bc - ad \neq 0).$$

18. $\displaystyle\int \frac{u}{(au + b)^2(cu + d)}\, du$

$$= -\frac{1}{bc - ad}\left\{\frac{b}{a(au + b)} + \frac{d}{bc - ad}\ln\left|\frac{cu + d}{au + b}\right|\right\} + C$$

$$(bc - ad \neq 0).$$

Irrational Integrands Involving $(a + bu)$

19. $\displaystyle\int u\sqrt{a + bu}\, du = -\frac{2(2u - 3bu)(a + bu)^{3/2}}{15b^2}.$

20. $\displaystyle\int u^m\sqrt{a + bu}\, du = \frac{2u^m(a + bu)^{3/2}}{b(2m + 3)}$

$$-\frac{2am}{b(2m + 3)}\int u^{m-1}\sqrt{a + bu}\, du.$$

21. $\displaystyle\int \frac{u\, du}{\sqrt{a + bu}} = \frac{2(bu - 2a)\sqrt{a + bu}}{3b^2}.$

22. $\displaystyle\int \frac{u^2\, du}{\sqrt{a + bu}} = \frac{2(3b^2u^2 - 4abu + 8a^2)\sqrt{a + bu}}{15b^3}.$

23. $\displaystyle\int \frac{u^m\, du}{\sqrt{a + bu}} = \frac{2u^m\sqrt{a + bu}}{b(2m + 1)} - \frac{2am}{b(2m + 1)}\int \frac{u^{m-1}\, du}{\sqrt{a + bu}}.$

24. $\displaystyle\int \frac{du}{u\sqrt{a + bu}} = \frac{1}{\sqrt{a}}\ln\left|\frac{\sqrt{a + bu} - \sqrt{a}}{\sqrt{a + bu} + \sqrt{a}}\right|,$ if $a > 0$.

25. $\displaystyle\int \frac{du}{u\sqrt{a + bu}} = \frac{2}{\sqrt{-a}}\arctan\sqrt{\frac{a + bu}{-a}},$ if $a < 0$.

26. $\displaystyle\int \frac{du}{u^m\sqrt{a + bu}} = -\frac{\sqrt{a + bu}}{a(m - 1)u^{m-1}}$

$$-\frac{b(2m - 3)}{2a(m - 1)}\int \frac{du}{u^{m-1}\sqrt{a + bu}}.$$

27. $\displaystyle\int \frac{\sqrt{a + bu}}{u}\, du = 2\sqrt{a + bu} + a\int \frac{du}{u\sqrt{a + bu}}.$

28. $\displaystyle\int \frac{\sqrt{a + bu}}{u^m}\, du = -\frac{(a + bu)^{3/2}}{a(m - 1)u^{m-1}}$

$$-\frac{b(2m - 5)}{2a(m - 1)}\int \frac{\sqrt{a + bu}\, du}{u^{m-1}}.$$

Rational Forms
Involving* $(a^2 \pm u^2)$

29. $\displaystyle\int \frac{du}{a^2 + u^2} = \frac{1}{a} \arctan \frac{u}{a}.$

30. $\displaystyle\int \frac{du}{u^2 - a^2} = \frac{1}{2a} \ln \left| \frac{u - a}{u + a} \right|.$

31₁. $\displaystyle\int \frac{du}{a^2 - u^2} = \frac{1}{a} \tanh^{-1} \frac{u}{a}. \qquad (u^2 < a^2)$

31₂. $\displaystyle\int \frac{du}{u^2 - a^2} = -\frac{1}{a} \coth^{-1} \frac{u}{a}. \qquad (u^2 > a^2)$

32. $\displaystyle\int \frac{du}{(a^2 + u^2)^n} = \frac{u}{2(n-1)a^2(a^2 + u^2)^{n-1}}$

$$+ \frac{2n - 3}{(2n - 2)a^2} \int \frac{du}{(a^2 + u^2)^{n-1}}.$$

Irrational Forms
Involving $\sqrt{a^2 - u^2}$

33. $\displaystyle\int \sqrt{a^2 - u^2}\, du = \frac{u}{2} \sqrt{a^2 - u^2} + \frac{a^2}{2} \arcsin \frac{u}{a}.$

34. $\displaystyle\int \frac{du}{\sqrt{a^2 - u^2}} = \arcsin \frac{u}{a}.$

35. $\displaystyle\int \frac{du}{(a^2 - u^2)^{3/2}} = \frac{u}{a^2 \sqrt{a^2 - u^2}}.$

36. $\displaystyle\int u^2 \sqrt{a^2 - u^2}\, du = -\frac{1}{4} u(a^2 - u^2)^{3/2} + \frac{1}{4} a^2 \int \sqrt{a^2 - u^2}\, du.$

37. $\displaystyle\int \frac{u^2\, du}{\sqrt{a^2 - u^2}} = -\frac{u}{2} \sqrt{a^2 - u^2} + \frac{a^2}{2} \arcsin \frac{u}{a}.$

38. $\displaystyle\int \frac{u^2\, du}{(a^2 - u^2)^{3/2}} = \frac{u}{\sqrt{a^2 - u^2}} - \arcsin \frac{u}{a}.$

39. $\displaystyle\int \frac{du}{u \sqrt{a^2 - u^2}} = -\frac{1}{a} \ln \left| \frac{a + \sqrt{a^2 - u^2}}{u} \right|$

$$= -\frac{1}{a} \cosh^{-1} \frac{a}{u}, \qquad \text{if } 0 < u < a.$$

40. $\displaystyle\int \frac{du}{u^2 \sqrt{a^2 - u^2}} = -\frac{\sqrt{a^2 - u^2}}{a^2 u}.$

41. $\displaystyle\int \frac{\sqrt{a^2 - u^2}\, du}{u} = \sqrt{a^2 - u^2} - a \ln \left| \frac{a + \sqrt{a^2 - u^2}}{u} \right|$

42. $\displaystyle\qquad\qquad = \sqrt{a^2 - u^2} - a \cosh^{-1} \frac{a}{u}, \qquad \text{if } 0 < u < a.$

43. $\displaystyle\int \frac{\sqrt{a^2 - u^2}}{u^2}\, du = -\frac{\sqrt{a^2 - u^2}}{u} - \arcsin \frac{u}{a}.$

*Wherever a constant a^2 enters, infer that $a > 0$.

44. $\displaystyle\int u^m (a^2 - u^2)^{n/2}\, du \qquad (m \geq 0 \quad \text{or} \quad m < 0)$

$$= \frac{u^{m+1}(a^2 - u^2)^{n/2}}{n + m + 1} + \frac{a^2 n}{n + m + 1} \int u^m (a^2 - u^2)^{(n/2) - 1}\, du.$$

45. $\displaystyle\int u^m (a^2 - u^2)^{n/2}\, du \qquad (n > 0 \quad \text{or} \quad n < 0)$

$$= -\frac{u^{m-1}(a^2 - u^2)^{(n/2)+1}}{n + m + 1} + \frac{a^2(m - 1)}{n + m + 1} \int u^{m-2} (a^2 - u^2)^{n/2}\, du.$$

46. $\displaystyle\int \frac{(a^2 - u^2)^{n/2}}{u^m}\, du \qquad (n > 0 \quad \text{or} \quad n < 0)$

$$= \frac{(a^2 - u^2)^{(n/2)+1}}{a^2(m - 1)u^{m-1}} + \frac{m - n - 3}{a^2(m - 1)} \int \frac{(a^2 - u^2)^{n/2}}{u^{m-2}}\, du.$$

47. $\displaystyle\int \frac{u^m\, du}{(a^2 - u^2)^{n/2}} \qquad (m \geq 0 \quad \text{or} \quad m < 0)$

$$= \frac{u^{m+1}}{a^2(n - 2)(a^2 - u^2)^{(n/2)-1}} - \frac{m - n + 3}{a^2(n - 2)} \int \frac{u^m\, du}{(a^2 - u^2)^{(n/2)-1}}.$$

Irrational Forms Involving $\sqrt{u^2 \pm a^2}$

In any formula of the present types, it can be shown that we may replace

$$\ln(u + \sqrt{u^2 + a^2}) \qquad by \qquad \sinh^{-1}\frac{u}{a}; \qquad \ln\left|\frac{a + \sqrt{u^2 + a^2}}{u}\right|$$

$$by \qquad \sinh^{-1}\left|\frac{a}{u}\right|$$

$$\ln|u + \sqrt{u^2 - a^2}| \qquad by \qquad \begin{cases} \cosh^{-1}\dfrac{u}{a}, & \text{if } u \geq a \\[2mm] -\cosh^{-1}\left|\dfrac{u}{a}\right|, & \text{if } u \leq -a \end{cases}$$

48. $\displaystyle\int \sqrt{u^2 \pm a^2}\, du = \tfrac{1}{2}[u\sqrt{u^2 \pm a^2} \pm a^2 \ln|u + \sqrt{u^2 \pm a^2}|].$

49. $\displaystyle\int u^2\sqrt{u^2 \pm a^2}\, du$

$$= \tfrac{1}{8}u(2u^2 \pm a^2)\sqrt{u^2 \pm a^2} - \tfrac{1}{8}a^4 \ln|u + \sqrt{u^2 \pm a^2}|.$$

50. $\displaystyle\int \frac{\sqrt{u^2 + a^2}}{u}\, du = \sqrt{u^2 + a^2} - a\ln\left|\frac{a + \sqrt{u^2 + a^2}}{u}\right|.$

51. $\displaystyle\int \frac{\sqrt{u^2 - a^2}}{u}\, du = \sqrt{u^2 - a^2} - a\operatorname{arcsec}\frac{u}{a}.$

52. $\displaystyle\int \frac{\sqrt{u^2 \pm a^2}}{u^2}\, du = -\frac{\sqrt{u^2 \pm a^2}}{u} + \ln|u + \sqrt{u^2 \pm a^2}|.$

$53_1.$ $\displaystyle\int \frac{du}{\sqrt{u^2 \pm a^2}} = \ln |u + \sqrt{u^2 \pm a^2}|.$

$53_2.$ $\displaystyle\int \frac{du}{\sqrt{u^2 + a^2}} = \sinh^{-1} \frac{u}{a}.$

$53_3.$ $\displaystyle\int \frac{du}{\sqrt{u^2 - a^2}} = \cosh^{-1} \frac{u}{a}, \qquad (u > a);$

$$= -\cosh^{-1} \left|\frac{u}{a}\right|, \qquad (u < -a).$$

$54.$ $\displaystyle\int \frac{du}{u\sqrt{u^2 - a^2}} = \frac{1}{a} \operatorname{arcsec} \frac{u}{a}.$

$55.$ $\displaystyle\int \frac{du}{u\sqrt{u^2 + a^2}} = \frac{1}{a} \ln \left|\frac{u}{a + \sqrt{u^2 + a^2}}\right|.$

$56.$ $\displaystyle\int \frac{u^2\, du}{\sqrt{u^2 \pm a^2}} = \frac{1}{2} \left(u\sqrt{u^2 \pm a^2} \mp a^2 \ln |u + \sqrt{u^2 \pm a^2}|\right).$

$57.$ $\displaystyle\int \frac{du}{u^2\sqrt{u^2 \pm a^2}} = \mp \frac{\sqrt{u^2 \pm a^2}}{a^2 u}.$

$58.$ $\displaystyle\int \frac{du}{(u^2 \pm a^2)^{3/2}} = \frac{\pm u}{a^2\sqrt{u^2 \pm a^2}}.$

$59.$ $\displaystyle\int \frac{u^2\, du}{(u^2 \pm a^2)^{3/2}} = \frac{-u}{\sqrt{u^2 \pm a^2}} + \ln |u + \sqrt{u^2 \pm a^2}|.$

$60.$ $\displaystyle\int u^m(u^2 \pm a^2)^{n/2}\, du \qquad (n > 0 \quad \text{or} \quad n < 0)$

$$= \frac{u^{m-1}(u^2 \pm a^2)^{(n/2)+1}}{n + m + 1} \mp \frac{a^2(m - 1)}{n + m + 1} \int u^{m-2}(u^2 \pm a^2)^{n/2}\, du.$$

$61.$ $\displaystyle\int \frac{(u^2 \pm a^2)^{n/2}\, du}{u^m} \qquad (n > 0 \quad \text{or} \quad n < 0)$

$$= \frac{\mp(u^2 \pm a^2)^{(n/2)+1}}{a^2(m - 1)u^{m-1}} \mp \frac{m - n - 3}{a^2(m - 1)} \int \frac{(u^2 \pm a^2)^{n/2}}{u^{m-2}}\, du.$$

$62.$ $\displaystyle\int \frac{u^m\, du}{(u^2 \pm a^2)^{n/2}} \qquad (m \geq 0 \quad \text{or} \quad m < 0)$

$$= \frac{\pm u^{m+1}}{a^2(n - 2)(u^2 \pm a^2)^{(n/2)-1}} \mp \frac{m - n + 3}{a^2(n - 2)} \int \frac{u^m\, du}{(u^2 \pm a^2)^{(n/2)-1}}.$$

$63.$ $\displaystyle\int u^m(u^2 \pm a^2)^{n/2}\, du \qquad (m \geq 0 \quad \text{or} \quad m < 0)$

$$= \frac{(u^2 \pm a^2)^{n/2}u^{m+1}}{n + m + 1} \pm \frac{a^2 n}{n + m + 1} \int u^m(u^2 \pm a^2)^{(n/2)-1}\, du.$$

**Irrational Forms
Involving $\sqrt{2au \pm u^2}$**

Note that $\sqrt{2au + u^2} = \sqrt{(u + a)^2 - a^2}$; $\sqrt{2au - u^2}$

$$= \sqrt{a^2 - (u - a)^2}.$$

64. $\displaystyle\int (2au + u^2)^{n/2}\, du.$ Use $\displaystyle\int (v^2 - a^2)^{n/2}\, dv$ with $v = u + a$. .

65. $\displaystyle\int (2au - u^2)^{n/2}\, du.$ Use $\displaystyle\int (a^2 - v^2)^{n/2}\, dv$ with $v = u - a$.

66. $\displaystyle\int \sqrt{2au - u^2}\, du = \frac{u - a}{2}\sqrt{2au - u^2} + \frac{a^2}{2}\arcsin\frac{u - a}{a}.$

67. $\displaystyle\int \sqrt{2au + u^2}\, du$

$$= \frac{u + a}{2}\sqrt{2au + u^2} - \frac{a^2}{2}\ln|u + a + \sqrt{2au + u^2}|.$$

68. $\displaystyle\int u\sqrt{2au + u^2}\, du$

$$= \frac{2u^2 + au - 3a^2}{6}\sqrt{2au + u^2} + \frac{a^3}{2}\ln|u + a + \sqrt{2au + u^2}|.$$

69. $\displaystyle\int u\sqrt{2au - u^2}\, du$

$$= \frac{2u^2 - au - 3a^2}{6}\sqrt{2au - u^2} + \frac{a^3}{2}\arcsin\frac{u - a}{a}.$$

70. $\displaystyle\int \frac{du}{\sqrt{2au - u^2}} = \arcsin\frac{u - a}{a}.$

71. $\displaystyle\int \frac{du}{\sqrt{2au + u^2}} = \ln|u + a + \sqrt{u^2 + 2au}|.$

72. $\displaystyle\int \frac{u\, du}{\sqrt{2au - u^2}} = -\sqrt{2au - u^2} + a\arcsin\frac{u - a}{a}.$

73. $\displaystyle\int \frac{u\, du}{\sqrt{2au + u^2}} = \sqrt{2au + u^2} - a\ln|u + a + \sqrt{2au + u^2}|.$

74. $\displaystyle\int \frac{u^n\, du}{\sqrt{2au - u^2}}$

$$= -\frac{u^{n-1}\sqrt{2au - u^2}}{n} + \frac{a(2n - 1)}{n}\int \frac{u^{n-1}\, du}{\sqrt{2au - u^2}}.$$

75. $\displaystyle\int \frac{du}{u^n\sqrt{2au - u^2}}$

$$= \frac{\sqrt{2au - u^2}}{a(1 - 2n)u^n} + \frac{n - 1}{(2n - 1)a}\int \frac{du}{u^{n-1}\sqrt{2au - u^2}}.$$

76. $\displaystyle\int u^n \sqrt{2au - u^2}\, du$

$$= -\frac{u^{n-1}(2au - u^2)^{3/2}}{n + 2} + \frac{(2n + 1)a}{n + 2} \int u^{n-1}\sqrt{2au - u^2}\, du.$$

77. $\displaystyle\int \frac{\sqrt{2au - u^2}}{u^n}\, du$

$$= \frac{(2au - u^2)^{3/2}}{a(3 - 2n)u^n} + \frac{n - 3}{(2n - 3)a} \int \frac{\sqrt{2au - u^2}}{u^{n-1}}\, du.$$

Forms Involving
$f(u) = a + bu \pm cu^2,\ c > 0$

By completing a square with the terms in u, $f(u)$ assumes the form $\pm c(A^2 + v^2)$, $\pm c(v^2 - A^2)$, or $\pm cv^2$, with $v = u \pm b/2c$. Then, substitution of $v = (2cu \pm b)/2c$ gives an integrand $F(v)$ involving $(A^2 \pm v^2)$, $(v^2 - A^2)$, or v^2. Any radicand $(4ac + b^2)$ is assumed to be positive.

78. $\displaystyle\int \frac{du}{a + bu - cu^2} = \frac{1}{\sqrt{b^2 + 4ac}} \ln \left| \frac{\sqrt{b^2 + 4ac} - b + 2cu}{\sqrt{b^2 + 4ac} + b - 2cu} \right|.$

79. $\displaystyle\int \frac{du}{\sqrt{a + bu + cu^2}} = \frac{1}{\sqrt{c}} \ln \left| 2cu + b + 2\sqrt{c}\sqrt{a + bu + cu^2} \right|$

80. $\displaystyle\int \frac{du}{\sqrt{a + bu - cu^2}} = \frac{1}{\sqrt{c}} \arcsin \frac{2cu - b}{\sqrt{b^2 + 4ac}}.$

81. $\displaystyle\int \frac{du}{(a + bu \pm cu^2)^{3/2}} = \frac{2(\pm 2cu + b)}{(\pm 4ac - b^2)\sqrt{a + bu \pm cu^2}}.$

82. $\displaystyle\int \sqrt{a + bu + cu^2}\, du$

$$= \frac{2cu + b}{4c} \sqrt{a + bu + cu^2}$$

$$- \frac{b^2 - 4ac}{8c^{3/2}} \ln \left| 2cu + b + 2\sqrt{c}\sqrt{a + bu + cu^2} \right|.$$

83. $\displaystyle\int \sqrt{a + bu - cu^2}\, du = \frac{2cu - b}{4c} \sqrt{a + bu - cu^2}$

$$+ \frac{b^2 + 4ac}{8c^{3/2}} \arcsin \frac{2cu - b}{\sqrt{b^2 + 4ac}}.$$

84. $\displaystyle\int \frac{u\, du}{\sqrt{a + bu + cu^2}}$

$$= \frac{\sqrt{a + bu + cu^2}}{c} - \frac{b}{2c^{3/2}} \ln \left| 2cu + b + 2\sqrt{c}\sqrt{a + bu + cu^2} \right|.$$

85. $\displaystyle \int \frac{u\,du}{\sqrt{a + bu - cu^2}} = -\frac{\sqrt{a + bu - cu^2}}{c}$

$$+ \frac{b}{2c^{3/2}} \arcsin \frac{2cu - b}{\sqrt{b^2 + 4ac}}.$$

86. $\displaystyle \int u\,\sqrt{a + bu \pm cu^2}\,du = \frac{(a + bu \pm cu^2)^{3/2}}{\pm 3c}$

$$\mp \frac{b}{2c} \int \sqrt{a + bu \pm cu^2}\,du.$$

Binomial Reduction Formulas

$\displaystyle \int u^m(a + bu^n)^p du:$ (m and p, positive or negative)

87. $\displaystyle = \frac{u^{m-n+1}(a + bu^n)^{p+1}}{b(np + m + 1)} - \frac{a(m - n + 1)}{b(np + m + 1)} \int u^{m-n}(a + bu^n)^p\,du.$

88. $\displaystyle = \frac{u^{m+1}(a + bu^n)^p}{np + m + 1} + \frac{npa}{np + m + 1} \int u^m(a + bu^n)^{p-1}\,du.$

89. $\displaystyle = \frac{u^{m+1}(a + bu^n)^{p+1}}{a(m + 1)}$

$$- \frac{b(np + n + m + 1)}{a(m + 1)} \int u^{m+n}(a + bu^n)^p\,du.$$

90. $\displaystyle = -\frac{u^{m+1}(a + bu^n)^{p+1}}{na(p + 1)}$

$$+ \frac{np + n + m + 1}{na(p + 1)} \int u^m(a + bu^n)^{p+1}\,du.$$

Trigonometric Forms

Hereafter any constant a, b, h, k may have any real value consistent with the domains of the given functions.

91. $\displaystyle \int \sin u\,du = \cos u.$

92. $\displaystyle \int \cos u\,du = -\sin u.$

93. $\displaystyle \int \tan u\,du = -\ln|\cos u| = \ln|\sec u|.$

94. $\displaystyle \int \cot u\,du = \ln|\sin u|.$

95. $\displaystyle \int \sec u\,du = \ln|\sec u + \tan u|.$

96. $\displaystyle \int \csc u\,du = \ln|\csc u - \cot u| = \ln\left|\tan\frac{u}{2}\right|.$

97. $\int \sec^2 u \, du = \tan u.$

98. $\int \csc^2 u \, du = -\cot u.$

99. $\int \sec u \tan u \, du = \sec u.$

100. $\int \csc u \cot u \, du = -\csc u.$

101. $\int \sin^2 au \, du = \dfrac{1}{2} u - \dfrac{1}{4a} \sin 2au.$

102. $\int \cos^2 au \, du = \dfrac{1}{2} u + \dfrac{1}{4a} \sin 2au.$

103. $\int \sin hu \sin ku \, du = -\dfrac{\sin(k+h)u}{2(k+h)} + \dfrac{\sin(k-h)u}{2(k-h)}.$

104. $\int \cos hu \cos ku \, du = \dfrac{\sin(k+h)u}{2(k+h)} + \dfrac{\sin(k-h)u}{2(k-h)}.$

105. $\int \sin hu \cos ku \, du = -\dfrac{\cos(k+h)u}{2(k+h)} - \dfrac{\cos(k-h)u}{2(k-h)}.$

106. $\int \dfrac{du}{a + b \cos u} = \dfrac{2}{\sqrt{a^2 - b^2}} \arctan \dfrac{\sqrt{a^2 - b^2} \tan \frac{1}{2}u}{a + b}$

$$(0 < b, a^2 > b^2).$$

$$= \dfrac{1}{\sqrt{b^2 - a^2}} \ln \left| \dfrac{a + b + \sqrt{b^2 - a^2} \tan \frac{1}{2}u}{a + b - \sqrt{b^2 - a^2} \tan \frac{1}{2}u} \right|$$

$$(0 < a, b^2 > a^2).$$

107. $\int \dfrac{du}{a + b \sin u} = \dfrac{2}{\sqrt{a^2 - b^2}} \arctan \dfrac{a \tan \frac{1}{2}u + b}{\sqrt{a^2 - b^2}}$

$$(0 < a, b^2 > a^2).$$

$$= \dfrac{1}{\sqrt{b^2 - a^2}} \ln \left| \dfrac{a \tan \frac{1}{2}u + b - \sqrt{b^2 - a^2}}{a \tan \frac{1}{2}u + b + \sqrt{b^2 - a^2}} \right|$$

$$(0 < a, b^2 > a^2).$$

108. $\int \dfrac{du}{a \sin u + b \cos u} = \dfrac{1}{\sqrt{a^2 + b^2}} \ln |\csc(u + \alpha) - \cot(u + \alpha)|$

$$\left(\tan \alpha = \dfrac{b}{a} \right)$$

109. $\int \sec^3 u \, du = \frac{1}{2}[\sec u \tan u + \ln|\sec u + \tan u|].$

110. $\displaystyle\int \csc^3 u\, du = -\tfrac{1}{2}[\cot u \csc u - \ln|\csc u - \cot u|].$

111. $\displaystyle\int \arcsin u\, du = u \arcsin u + \sqrt{1 - u^2}.$

112. $\displaystyle\int \arctan u\, du = u \arctan u - \ln\sqrt{1 + u^2}.$

Trigonometric Reduction Formulas

113. $\displaystyle\int \sin^n u\, du = -\frac{\sin^{n-1} u \cos u}{n} + \frac{n-1}{n}\int \sin^{n-2} u\, du.$

114. $\displaystyle\int \cos^n u\, du = \frac{\cos^{n-1} u \sin u}{n} + \frac{n-1}{n}\int \cos^{n-2} u\, du.$

115. $\displaystyle\int \sec^n u\, du = \frac{\tan u \sec^{n-2} u}{n-1} + \frac{n-2}{n-1}\int \sec^{n-2} u\, du.$

116. $\displaystyle\int \csc^n u\, du = -\frac{\cot u \csc^{n-2} u}{n-1} + \frac{n-2}{n-1}\int \csc^{n-2} u\, du.$

117. $\displaystyle\int \tan^n u\, du = \frac{\tan^{n-1} u}{n-1} - \int \tan^{n-2} u\, du.$

118. $\displaystyle\int \cot^n u\, du = -\frac{\cot^{n-1} u}{n-1} - \int \cot^{n-2} u\, du.$

119. $\displaystyle\int \cos^m u \sin^n u\, du \qquad (m \text{ and } n, \text{ positive or negative})$

$$= \frac{\cos^{m-1} u \sin^{n+1} u}{m+n} + \frac{m-1}{m+n}\int \cos^{m-2} u \sin^n u\, du.$$

120. $$= -\frac{\sin^{n-1} u \cos^{m+1} u}{m+n} + \frac{n-1}{m+n}\int \cos^m u \sin^{n-2} u\, du.$$

121. $$= -\frac{\sin^{n+1} u \cos^{m+1} u}{m+1} + \frac{m+n+2}{m+1}\int \cos^{m+2} u \sin^n u\, du.$$

122. $$= \frac{\sin^{n+1} u \cos^{m+1} u}{n+1} + \frac{m+n+2}{n+1}\int \cos^m u \sin^{n+2} u\, du.$$

123. $\displaystyle\int u^n \sin au\, du = -\frac{u^n \cos au}{a} + \frac{n}{a}\int u^{n-1} \cos au\, du.$

124. $\displaystyle\int u^n \cos au\, du = \frac{u^n \sin au}{a} - \frac{n}{a}\int u^{n-1} \sin au\, du.$

Exponential and Logarithmic Forms

125. $\displaystyle\int e^u\, du = e^u.$

126. $\displaystyle\int a^u\, du = \frac{a^u}{\ln a}.$

127. $\displaystyle\int u^n e^{au}\, du = \frac{u^n e^{au}}{a} - \frac{n}{a}\int u^{n-1} e^{au}\, du.$

128. $\displaystyle\int \frac{e^{au}}{u^n}\, du = -\frac{e^{au}}{(n-1)u^{n-1}} + \frac{a}{n-1} \int \frac{e^{au}\, du}{u^{n-1}}.$

129. $\displaystyle\int \ln u\, du = u \ln u - u.$

130. $\displaystyle\int u^n \ln u\, du = \frac{u^{n+1} \ln u}{n+1} - \frac{u^{n+1}}{(n+1)^2}.$

131. $\displaystyle\int u^m \ln^n u\, du = \frac{u^{m+1} \ln^n u}{m+1} - \frac{n}{m+1} \int u^m \ln^{n-1} u\, du.$

132. $\displaystyle\int e^{au} \sin nu\, du = \frac{e^{au}(a \sin nu - n \cos nu)}{a^2 + n^2}.$

133. $\displaystyle\int e^{au} \cos nu\, du = \frac{e^{au}(n \sin nu + a \cos nu)}{a^2 + n^2}.$

Forms Involving Hyperbolic Functions

134. $\displaystyle\int \sinh u\, du = \cosh u.$

135. $\displaystyle\int \cosh u\, du = \sinh u.$

136. $\displaystyle\int \tanh u\, du = \ln \cosh u.$

137. $\displaystyle\int \coth u\, du = \ln |\sinh u|.$

138. $\displaystyle\int \operatorname{sech} u\, du = \arctan (\sinh u).$

139. $\displaystyle\int \operatorname{csch} u\, du = \ln |\tanh \tfrac{1}{2}u|.$

140. $\displaystyle\int \operatorname{sech}^2 u\, du = \tanh u.$

141. $\displaystyle\int \operatorname{csch}^2 u\, du = -\coth u.$

142. $\displaystyle\int \operatorname{sech} u \tanh u = -\operatorname{sech} u.$

143. $\displaystyle\int \operatorname{csch} u \coth u = -\operatorname{csch} u.$

Tables of Trigonometric Functions

The Trigonometric Functions (at Angles, in Radians)

t	sin t	cos t	tan t	cot t	sec t	csc t
.00	.0000	1.0000	.0000	. . . . .	1.000	. . . . .
.01	.0100	1.0000	.0100	99.997	1.000	100.00
.02	.0200	.9998	.0200	49.993	1.000	50.00
.03	.0300	.9996	.0300	33.323	1.000	33.34
.04	.0400	.9992	.0400	24.987	1.001	25.01
.05	.0500	.9988	.0500	19.983	1.001	20.01
.06	.0600	.9982	.0601	16.647	1.002	16.68
.07	.0699	.9976	.0701	14.262	1.002	14.30
.08	.0799	.9968	.0802	12.473	1.003	12.51
.09	.0899	.9960	.0902	11.081	1.004	11.13
.10	.0998	.9950	.1003	9.967	1.005	10.02
.11	.1098	.9940	.1104	9.054	1.006	9.109
.12	.1197	.9928	.1206	8.293	1.007	8.353
.13	.1296	.9916	.1307	7.649	1.009	7.714
.14	.1395	.9902	.1409	7.096	1.010	7.166
.15	.1494	.9888	.1511	6.617	1.011	6.692
.16	.1593	.9872	.1614	6.197	1.013	6.277
.17	.1692	.9856	.1717	5.826	1.015	5.911
.18	.1790	.9838	.1820	5.495	1.016	5.586
.19	.1889	.9820	.1923	5.200	1.018	5.295
.20	.1987	.9801	.2027	4.933	1.020	5.033
.21	.2085	.9780	.2131	4.692	1.022	4.797
.22	.2182	.9759	.2236	4.472	1.025	4.582
.23	.2280	.9737	.2341	4.271	1.027	4.386
.24	.2377	.9713	.2447	4.086	1.030	4.207
.25	.2474	.9689	.2553	3.916	1.032	4.042
.26	.2571	.9664	.2660	3.759	1.035	3.890
.27	.2667	.9638	.2768	3.613	1.038	3.749
.28	.2764	.9611	.2876	3.478	1.041	3.619
.29	.2860	.9582	.2984	3.351	1.044	3.497
.30	.2955	.9553	.3093	3.233	1.047	3.384
.31	.3051	.9523	.3203	3.122	1.050	3.278
.32	.3146	.9492	.3314	3.018	1.053	3.179
.33	.3240	.9460	.3425	2.920	1.057	3.086
.34	.3335	.9428	.3537	2.827	1.061	2.999
.35	.3429	.9394	.3650	2.740	1.065	2.916
.36	.3523	.9359	.3764	2.657	1.068	2.839
.37	.3616	.9323	.3879	2.578	1.073	2.765
.38	.3709	.9287	.3994	2.504	1.077	2.696
.39	.3802	.9249	.4111	2.433	1.081	2.630
.40	.3894	.9211	.4228	2.365	1.086	2.568
.41	.3986	.9171	.4346	2.301	1.090	2.509
.42	.4078	.9131	.4466	2.239	1.095	2.452
.43	.4169	.9090	.4586	2.180	1.100	2.399
.44	.4259	.9048	.4708	2.124	1.105	2.348

The Trigonometric Functions (Cont.)

t	$\sin t$	$\cos t$	$\tan t$	$\cot t$	$\sec t$	$\csc t$
.45	.4350	.9004	.4831	2.070	1.111	2.299
.46	.4439	.8961	.4954	2.018	1.116	2.253
.47	.4529	.8916	.5080	1.969	1.122	2.208
.48	.4618	.8870	.5206	1.921	1.127	2.166
.49	.4706	.8823	.5334	1.875	1.133	2.125
.50	.4794	.8776	.5463	1.830	1.139	2.086
.51	.4882	.8727	.5594	1.788	1.146	2.048
.52	.4969	.8678	.5726	1.747	1.152	2.013
.53	.5055	.8628	.5859	1.707	1.159	1.978
.54	.5141	.8577	.5994	1.668	1.166	1.945
.55	.5227	.8525	.6131	1.631	1.173	1.913
.56	.5312	.8473	.6269	1.595	1.180	1.883
.57	.5396	.8419	.6310	1.560	1.188	1.853
.58	.5480	.8365	.6552	1.526	1.196	1.825
.59	.5564	.8309	.6696	1.494	1.203	1.797
.60	.5646	.8253	.6841	1.462	1.212	1.771
.61	.5729	.8196	.6989	1.431	1.220	1.746
.62	.5810	.8139	.7139	1.401	1.229	1.721
.63	.5891	.8080	.7291	1.372	1.238	1.697
.64	.5972	.8021	.7445	1.343	1.247	1.674
.65	.6052	.7961	.7602	1.315	1.256	1.652
.66	.6131	.7900	.7761	1.288	1.266	1.631
.67	.6210	.7838	.7923	1.262	1.276	1.610
.68	.6288	.7776	.8087	1.237	1.286	1.590
.69	.6365	.7712	.8253	1.212	1.297	1.571
.70	.6442	.7648	.8423	1.187	1.307	1.552
.71	.6518	.7584	.8595	1.163	1.319	1.534
.72	.6594	.7518	.8771	1.140	1.330	1.517
.73	.6669	.7452	.8949	1.117	1.342	1.500
.74	.6743	.7358	.9131	1.095	1.354	1.483
.75	.6816	.7317	.9316	1.073	1.367	1.467
.76	.6889	.7248	.9505	1.052	1.380	1.452
.77	.6961	.7179	.9697	1.031	1.393	1.437
.78	.7033	.7109	.9893	1.011	1.407	1.422
.79	.7104	.7038	1.009	.9908	1.421	1.408
.80	.7174	6967	1.030	.9712	1.435	1.394
.81	.7243	.6895	1.050	.9520	1.450	1.381
.82	.7311	.6822	1.072	.9331	1.466	1.368
.83	.7379	.6749	1.093	.9146	1.482	1.355
.84	.7446	.6675	1.116	.8964	1.498	1.343
.85	.7513	.6600	1.138	.8785	1.515	1.331
.86	.7578	.6524	1.162	.8609	1.533	1.320
.87	.7643	.6448	1.185	.8437	1.551	1.308
.88	.7707	.6372	1.210	.8267	1.569	1.297
.89	.7771	.6294	1.235	.8100	1.589	1.287

The Trigonometric Functions (Cont.)

t	sin t	cos t	tan t	cot t	sec t	csc t
.90	.7833	.6216	1.260	.7936	1.609	1.277
.91	.7895	.6137	1.286	.7774	1.629	1.267
.92	.7956	.6058	1.313	.7615	1.651	1.257
.93	.8016	.5978	1.341	.7458	1.673	1.247
.94	.8076	.5898	1.369	.7303	1.696	1.238
.95	.8134	.5817	1.398	.7151	1.719	1.229
.96	.8192	.5735	1.428	.7001	1.744	1.221
.97	.8249	.5653	1.459	.6853	1.769	1.212
.98	.8305	.5570	1.491	.6707	1.795	1.204
.99	.8360	.5487	1.524	.6563	1.823	1.196
1.00	.8415	.5403	1.557	.6421	1.851	1.188
1.01	.8468	.5319	1.592	.6281	1.880	1.181
1.02	.8521	.5234	1.628	.6142	1.911	1.174
1.03	.8573	.5148	1.665	.6005	1.942	1.166
1.04	.8624	.5062	1.704	.5870	1.975	1.160
1.05	.8674	.4976	1.743	.5736	2.010	1.153
1.06	.8724	.4889	1.784	.5604	2.046	1.146
1.07	.8772	.4801	1.827	.5473	2.083	1.140
1.08	.8820	.4713	1.871	.5344	2.122	1.134
1.09	.8866	.4625	1.917	.5216	2.162	1.128
1.10	.8912	.4536	1.965	.5090	2.205	1.122
1.11	.8957	.4447	2.014	.4964	2.249	1.116
1.12	.9001	.4357	2.066	.4840	2.295	1.111
1.13	.9044	.4267	2.120	.4718	2.344	1.106
1.14	.9086	.4176	2.176	.4596	2.395	1.101
1.15	.9128	.4085	2.234	.4475	2.448	1.096
1.16	.9168	.3993	2.296	.4356	2.504	1.091
1.17	.9208	.3902	2.360	.4237	2.563	1.086
1.18	.9246	.3809	2.427	.4120	2.625	1.082
1.19	.9284	.3717	2.498	.4003	2.691	1.077
1.20	.9320	.3624	2.572	.3888	2.760	1.073
1.21	.9356	.3530	2.650	.3773	2.833	1.069
1.22	.9391	.3436	2.733	.3659	2.910	1.065
1.23	.9425	.3342	2.820	.3546	2.992	1.061
1.24	.9458	.3248	2.912	.3434	3.079	1.057
1.25	.9490	.3153	3.010	.3323	3.171	1.054
1.26	.9521	.3058	3.113	.3212	3.270	1.050
1.27	.9551	.2963	3.224	.3102	3.375	1.047
1.28	.9580	.2867	3.341	.2993	3.488	1.044
1.29	.9608	.2771	3.467	.2884	3.609	1.041
1.30	.9636	.2675	3.602	.2776	3.738	1.038
1.31	.9662	.2579	3.747	.2669	3.878	1.035
1.32	.9687	.2482	3.903	.2562	4.029	1.032
1.33	.9711	.2385	4.072	.2456	4.193	1.030
1.34	.9735	.2288	4.256	.2350	4.372	1.027

The Trigonometric Functions (Cont.)

t	sin t	cos t	tan t	cot t	sec t	csc t
1.35	.9757	.2190	4.455	.2245	4.566	1.025
1.36	.9779	.2092	4.673	.2140	4.779	1.023
1.37	.9799	.1994	4.913	.2035	5.014	1.021
1.38	.9819	.1896	5.177	.1931	5.273	1.018
1.39	.9837	.1798	5.471	.1828	5.561	1.017
1.40	.9854	.1700	5.798	.1725	5.883	1.015
1.41	.9871	.1601	6.165	.1622	6.246	1.013
1.42	.9887	.1502	6.581	.1519	6.657	1.011
1.43	.9901	.1403	7.055	.1417	7.126	1.010
1.44	.9915	.1304	7.602	.1315	7.667	1.009
1.45	.9927	.1205	8.238	.1214	8.299	1.007
1.46	.9939	.1106	8.989	.1113	9.044	1.006
1.47	.9949	.1006	9.887	.1011	9.938	1.005
1.48	.9959	.0907	10.938	.0910	11.029	1.004
1.49	.9967	.0807	12.350	.0810	12.390	1.003
1.50	.9975	.0707	14.101	.0709	14.137	1.003
1.51	.9982	.0608	16.428	.0609	16.458	1.002
1.52	.9987	.0508	19.670	.0508	19.965	1.001
1.53	.9992	.0408	24.498	.0408	24.519	1.001
1.54	.9995	.0308	32.461	.0308	32.476	1.000
1.55	.9998	.0208	48.078	.0208	48.089	1.000
1.56	.9999	.0108	92.620	.0108	92.626	1.000
1.57	1.0000	.0008	1255.8	.0008	1255.8	1.000
1.58	1.0000	−.0092	−108.65	−.0092	−108.65	1.000
1.59	.9998	−.0192	−52.067	−.0192	−52.08	1.000
1.60	.9996	−.0292	−34.233	−.0292	−34.25	1.000

Values of Trigonometric Functions at Angles

θ	$\sin\theta$	$\tan\theta$	$\cot\theta$	$\cos\theta$	
0°	.0000	.0000	.....	1.0000	90°
1°	.0175	.0175	57.290	.9998	89°
2°	.0349	.0349	28.636	.9994	88°
3°	.0523	.0524	19.081	.9986	87°
4°	.0698	.0699	14.301	.9976	86°
5°	.0872	.0875	11.430	.9962	85°
6°	.1045	.1051	9.5144	.9945	84°
7°	.1219	.1228	8.1443	.9925	83°
8°	.1392	.1405	7.1154	.9903	82°
9°	.1564	.1584	6.3138	.9877	81°
10°	.1736	.1763	5.6713	.9848	80°
11°	.1908	.1944	5.1446	.9816	79°
12°	.2079	.2126	4.7046	.9781	78°
13°	.2250	.2309	4.3315	.9744	77°
14°	.2419	.2493	4.0108	.9703	76°
15°	.2588	.2679	3.7321	.9659	75°
16°	.2756	.2867	3.4874	.9613	74°
17°	.2924	.3057	3.2709	.9563	73°
18°	.3090	.3249	3.0777	.9511	72°
19°	.3256	.3443	2.9042	.9455	71°
20°	.3420	.3640	2.7475	.9397	70°
21°	.3584	.3839	2.6051	.9336	69°
22°	.3746	.4040	2.4751	.9272	68°
23°	.3907	.4245	2.3559	.9205	67°
24°	.4067	.4452	2.2460	.9135	66°
25°	.4226	.4663	2.1445	.9063	65°
26°	.4384	.4877	2.0503	.8988	64°
27°	.4540	.5095	1.9626	.8910	63°
28°	.4695	.5317	1.8807	.8829	62°
29°	.4848	.5543	1.8040	.8746	61°
30°	.5000	.5774	1.7321	.8660	60°
31°	.5150	.6009	1.6643	.8572	59°
32°	.5299	.6249	1.6003	.8480	58°
33°	.5446	.6494	1.5399	.8387	57°
34°	.5592	.6745	1.4826	.8290	56°
35°	.5736	.7002	1.4281	.8192	55°
36°	.5878	.7265	1.3764	.8090	54°
37°	.6018	.7536	1.3270	.7986	53°
38°	.6157	.7813	1.2799	.7880	52°
39°	.6293	.8098	1.2349	.7771	51°
40°	.6428	.8391	1.1918	.7660	50°
41°	.6561	.8693	1.1504	.7547	49°
42°	.6691	.9004	1.1106	.7431	48°
43°	.6820	.9325	1.0724	.7314	47°
44°	.6947	.9657	1.0355	.7193	46°
45°	.7071	1.0000	1.0000	.7071	45°
	$\cos\theta$	$\cot\theta$	$\tan\theta$	$\sin\theta$	θ

SOLUTIONS
TO PROBLEMS

Chapter 1

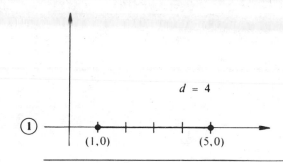

① $d = 4$

$(1,0)$ $(5,0)$

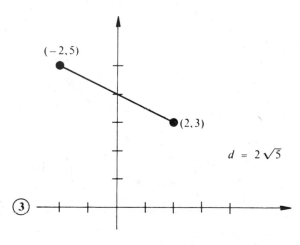

③ $(-2,5)$ $(2,3)$ $d = 2\sqrt{5}$

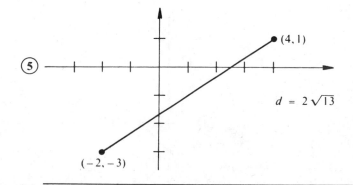

⑤ $(4,1)$ $d = 2\sqrt{13}$ $(-2,-3)$

⑦ a. $\overline{AB} = 5,$ $\overline{BC} = 5\sqrt{2},$ $\overline{CA} = 5$
   b. $\overline{AB} = \sqrt{8},$ $\overline{BC} = \sqrt{8},$ $\overline{CA} = 4$
   The Pythagorean Theorem states that if the sum of the squares
   of the lengths of two sides of a triangle is equal to the square of
   the length of the third side, then the triangle is a right triangle
   (that is, if the sides of a triangle have lengths A, B, and C, and
   $A^2 + B^2 = C^2$, then the triangle is a right triangle).
   Part (a) $5^2 + 5^2 = (5\sqrt{2})^2$
   Part (b) $(\sqrt{8})^2 + (\sqrt{8})^2 = 4^2$; both are right triangles
⑨ $\sqrt{(x-1)^2 + (y-3)^2} = \sqrt{(x+2)^2 + (y-4)^2}$
   reduces to $y = 3x + 5$

SECTION 2

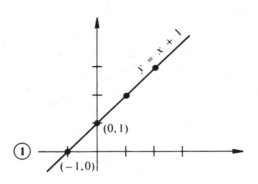

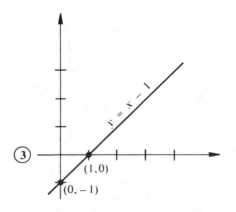

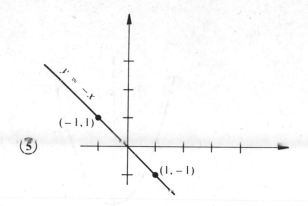

⑤

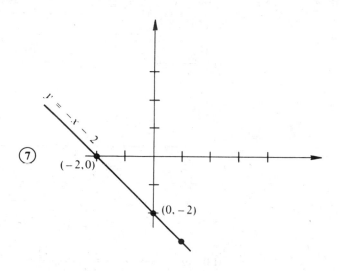

⑦

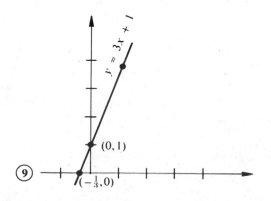

⑨

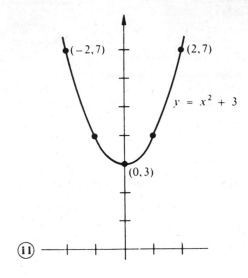

(11)

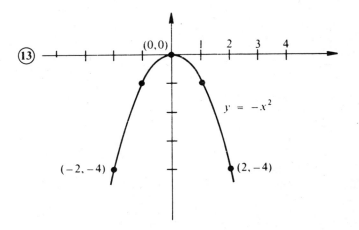

(13)

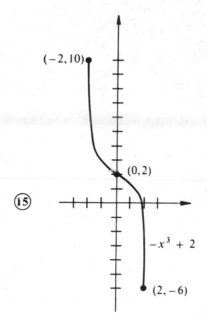

⑮

$-x^3 + 2$

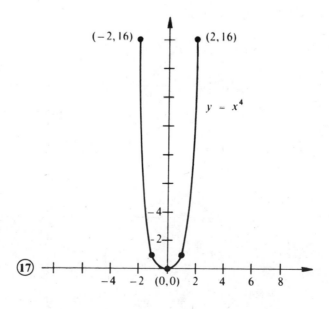

⑰

$y - x^4$

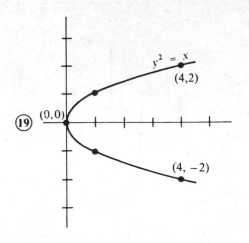

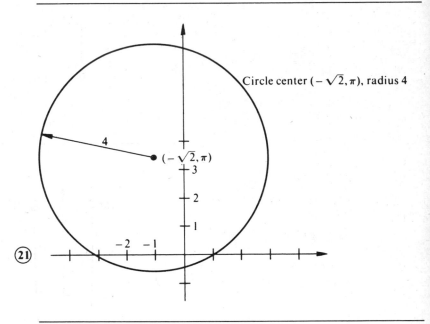

Circle center $(-\sqrt{2}, \pi)$, radius 4

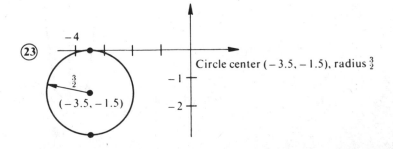

Circle center $(-3.5, -1.5)$, radius $\frac{3}{2}$

25 $x^2 + y^2 - 4x + 4y + 3 = 0$
$(x^2 - 4x + 4) + (y^2 + 4y + 4) = -3 + 4 + 4$
$(x - 2)^2 + (y + 2)^2 = 5$
Circle: center $(2, -2)$,   radius $\sqrt{5}$

27 $x^2 + y^2 - 4x + 6y + 11 = 0$
$(x^2 - 4x + 4) + (y^2 + 6y + 9) = -11 + 4 + 9$
$(x - 2)^2 + (y + 3)^2 = 2$
Circle: center $(2, -3)$,   radius $\sqrt{2}$

SECTION 3

1 $y = -3x + 7$

3 $y = \frac{2}{3}x + \frac{4}{3}$

5 $y = \left(\frac{d - b}{-a}\right)x + d$

7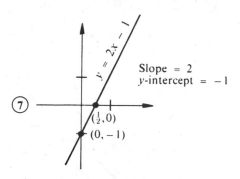

Slope = 2
y-intercept = -1

9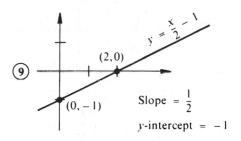

Slope = $\frac{1}{2}$
y-intercept = -1

11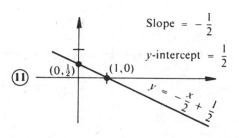

Slope = $-\frac{1}{2}$
y-intercept = $\frac{1}{2}$

(13) $y = -x + 4$, slope $= -1$, $x -$ intercept $= (4, 0)$, $y -$ intercept $=$ $(0, 4)$

(15) $y = \frac{2}{3}x - \frac{1}{3}$, slope $= \frac{2}{3}$, $x -$ intercept $= (\frac{1}{2}, 0)$ $y -$ intercept $=$ $(0, -\frac{1}{3})$

(17) $y = 3x - 1$

SECTION 4

(1) $f(-1) = 2$
$f(0) = 1$
$f(1) = 2$
$f(2) = 17$
$f(\frac{1}{2}) = \frac{17}{16}$
$f(-\frac{1}{2}) = \frac{17}{16}$
$f\left(\frac{1}{a}\right) = \frac{a^4 + 1}{a^4}$

(3) $f(0) = 1$
$f(1) = 2$
$f(-1) = -2$
$f(2) = 7$
$f(-2) = -13$

(5) $-2 \leq x \leq 1$

(7) All x such that $x \neq -1, 1$

(9)
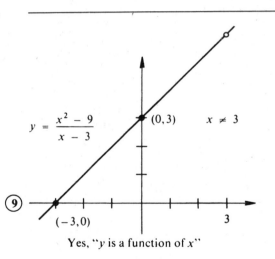
$$y = \frac{x^2 - 9}{x - 3}$$
$(0, 3)$ $x \neq 3$
$(-3, 0)$ 3
Yes, "y is a function of x"

(11)
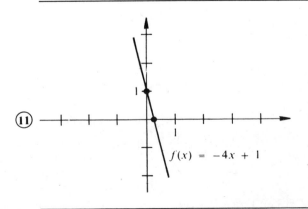
1
1
$f(x) = -4x + 1$

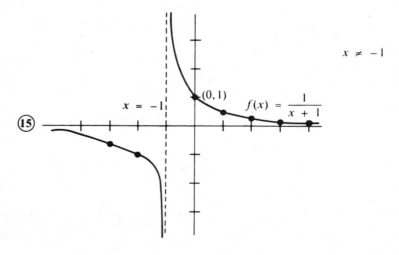

(13)

$(-\sqrt{3},0)$ $(\sqrt{3},0)$ $f(x) = x^2 - 3$

$(0,-3)$

$x \neq -1$

$(0,1)$

$x = -1$ $f(x) = \dfrac{1}{x+1}$

(15)

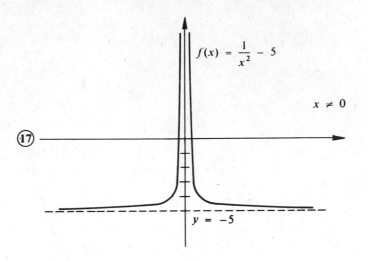

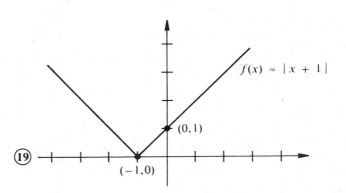

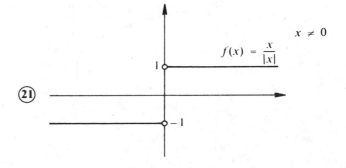

㉓ $f(-3) = -1$
     $f(-2) = -1$
     $f(0) = 1$
     $f(1) = 1$
     $f(2) = 1$

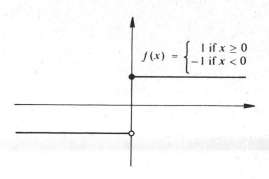

$$f(x) = \begin{cases} 1 \text{ if } x \geq 0 \\ -1 \text{ if } x < 0 \end{cases}$$

㉕ $x^2 + y^2 = 1$
     $y^2 = 1 - x^2$
     $y = \pm\sqrt{1 - x^2}$

㉗ a. $x = 0$

     b. $x = \dfrac{10}{3}$

     c. $x = \dfrac{10n - 10}{5 - n}, \qquad n < 5$

SECTION 5

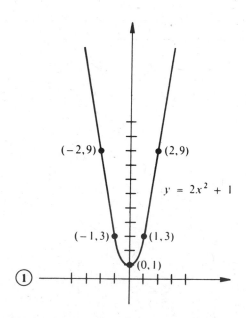

$(-2, 9)$ $(2, 9)$

$y = 2x^2 + 1$

$(-1, 3)$ $(1, 3)$

$(0, 1)$

①

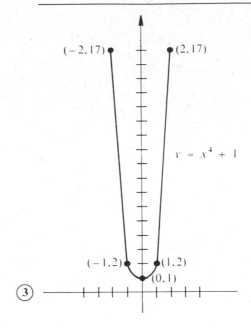

(-2,17) •   • (2,17)

$y = x^4 + 1$

(-1,2) •   • (1,2)
        (0,1)

③

⑤ $(f + g)(x) = \dfrac{2x + 1}{x^2 - 1}$, $x \neq -1, 1$. Therefore, the graph crosses the **x**-axis at $\left(-\tfrac{1}{2}, 0\right)$

⑦ $f(x) = 4x^2 + 1$

⑨ $f(x) = 27x^3$

⑪ $f(x) = (x + 1)^4 = x^4 + 4x^3 + 6x^2 + 4x + 1$

Chapter 2

SECTION 1

① $\dfrac{-2}{3}$ ③ 11   ⑤ $\dfrac{-3}{-4}$ ⑦ 0

⑨ 0

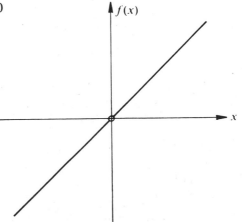

$f(x)$

x

⑪ 0

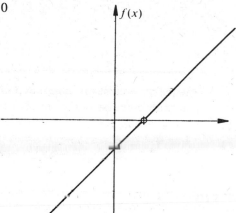

⑬ −1

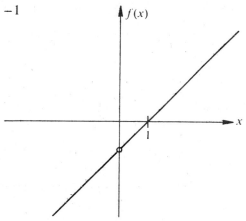

⑮ 0

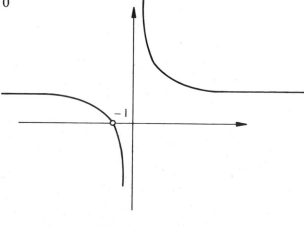

⑰ 2   ⑲ No limit   ㉑ a. 0   ㉓ 10
                              b. −1
                              c. 3
                              d. No limit

㉕ 0   ㉗ No limit   ㉙ 0 if $n = 0$, no limit elsewhere

SECTION 2

① 0   ③ 0   ⑤ No limit   ⑦ $-\frac{2}{3}$

⑨ 4   ⑪ 8   ⑬ No limit

⑮ $1/x^2 - 1/x = (1 - x)/x^2$ by algebraic manipulation. As $x \to 0$,
$1 - x \to 1$, and $x^2 \to 0$; hence, since the denominator tends to 0,
$(1 - x)/x^2$ becomes large. Because $1/x^2$ goes to ∞ "faster than"
$1/x$ does, the difference is not zero.

⑰ $\lim_{x \to -4} \sqrt{x^2} = 4$, $\lim_{x \to 0} \sqrt{x^2} = 0$

SECTION 3

① Vertical: $x = -1, -\frac{1}{2}$
   Horizontal: $y = 0$
③ Vertical: $x = -4, 5$
   Horizontal: $y = 2$
⑤ Vertical: $x = -\frac{1}{2}, 3$
   Horizontal: $y = \frac{1}{2}$
⑦ Vertical: $x = -2, 2$
   No horizontal
⑨ Vertical: $x = 2$
   Horizontal: $y = 0$

⑪ $-\frac{1}{2}$ ⑬ $\frac{1}{2}$
⑮ 1   ⑰ 0
⑲ 0   ㉑ No limit

㉓ $\lim_{n \to \infty} \dfrac{6x^2 + 7x}{5x^2 + 1} = \dfrac{6}{5}$

SECTION 4

① Continuous, continuous, discontinuous
③ Continuous, continuous, continuous
⑤ Continuous
⑦ Continuous

⑨ $G(x) = \begin{cases} \dfrac{x^2 - 9}{x - 3}, & x \neq 3 \\ 6, & x = 3 \end{cases}$

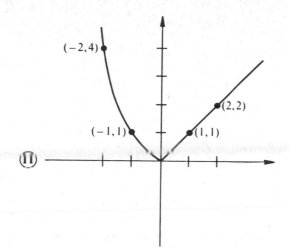

(-2,4)

(2,2)

(-1,1)   (1,1)

(11)

SECTION 5

(1) a. $\bar{v} = 16$ ft/sec
b. $\bar{v} = 32a + 16h$ ft/sec
c. $32a$ ft/sec

(3) a. 12 ft/sec
b. $4t$ ft/sec

(5) a. 12
b. $6t$

(7) a. $\bar{v} = (P(t + h) - P(t))/h$
b. $v = \lim\limits_{h \to 0} (P(t + h) - P(t))/h$
c. $8t + 2$

SECTION 6

(1) $800x^{99}$

(3) $x^{-3/4}$

(5) $-\dfrac{1}{5} x^{-4/5}$

(7) $\dfrac{-5}{x^2}$

(9) $x^{-4/3}$

(11) $s'(t) = \dfrac{3}{2} t^{1/2} = \dfrac{3\sqrt{t}}{2} =$ velocity

$s''(t) = \dfrac{3}{4} t^{-1/2} = \dfrac{3}{4\sqrt{t}} =$ acceleration

(13) Area $= \pi r^2$ Circumference $= 2\pi r$
$A(r) = \pi r^2$
$A'(r) = 2\pi r$
Therefore, the rate of change equals the circumference.

(15) $-1500, -1000, -1000, 0$ gal/min

(17) Average cost $= 20,000$
Marginal cost $= 60,000$

(19) $Dx^{-n} = D\left(\dfrac{1}{x^n}\right) = \lim\limits_{h\to 0} \dfrac{\dfrac{1}{(x+h)^n} - \dfrac{1}{x^n}}{h} =$ (put over a common

denominator) $\lim\limits_{h\to 0} \dfrac{\dfrac{x^n - (x+h)^n}{(x+h)^n(x^n)}}{h} = \lim\limits_{h\to 0} \dfrac{x^n - (x+h)^n}{h} \cdot$

$\dfrac{1}{(x+h)^n(x^n)} =$ (by the definition of the derivative)

$-(Dx^n) \cdot \lim\limits_{h\to 0} \dfrac{1}{(x+h)^n(x^n)} = -(Dx^n) \cdot \dfrac{1}{x^{2n}} =$ (take the

derivative) $\dfrac{-nx^{n-1}}{x^{2n}} = -nx^{-n-1}$

SECTION 7

(1) At $x = 1$, $y = \dfrac{x}{2} + \dfrac{1}{2}$

At $x = 2$, $y = \dfrac{x}{2\sqrt{2}} - \dfrac{1}{\sqrt{2}} + \sqrt{2}$

At $x = 4$, $y = \dfrac{x}{4} + 1$

At $x = 9$, $y = \dfrac{x}{6} + \dfrac{3}{2}$

A tangent line at $(0,0)$ is the **y**-axis.

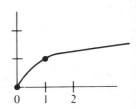

(3) $\left[\left(\dfrac{8}{27}\right)^3, \left(\dfrac{8}{27}\right)^2\right]$

(5) $y = \dfrac{3}{8}x + \dfrac{3}{2}$

(7) $f' = g'$ since the graph of g is simply the graph of f moved up one unit.

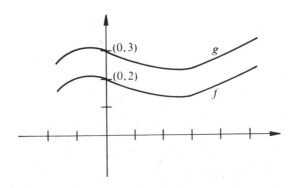

SECTION 8

① $f'(1) = 4$

③ $f'(x) = 10x^9 - 42x^5$
$f''(x) = 90x^8 - 210x^4$

⑤ $f'(x) = 2x + 4$
$f''(x) = 2$

⑦ $f'(x) = 2x + 2$
$f''(x) = 2$
$f'''(x) = 0$

⑨ $f'(x) = \dfrac{5}{2} x^{3/2} + 3x^{1/2} + \dfrac{1}{2} x^{-1/2}$

$= \dfrac{5x \sqrt{x}}{2} + 3\sqrt{x} + \dfrac{1}{2\sqrt{x}}$

⑪ $(2,0)$ $(-1,27)$

⑬ $\left(-2, \dfrac{19}{3}\right)$ $\left(1, \dfrac{-7}{6}\right)$

⑮ $c = \dfrac{3}{4\sqrt[3]{4}}$; observe that x must satisfy the equations $x = x^4 + c$

and $f'(x) = 4x^3 = 1$. (Why?)

⑰ $B'(t) = (4t^{-1/2} + 4 + 6t) 10^5$
$B\left(\tfrac{1}{4}\right) = 13.5 \times 10^5$
$B(2) = (16 + 2\sqrt{2}) 10^5$

⑲ $f'(t) = t^2 - 2t$
$15 = t^2 - 2t, \ t = 5$ (that is, turn off the water after 5 minutes)

SECTION 9

① $f'(x) = -2(2x + 3)^{-2} = \dfrac{-2}{(2x + 3)^2}$

③ $f'(x) = \dfrac{1}{2} (x + 1)^{-1/2} = \dfrac{1}{2\sqrt{x + 1}}$

⑤ $f'(x) = \dfrac{1}{2} (2x)^{-1/2} \cdot 2 = \dfrac{1}{\sqrt{2x}}$

⑦ $f'(x) = \dfrac{-1}{9x^2 + 6x + 1}$ ⑨ $f'(x) = \dfrac{x^2 + 2x}{(x^2 + x + 1)^2}$

⑪ $f'(x) = \dfrac{34x}{(3x^2 + 1)^2}$ ⑬ $f'(x) = -30x^2(-x^3 + 5)^9$

⑮ $f'(x) = 7(4x^6 + 2x^3)^6(24x^5 + 6x^2)$

⑰ $f'(x) = \dfrac{3}{2} x^{1/2} + \dfrac{1}{2} x^{-1/2} = \dfrac{3}{2} \sqrt{x} + \dfrac{1}{2\sqrt{x}}$

⑲ $f'(x) = \dfrac{3(x + 1)^4(x - 1)^2 - 4(x - 1)^3(x + 1)^3}{(x + 1)^8}$

$= \dfrac{3(x + 1)(x - 1)^2 - 4(x - 1)^3}{(x + 1)^5} = \dfrac{(7 - x)(x - 1)^2}{(x + 1)^5}$

㉑ $f'(x) = \dfrac{2x(x^3 - 1)^5}{3(x^2 + 2)^{2/3}} + 15x^2(x^2 + 2)^{1/3}(x^3 - 1)^4$

㉓ $f'(x) = 8(x + 2)(x - 5)^7 + (x - 5)^8 = (x - 5)^7(9x + 11)$

㉕ $f'(x) = x^{1/2}(\pi + 1)(x + 1)^\pi + (x + 1)^{\pi+1}(\frac{1}{2})x^{-1/2}$

$= (\pi + 1)\sqrt{x}(x + 1)^\pi + \dfrac{(x + 1)^{\pi+1}}{2\sqrt{x}}$

㉗ $f'(x) = (x - 1)^3 5(x + 2)^4 + (x + 2)^5 3(x - 1)^2$

$= 5(x - 1)^3(x + 2)^4 + 3(x + 2)^5(x - 1)^2$

㉙ $f'(x) = (x^3 - 1)^5 \frac{1}{3}(x^2 + 2)^{-2/3}(2x) + (x^2 + 2)^{1/3} 5(x^3 - 1)^4 \cdot 3x^2$

$= \frac{2}{3}x(x^3 - 1)^5(x^2 + 2)^{-2/3} + 15x^2(x^2 + 2)^{1/3}(x^3 - 1)^4$

㉛ $f'(x) = \dfrac{1}{2}(x + x^{1/2})^{-1/2}\left(1 + \dfrac{1}{2}x^{-1/2}\right) = \dfrac{1 + \dfrac{1}{2\sqrt{x}}}{2\sqrt{x + \sqrt{x}}}$

㉝ $f'(x) = 3(x^{1/2} + x)^2\left(\dfrac{1}{2}x^{-1/2} + 1\right) = 3(\sqrt{x} + x)^2\left(\dfrac{1}{2\sqrt{x}} + 1\right)$

㉟ Total revenue $= x \cdot p(x)$

$x \cdot \dfrac{8}{2 + x} = \dfrac{8x}{2 + x}$

Marginal revenue $= xp'(x) + p(x)$

$= 8x(-1)(2 + x)^2 + (2 + x)^{-1}8$

$= \dfrac{-8x}{(2 + x)^2} + \dfrac{8}{2 + x} = \dfrac{16}{(2 + x)^2}$

㊲ $N'(t) = 6 - (t^2 + 1)^{-2}(2t) = 6 - \dfrac{2t}{(t^2 + 1)^2}$

SECTION 10

① a. $y = \pm\sqrt{\dfrac{x}{1 + x}}$. The derivative of the positive function is

$\dfrac{1}{2(x)^{1/2}(1 + x)^{3/2}}$; the derivative of the negative function is

$-\dfrac{1}{2(x)^{1/2}(1 + x)^{3/2}}$.

b. $F'(x) = \dfrac{1 - (F(x))^2}{2F(x)(1 + x)}$

$F(1) = \pm\sqrt{\frac{1}{2}}$, depending upon whether the positive or negative function is used. Therefore, $F'(1) = \pm\dfrac{\sqrt{2}}{8}$.

$$y = +\sqrt{\frac{x}{1+x}} \qquad y = -\sqrt{\frac{x}{1+x}}$$

③ $y' = \dfrac{5x^4}{2y}$ 　　　　　⑤ $y' = \dfrac{1}{10y^4\sqrt{x}}$

⑦ $y' = \dfrac{-\sqrt{y}}{\sqrt{x}}$ 　　　　⑨ $y' = \dfrac{-3x^2}{2y}$

⑪ $y' = \dfrac{-2x + y}{2y - x}$ 　　　⑬ $y' = \dfrac{xy^2(2y)}{(x + 2y^2)(x^2)} = \dfrac{2y^3}{x(x + 2y^2)}$

⑮ $y' = \dfrac{-y^2(2x)}{x^2(2y) - 1}$ 　　$y' = \tfrac{2}{3}$ at $(1, -1)$ 　 $y' = -\tfrac{8}{3}$ at $(1, 2)$

⑰ $y' = \left(-y^{1/2} + \dfrac{y}{2x^{1/2}}\right)\left(\dfrac{2y^{1/2}}{x - 2y^{1/2}x^{1/2}}\right)$

　　$y' = 1$ at the point $(1, 1)$

⑲ $\dfrac{-8}{3}$ ft/sec 　　　　⑳ $\tfrac{4}{3}$ ft/sec

㉓ 50 mph 　　　　　　　㉕ -40 ft/sec

㉗ $x = 3$ ft/sec

SECTION 11

① $f(x) = 4x^2 + 10x + 3$ 　　③ $f(x) = x^4 + 4x^2 + 2$
　　$f'(x) = 8x + 10$ 　　　　　　$f'(x) = 4x^3 + 8x$

⑤ $f(x) = (4x^2 + x - 1)^{25}$
　　$f'(x) = 25(4x^2 + x - 1)^{24}(8x + 1)$

⑦ $f(x) = \sqrt{x^2 + 1}$
　　$f'(x) = \dfrac{x}{\sqrt{x^2 + 1}}$

⑨ $f(x) = \left(\dfrac{x}{x + 1}\right)^{30}$
　　$f'(x) = \dfrac{30}{(x + 1)^2}\left(\dfrac{x}{x + 1}\right)^{29}$, 　or 　$\dfrac{30x^{29}}{(x + 1)^{31}}$

⑪ $x = \dfrac{-12}{25}$

⑬ a. $f(x^2) = \dfrac{x^{12}}{6}$ b. $f(x^2 + 1) = \dfrac{(x^2 + 1)^6}{6}$

$f'(x^2) = 2x^{11}$ $f'(x^2 + 1) = (x^2 + 1)^5(2x)$

c. $f'(x^2 - 5x + 1) = (x^2 - 5x + 1)^5(2x - 5)$

Chapter 3

SECTION 1

① **S.I.** for $x > 0$, **S.D.** for $x < 0$; horizontal tangent at $x = 0$

③ **S.I.** for all x; horizontal tangent at $x = 0$

⑤ **S.I.** for $x < -1$, $x > \frac{1}{3}$, **S.D.** for $-1 < x < \frac{1}{3}$; horizontal tangents at $x = -1$, $x = \frac{1}{3}$

⑦ **S.I.** for $x < -2$, $x > 1$, **S.D.** for $-2 < x < 1$; horizontal tangents at $x = -2$, $x = 1$

⑨ **S.I.** for $-3 < x < 0$, $x > 2$, **S.D.** for $x < -3$, $0 < x < 2$; horizontal tangents at $x = -3$, $x = 0$, $x = 2$

⑪ **S.I.** for $-\frac{1}{2} < x < 0$, $x > 2$, **S.D.** for $x < -\frac{1}{2}$, $0 < x < 2$; horizontal tangents at $x = -\frac{1}{2}$, $x = 0$, $x = 2$

⑬ **S.I.** for all x; horizontal tangents at $x = 0$, $x = 1$

⑮ **S.I.** for $x > 0$; horizontal tangent at $x = 1$

⑰ **S.I.** for $x > 2$, **S.D.** for $x < -2$; no horizontal tangents

⑲ **S.I.** for $-2 < x < 0$, **S.D.** for $x < -2$, $x > 0$; horizontal tangents at $x = -2$, $x = 0$, $x = 2$

㉑ **S.I.** for $p < \frac{1}{2}$, $p > 2$, **S.D.** for $\frac{1}{2} < p < 2$

SECTION 2

① Max: $f(0) = f(3) = 0$
Min: $f(-1) = f(2) = -4$

③ Max: $f(3) = 17$
Min: $f(0) = 2$

⑤ Max: $f(2) = 2^{1/3}$
Min: $f(-1) = -1$

⑦ Max: $f(1) = 1$,
Min: $f(\frac{1}{2}) = \frac{1}{8}$

⑨ Max: $f(0) = 0$
Min: $f(2) = -4$

⑪ Max: $f(1) = 11$
Min: $f(0) = 7$

⑬ Max: $f(2) = 23$
Min: $f(-2) = -21$

⑮ *Hint:* Consider $(f(x))^3$. Max: $f(1) = f(4) = 4^{1/3}$
Min: $f(-1) = -2\,(2^{1/3})$

⑰ Max: $f(-3) = 313$
Min: $f(2) = 88$

⑲ Max: $f(1) = 48$
Min: $f(-1) = -28$

SECTION 3

① Max: $f(0) = 0$
Min: $f(2) = -4$

③ Max: $f(1) = 5$
Min: $f(\frac{4}{3}) = \frac{134}{27}$

⑤ There are no maxima or minima because $f'(x) \neq 0$ for any x.

⑦ $(0,0)$ is an inflection point and there are no maxima or minima.

⑨ Max: $f(0) = 0$
Min: $f(2) = -4$

⑪ $(0,7)$ is an inflection point; $f'(x) \neq 0$; therefore, there are no maxima or minima.

⑬ Max: $f(-1) = 5$ ⑰ No maxima
   Min: $f(1) = -3$ Min: $f(2) = 88$
⑲ Max: $f(-2) = -6$, $f(1) = 48$
   Min: $f(-1) = -28$, $f(2) = 26$

SECTION 4

① Remember: total revenue $= x \cdot p(x)$, marginal revenue $= xp'(x) + p(x)$; total revenue max: $\frac{5}{4}$, min: 0,5; marginal revenue max: 0, min: $\frac{5}{2}$

③ Average cost $= s(x)/x$, the minimum average cost is (1/4) and it occurs when $x = 5/2$; the minimum average cost if x is an integer is 0 and it occurs when $x = 2$ or $x = 3$.

⑤ Maximum total revenue is 2 and it occurs when $x = 2$. Maximum since second derivative of total revenue function is negative when $x = 2$.

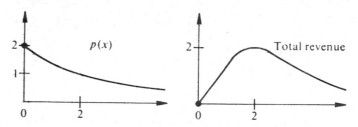

⑦ Let $p =$ price per inexpensive doll.

  a. $R(x) = 2yp + xp = \dfrac{20p(5 - x)}{10 - x} + xp$

  b. To maximize R, x must be 20 and y must be 15.

⑨ a. $P(x) = \left(20x - \dfrac{x^2}{20}\right) - \left(5x + \dfrac{x^2}{50}\right)$

  b. $x = 107$ (approx)
  c. $P = 14.7$ (approx) and max: $P = 804$ (approx)

⑪ $x = 20$, $y = 20$

⑬ $v = x(10 - 2x)(12 - 2x)$ where $x =$ length of a side of the square; $x = (11 - \sqrt{31})/3$.

⑮ Area $= 2x\sqrt{R^2 - x^2}$ Max area $= R^2$

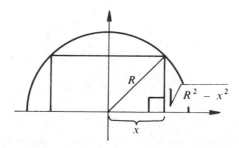

⑰ The line connecting R and Q has the equation: $y = -(R/Q)x + R$, the area of the triangle is $\frac{1}{2}(R)\,Q$; Q is 4 miles from P, R is 2 miles from P.

⑲ Find $F'(p)$, set it equal to zero and solve. The maximum likelihood estimate of the probability is the ratio of the occurrences of A to the number of observations.

㉑ Maximum work per second during the first 5 seconds is 136.5, which occurs at $n = \frac{1}{2}$.

SUPPLEMENTARY 1

① $N = \dfrac{-p}{\mathbf{x}(p)} \cdot \dfrac{d\mathbf{x}(p)}{dp} = \dfrac{-p}{3p^{-4}} \dfrac{d}{dp}(3p^{-4}) = -\dfrac{p}{3}\,p^{4} \cdot (-12)\,p^{-5} = 4.$

In general, $N = \dfrac{p}{ap^{-m}} \cdot -\mathrm{amp}^{-m-1} = m$

③ $N = \dfrac{-p(x)}{x} \cdot \dfrac{1}{p'(x)}$

$N = \dfrac{1 + x^{2}}{2x^{2}}$

⑤ $Q = \dfrac{-a}{\mathbf{x}(a)} \cdot \dfrac{\mathbf{x}(b) - \mathbf{x}(a)}{b - a}$

$Q = 2\left[1 - \left(\tfrac{27}{28}\right)^{3}\right]$

$N = \tfrac{3}{14}$

⑦ $N = 6$; therefore, every point is both a max and a min.

⑨ $N = 7$; therefore, he should lower the price.

$N = \frac{1}{2}$; therefore, raise the price. He should increase the price when $m < 1$, and he should decrease the price if $m > 1$.

SUPPLEMENTARY 2

① We must first solve the inequality $G'(x) < 0$; this gives us

$$\frac{y}{2} - \frac{sa}{x^{2}} < 0$$

keeping in mind that u and s are positive numbers. After some algebra, we obtain

$$x^{2} < \frac{2sa}{u}, \qquad \text{with } x \geq 0$$

thus, $x < \sqrt{\dfrac{2sa}{u}}$; the result for $G'(x) > 0$ can be obtained by similar computations.

Chapter 4

SECTION 1

(1) $x^2 + c$

(3) $\dfrac{t^4}{4} - t + c$

(5) $\dfrac{(x + 1)^3}{3} + c$ or $\dfrac{x^3}{3} + x^2 + x + k$, the two integrals differ by a constant

(7) $\dfrac{x^5}{5} - \dfrac{x^3}{3} + c$

(9) $\dfrac{5x^{6/5}}{6} + \dfrac{2x^{3/2}}{3} + c$

(11) $x - \dfrac{1}{x} + c$

(13) $\displaystyle\int a(t)\, dt = v(t) = \dfrac{t^2}{2} + \dfrac{3}{2}$

$\displaystyle\int v(t)\, dt = s(t) = \dfrac{t^3}{6} + \dfrac{3t}{2} - \dfrac{1}{3}$

(15) $\displaystyle\int a(t)\, dt = v(t) = t^3 - t^2 + 5$

$\displaystyle\int v(t)\, dt = s(t) = \dfrac{t^4}{4} - \dfrac{t^3}{3} + 5t - \dfrac{4}{3}$

(17) $f(x) = \dfrac{x^6}{6} - \dfrac{x^{1+\sqrt{2}}}{1 + \sqrt{2}} - 2x^{-1/2} + \dfrac{5x^{-24}}{24} + c$

$\quad = \dfrac{x^6}{6} - \dfrac{(x^{1+\sqrt{2}})(1 - \sqrt{2})}{3} - \dfrac{2\sqrt{x}}{x} + \dfrac{5}{24x^{24}} + c$

(19) $f''(x) = x^3 + x$

$f'(x) = \dfrac{x^4}{4} + \dfrac{x^2}{2} + \dfrac{1}{4}$

$f(x) = \dfrac{x^5}{20} + \dfrac{x^3}{6} + \dfrac{x}{4} + 2$

(21) $s(6) = 360$, $v(0) = 0$, $v(t) = at$. Therefore, $\int v(t)\, dt = s(t) = at^2/2$. When $t = 6$, $a = 20$; therefore, the constant acceleration is $A(t) = 20$ (feet per second per second).

(23) Total revenue $= (d^3/3) + d^2 + d$

SECTION 2

(1) $\frac{3}{2}$

(3) $\frac{11}{2} = 5\frac{1}{2}$

(5) 14

(7) $\dfrac{5x^2}{2}$

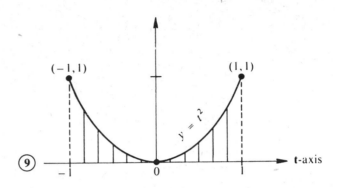

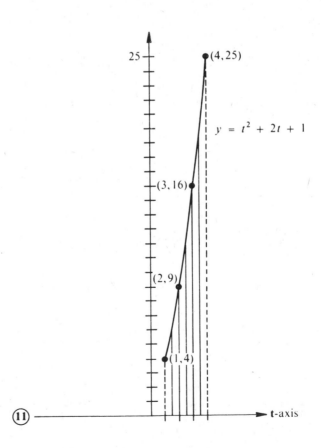

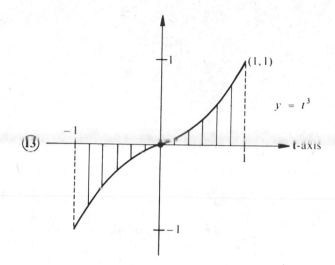

$y = t^3$

$(1,1)$

t-axis

⑬

SECTION 3

① a. $x^{15} - x^3 + x - 1$
   b. $\sqrt[3]{x^2 + x + 1}$
   c. $(x^2 + 1)^3 \sqrt{x^4 + 1}$

③ 140

⑤ $\frac{129}{88}$

⑦ $\frac{8}{3}$

⑨ $\frac{26}{27}$

⑪ $\frac{31}{30}$

⑬ $\frac{52}{5}$

⑮ $\frac{1}{6}$

⑰ 0

⑲ $\frac{5}{9}$

㉑ $\left(\dfrac{Ab^3}{3} + \dfrac{Bb^2}{2} + Cb \right) - \left(\dfrac{Aa^3}{3} + \dfrac{Ba^2}{2} + Ca \right)$

   $= \dfrac{A}{3}(b^3 - a^3) + \dfrac{B}{2}(b^2 - a^2) + C(b - a)$

㉓ $2\sqrt{2} - \frac{29}{24}$

㉕ 5

㉗ $\frac{2}{3}$

㉙ $\frac{3}{2} \cdot 2^{4/3}$, or equivalently, $3\sqrt[3]{2}$

SECTION 4

① $\frac{880}{21}$

③ Odd; therefore, 0

⑤ $\displaystyle\int_{-1}^{1} x^{17}\,dx - \int_{-1}^{1} x^2\,dx = 0 - (2) \left.\dfrac{x^3}{3}\right|_0^1 = -(2)\left[\dfrac{1}{3}\right] = \dfrac{2}{3}$

⑦ Odd; therefore, 0

⑨ Odd; therefore, 0

⑪ $\frac{32}{3}$

Chapter 5

SECTION 1

(1) $\frac{1}{2}, \frac{3}{4}, \frac{7}{8}, \frac{15}{16}, \frac{31}{32}, \ldots$

(3) $1, \frac{9}{2}, \frac{1}{3}, \frac{33}{4}, \frac{1}{5}, \ldots$

(5) $-1, \frac{1}{2}, -\frac{1}{3}, \frac{1}{4}, -\frac{1}{5}, \ldots$

(7) $0, 2, 0, 2, 0, \ldots$

(9) $2, 1, \frac{1}{2}, \frac{1}{4}, \frac{1}{8}, \ldots$

(11) $0, \frac{1}{3}, \frac{1}{2}, \frac{3}{5}, \frac{2}{3}, \ldots$

(13) $\frac{1}{2}$ (15) $-\frac{1}{2}$

(17) 0 (19) $\frac{5}{4}$

(21) $1, \frac{1}{2}, \frac{1}{4}, \frac{1}{8}, \frac{1}{16}$

(23) $-2, 1, -\frac{1}{2}, \frac{1}{4}, -\frac{1}{8}$

(25) 4 (27) $\frac{3}{2}$

(29)
$$\lim_{n \to \infty} n(\sqrt{n^2 + 1} - n) = \lim_{n \to \infty} \frac{n(\sqrt{n^2 + 1} - n)(\sqrt{n^2 + 1} + n)}{\sqrt{n^2 + 1} + n}$$

$$= \lim_{n \to \infty} \frac{n}{\sqrt{n^2 + 1} + n}$$

$$= \lim_{n \to \infty} \frac{1}{\sqrt{1 + (1/n^2)} + 1} = \frac{1}{2}$$

(31) To show: $\lim_{n \to \infty} \left(\frac{1}{a}\right) \cdot \frac{[a(n + b)]}{(n + c)} = 1$

$$\left| \lim_{n \to \infty} \frac{\cancel{a}}{\cancel{a}} \frac{(n + b)}{(n + c)} = 1 \text{ because } b \text{ and } c \text{ are constants} \right|$$

SECTION 2

(1) 10

(3) $\frac{1}{4} + \frac{1}{6} + \frac{1}{8} + \frac{1}{10} + \frac{1}{12} + \frac{1}{14} + \frac{1}{16}$

(5) $\frac{2}{3} + \frac{3}{4} + \frac{4}{5} + \frac{5}{6} = \frac{61}{20}$

(7) $1 + 2 + 4 + 8 + 16 + 32 + 64 + 128 + 256 + 512 = 1{,}023$

(9) 1 (11) 0

(13) a. -194

b. 314

(15) $\frac{1}{4}\left[1 - \left(\frac{1}{5}\right)^{12}\right]$ (17) $\frac{1}{4}\left[1 + \left(\frac{1}{3}\right)^{15}\right]$

(19) $\dfrac{8 - 8^{31}}{1 - 8^5}$ (21) $4 - \dfrac{1}{4^8}$

SECTION 3

(1) $\frac{3}{2}$ (3) $\frac{10}{3}$

(5) $\frac{1}{8}$ (7) $\frac{30}{65}$

(9) 2 (11) $\frac{17}{2}$

(13) $-\frac{6}{17}$

⑮ $r = a/5$. If $|a| \geq 5$, then the series has no sum (diverges). If $|a| < 5$, then $|r| < 1$ and the series $a/5 + (a/5)^2 + \cdots$ has a sum (converges).

⑰ $\frac{41}{333}$ ⑲ $\frac{1}{999}$

㉑ $\frac{101}{999}$

SECTION 4

① $f'(x) = -3e^{-3x}$ ③ $2 \exp(2x + 3)$

⑤ $f'(x) = (2x + 1) \exp(x^2 + x)$

⑦ $f'(x) = \frac{1}{2}(x - 1)^{-1/2} \exp(\sqrt{x - 1})$

⑨ $f'(x) = [\sqrt{x + 1} + \frac{1}{2}x(x + 1)^{-1/2}] \exp[x\sqrt{x + 1}]$

⑪ $f'(x) = 2x^{-3} \exp\left(-\frac{1}{x^2}\right)$

⑬ $f'(x) = 20x(x^2 + 1)^9 \exp(x^2 + 1)^{10}$

⑮ a. ae^{ax}

   b. $(1/a)e^{ax} + c$

   c. $-e^{-1} + \frac{2}{3}e^3 + \frac{1}{3}$

⑰ $\dfrac{dy}{dx} = \dfrac{2 - e^y - ye^x}{xe^y + e^x}$

⑲ 0 ㉑ -1

㉓ e^2 ㉕ Max total revenue $= 27/e$

SECTION 5

① $f'(x) = \dfrac{1}{2x + 2}$ ③ $f'(x) = \dfrac{2 \ln x}{x}$

⑤ $f'(x) = 3x^2 \ln 4x + x^2 = x^2(1 + 3 \ln 4x)$

⑦ $f'(x) = \dfrac{1}{x(1 + 3x)}$ ⑨ $f'(x) = \dfrac{3x^2 - 6x + 2}{x^3 - 3x^2 + 2x + 5}$

⑪ $f'(x) = \dfrac{6x + 1}{x(x + 1)}$ ⑬ $f'(x) = \dfrac{3x^2 + 6x + 2}{2(x)(x + 1)(x + 2)}$

⑮ $f'(x) = \dfrac{1}{x \ln x}$

⑰ $f'(x) = \dfrac{1}{7}\left[\left(\dfrac{2x}{x^2 + 1}\right) + \left(\dfrac{1}{(x - 3)}\right) - \left(\dfrac{4x^3}{x^4 + 2}\right)\right]$

⑲ $y' = (x^{x^2 + x - 2})\left[(2x + 1) \ln x + \left(\dfrac{x^2 + x - 2}{x}\right)\right]$

㉑ $y' = (\ln x^{\ln x})\left(\dfrac{\ln \ln x}{x} + \dfrac{1}{x}\right)$

㉓ $y' = \left(\dfrac{1 + \ln x}{2\sqrt{x}}\right)\sqrt{x}^{\sqrt{x}}$

㉕ $y' = \left(\sqrt[10]{\dfrac{x^8 + 3}{x^{10} - 5}}\right)\dfrac{(x^{10} - 5)4x^7 - [(x^8 + 3)5x^9]}{5(x^{10} - 5)(x^8 + 3)}$

㉗ $f(x)$ is **S.I.** for $0 < x < e$; $f(x)$ is max at $x = e$; $f(x)$ is **S.D.** for $x > e$.

㉙ Limit = 0

㉛ $\displaystyle\sum_{k=2}^{99} \ln\left(\frac{k+1}{k}\right) = \ln 100 - \ln 2 = \ln 50$

$\displaystyle\sum_{k=2}^{\infty} \ln\left(\frac{k+1}{k}\right) = \infty$

㉝ $f'(x) = \left(\dfrac{(\ln x)^x}{(x^{\ln x})}\right)\left[\ln \ln x + \dfrac{1}{\ln x} - \dfrac{2\ln x}{x}\right]$

SECTION 6

① $y(t) = 2e^t$ ③ $y(t) = 5e^{-t}$

⑤ $y_{2000} = e^{(-20/17)\ln 2}$ grams $= \left(\frac{1}{2}\right)^{20/17}$ grams

⑦ Time $= 10\,\dfrac{\ln 6}{\ln 4}$; population $= 16,000$ after 20 minutes; change $=$ $100 \ln 4 e^{(\ln 4/10)t}$ When $t = 20$, the change $= 1,600 \ln 4$ objects/minute.

⑨ Time $= \dfrac{20 \ln 10}{\ln 2}$,   (1,000 obj)

Time 2 $= \dfrac{20 \ln 100}{\ln 2}$,   (100 obj);

Decline $= -500 \ln 2 e^{-\ln 100}$

⑪ a. $y(t) = y_0 e^{-(\ln 2/5,550)t}$

b. $t = \frac{5550}{0.693}\ln \frac{25}{3} \sim 20,000$

Chapter 6

SECTION 1

① $\frac{2}{3}\sqrt{x^3 + 2} + c$

③ $\frac{3}{4}(x^2 - 6x + 15)^{2/3} + c$ ⑤ $\frac{1}{12}$

⑦ $\frac{1}{2}$ ⑨ $\frac{1}{2}\ln|2x + 3| + c$

⑪ $-\frac{1}{2}\ln|1 - x^2| + c$ ⑬ 1

⑮ $\frac{1}{2}$

⑰ $\frac{1}{4}[\ln(1 + x^2)]^2 + c$ ⑲ $\frac{1}{5}\ln(1 + e^{5x}) + c$

㉑ $x + 1 - \ln|x + 1| + c$ ㉓ $\frac{2}{5}(1 + x)^{5/2} - \frac{2}{3}(1 + x)^{3/2} + c$

㉕ $-e^{-x}(x + 1) + c$ ㉗ $\dfrac{x^{11}}{11}\left(\ln x - \dfrac{1}{11}\right) + c$

㉙ $\frac{2}{3}x^{3/2}\left(\ln x - \frac{2}{3}\right) + c$ ㉛ $e^x[x^2 - 2x + 2] + c$

㉝ $f(x) = P(x)$ $f'(x) = P'(x)$
$g'(x) = e^x$ $g(x) = e^x$

$\displaystyle\int e^x P(x)\,dx = e^x P(x) - \int e^x P'(x)\,dx$

Next:
$$f(x) = P'(x) \qquad f'(x) = P''(x)$$
$$g'(x) = e^x \qquad g(x) = e^x$$
Thus,

$$\int e^x P(x)\,dx = e^x P(x) - e^x P'(x) - \int e^x P''(x)\,dx \quad \text{(etc.)}$$

(35) Notice $\int \dfrac{1}{x}\,dx = 1 + \int \dfrac{1}{x}\,dx + c$

$1 + c = 0$; there are c for which $1 + c = 0$

For definite integrals, note that what we have is

$$\int_a^b \frac{1}{x}\,dx - 1\Big|_a^b + \int_a^b \frac{1}{x}\,dx$$

which simply asserts that $1\Big|_a^b = 0$.

SECTION 2

(1) $\dfrac{1}{2\sqrt{5}} \ln \left| \dfrac{x - \sqrt{5}}{x + \sqrt{5}} \right| + c$

(3) $-\dfrac{1}{(x + 1)} + \ln \left| \dfrac{x + 2}{x + 1} \right| + c$

(5) $\dfrac{\sqrt{3}}{4}$

(7) $\frac{9}{20} \ln |x + 10| + \frac{11}{20} \ln |x - 10| + c$

(9) $-\dfrac{1}{x^2} + \dfrac{2}{x} + 3 \ln |x| + \dfrac{1}{(x + 1)} - 3 \ln (x + 1) + c$

(11) $\dfrac{1}{(1 + x)} + \ln |1 + x| + c$

SECTION 3

(1) 0.771 (3) -0.16 (5) 11.209

SECTION 4

(1) $A = \dfrac{9}{2}$ (3) $A = \dfrac{1}{6}$

(5) $A = \dfrac{1}{4}$ (7) $162 - \dfrac{2 \cdot 3^5}{5}$

(9) $A = 16$ (11) $A = \dfrac{1}{6}$

SECTION 5

(1) $v = \dfrac{1}{3} \pi R^2 H$

(3) a. $\dfrac{\pi}{2}$

b. $\left(\dfrac{49}{30}\right)\pi$

c. $\left(\dfrac{\pi}{3}\right)\left(2\sqrt{2}-1\right)$

d. $\dfrac{\pi}{2}$

e. $\pi(e-2)$

f. $\dfrac{\pi}{6\pi+3}\,[6^{2\pi+1}-5^{2\pi+1}]$

(5) a. $\dfrac{2\pi}{35}$ about **x**-axis; $\dfrac{\pi}{10}$ about **y**-axis

b. $\dfrac{72\pi\sqrt{3}}{5}$

c. $\dfrac{512\pi}{15}$

d. 25π

e. $\dfrac{2\pi}{5}$ about **x**-axis; $\dfrac{5\pi}{14}$ about **y**-axis

f. $\dfrac{144\pi}{35}$ about **x**-axis; $\dfrac{8\pi}{5}$ about **y**-axis

---

SECTION 6

(1) $\dfrac{5}{4}$

(3) 0

(5) Doesn't exist

(7) $\dfrac{3}{2}\,4^{2/3}$

(9) $-\dfrac{1}{4}$

(11) 1

(13) Doesn't exist

(15) $\dfrac{1}{(1-p)}$ for $p<1$

SECTION 7

(1) $\dfrac{1}{2}$

(3) 1

(5) -1

(7) Does not exist

(9) 0

(11) Does not exist

(13) Does not exist

(15) $\dfrac{1}{2}$

(17) $\dfrac{1}{-p+1}$

(19) ∞ ; $\dfrac{1}{\ln 2}$; for $p > 1$, $\dfrac{1}{(p-1)(\ln 2)^{p-1}}$

(21) 1

SECTION 8

(1) $y = ce^{(1/2)x^2}$

(3) $y^{3/2} = x^{3/2} + 8$

(5) $y = \frac{1}{2}x^2 + \frac{1}{2}$

(7) $y = ce^{(2/3)\sqrt{x^3+1}}$

(9) $y = (2e^x(x-1) + 2x + k)^{1/2}$ (11) $5x$

(13) $\frac{9}{5}(y+2)^{5/2} - \frac{4}{3}(y+2)^{3/2} + c = \ln x$

(15) $\varepsilon = \dfrac{-ce^{-bt} + av}{b}$

(17) $\dfrac{y}{1-y} = e^{ct+d}$

Since $y(0) = \dfrac{1}{e+1}$, we obtain $d = -1$ so that $y/(1-y) = c^{ct-1}$,

or $y = e^{ct-1} - ye^{ct-1}$. Solving for y (and after some simplification), we obtain $y = e^{ct}/(e + e^{ct})$.

Supplementary Readings

SECTION SR-1

(1) $k + (10{,}000 - k)e^{-12}$

(3) $(9999/e^4) + 1$

SECTION SR-2

(1) $\left(\frac{9}{10}\right) - (\ln 10)\frac{1}{10}$

(3) $\dfrac{1}{\ln 2} - \dfrac{1}{\ln 5} + \ln(\ln 5) - \ln(\ln 2) = u(x)$

Chapter 7

SECTION 1

(1) $\sqrt{x+1} \approx 1 + \frac{1}{2}x - \frac{1}{8}x^2 + \frac{1}{16}x^3$
$\sqrt{1.2} \approx 1.095$

(3) $\ln(x+1) \approx x - \frac{1}{2}x^2 + \frac{1}{3}x^3 - \frac{1}{4}x^4 + \frac{1}{5}x^5$
$\ln(1.1) \approx 0.095$

(5) $x \ln x + 1 \approx x^2 - \frac{1}{2}x^3$

(7) $e^x \approx 1 + x + \dfrac{1}{2!}x^2 + \dfrac{1}{3!}x^3$

$= 1 + x + \dfrac{1}{2}x^2 + \dfrac{1}{6}x^3$

a. $\left(x = \dfrac{1}{2}\right)$ $\sqrt{e} \approx 1 + \dfrac{1}{2} + \dfrac{1}{8} + \dfrac{1}{48} \approx 63$

b. $\left(x = \dfrac{1}{3}\right)$ $\sqrt[3]{e} \approx 1 + \dfrac{1}{3} + \dfrac{1}{18} + \dfrac{1}{27 \cdot 3!} \approx 1.39$

SECTION 2

(1) $1 - x + \dfrac{x^2}{2} - \dfrac{x^3}{3!} + \dfrac{x^4}{4!} - \dfrac{x^5}{5!}$

(3) $P_6(x) = 0 + 0 + 0 + 0 + 0 + \dfrac{120x^5}{5!} + 0$ (with $c = 0$); with $c = 1$,

$$P_6(x) = 1 + 5(x - 1) + \frac{20(x - 1)^2}{2!} + \frac{60(x - 1)^3}{3!}$$

$$+ \frac{120(x - 1)^4}{4!} + \frac{120(x - 1)^5}{5!} + 0$$

$$= 1 + 5(x - 1) + 10(x - 1)^2 + 10(x - 1)^3$$

$$+ 5(x - 1)^4 + (x - 1)^5$$

(5) $1 + (x - 1) \cdot \dfrac{1}{2} - \dfrac{(x - 1)^2}{2!} \dfrac{1}{4} + \dfrac{(x - 1)^3}{3!} \dfrac{3}{8} - \dfrac{(x - 1)^4}{4!} \dfrac{15}{16}$

(7) $1 + \frac{1}{2}x + \frac{3}{8}x^2 + \frac{5}{16}x^3 + \frac{105}{384}x^4$

(9) $e^{-1} + e^{-1}(x + 1) + \dfrac{e^{-1}(x + 1)^2}{2!} + \dfrac{e^{-1}(x + 1)^3}{3!} + \dfrac{e^{-1}(x + 1)^4}{4!}$

(11) $\ln 5 + \dfrac{x}{5} - \dfrac{x^2}{5^2 2} + \dfrac{x^3}{5^3 3} - \dfrac{x^4}{5^4 4} + \dfrac{x^5}{5^5 5} - \dfrac{x^6}{5^6 6}$

(13) $e + \frac{1}{2}e(x - 1)$

SECTION 3

(1) $\dfrac{1}{1 + x^2} = 1 - x^2 + x^4 - x^6 \cdots$ (3) $\displaystyle\int_0^1 xe^x \, dx = 1$

(5) $D\left(\dfrac{1}{1 - x}\right) = D(1 + x + x^2 + x^3 + \cdots)$

$$= D1 + Dx + Dx^2 + Dx^3 + \cdots$$

Thus, $\dfrac{1}{(1 - x)^2} = 1 + 2x + 3x^2 + \cdots$ (for $|x| < 1$)

(7) Differentiate term by term

$$D \sin x = D\left(x - \frac{x^3}{3!} + \frac{x^5}{5!} - \frac{x^7}{7!} + \cdots\right)$$

$$= 1 - \frac{x^2}{2!} + \frac{x^4}{4!} - \frac{x^6}{6!} + \cdots = \cos x$$

and

$$D \cos x = D\left(1 - \frac{x^2}{2!} + \frac{x^4}{4!} - \frac{x^6}{6!} + \frac{x^8}{8!} - \cdots\right)$$

$$= -x + \frac{x^3}{3!} - \frac{x^5}{5!} + \frac{x^7}{7!} - \cdots = -\sin x$$

(9) a. $e^{-x^2} = 1 - x^2 + \dfrac{x^4}{2!} - \dfrac{x^6}{3!} \cdots$

b. $\displaystyle\int_0^x e^{-t^2}\,dt = x - \frac{x^3}{3} + \frac{x^5}{5\cdot 2!} - \frac{x^7}{3!\cdot 7}\cdots$

c. $1 - \frac{1}{3} + \frac{1}{10} - \frac{1}{42} \cong 0.74277$

Chapter 8

SECTION 1

(1) $f(0,0) = 0$ $f(-1,0) = 0$ $f(0,-1) = 0$
$f(1,1) = 2$ $f(2,4) = 48$ $f(t,t) = 2t^3$
        $f(1-t,t) = t - t^2$ $f(t,t^2) = t^4 + t^5$

(3) $f(0,0,0) = 0$ $f(1,-1,1) = 1 - e$ $f(-1,1,-1) = e + 1$

$\dfrac{df}{dx}(x,x,x) = 2x^2 e^{x^2} + x^4 e^{x^2} + 2x$

$\dfrac{df}{dy}(1,y,1) = e + 2y$ $\dfrac{df}{dz}(1,1,z^7) = e^{z^2} - 4 + 2z^7$

(5) a. $\sqrt{3}$ b. $\sqrt{10}$

(7)

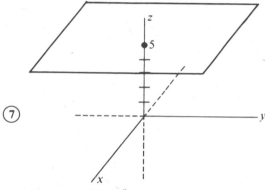

Plane $\parallel$ to xy, at $z = 5$

(9)

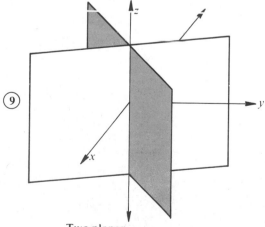

Two planes
in z direction: intersection
at **z** axis

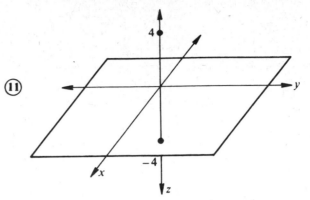

⑪

Plane ∥ to xy plane, at $z = -4$

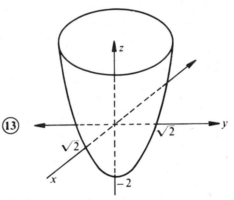

⑬

Paraboloid of revolution about z-axis
Vertex: $z = -2$

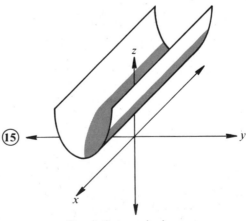

⑮

Parabola trough along
x direction

Hyperbolic paraboloid

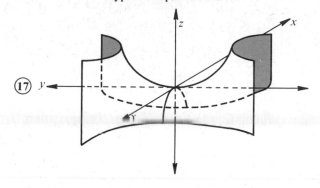

⑰

⑲

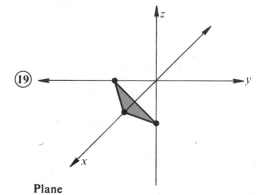

Plane

$y + z = 1$

⑳

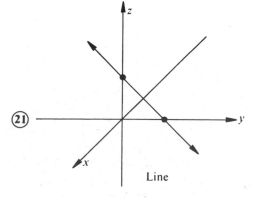

Line

㉓ $(x + 1)^2 + (y - 2)^2 + (z - 3)^2 = 4$
Sphere: radius $= 2$,
Center $= (-1,2,3)$

㉕ $x^2 + 4y^2 + 16z^2 = 16$

$$\frac{x^2}{4^2} + \frac{y}{2^2} + z^2 = 1$$

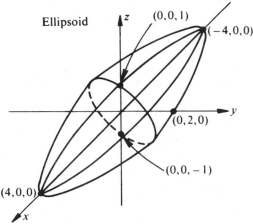

Ellipsoid

$(0,0,1)$

$(-4,0,0)$

$(0,2,0)$

$(0,0,-1)$

$(4,0,0)$

㉗ $y^2 - 4x^2 = 4z$
Hyperbolic paraboloid

SECTION 2

① a. 5
   b. 5
   c. 5
   d. 5

③ -1

⑤ 50

⑦ $2a$

⑨ $3x^2$

⑪ Let $x = y^2 \to 0$,   then $\dfrac{x^2}{x^2 + x^2} \to \dfrac{1}{2}$

Let $y = 0$, $x \to 0$: lim $= 0$
Let $x = 0$, $y \to 0$: lim $= 0$
Limit does not exist.

⑬ No limit: Note that $\lim_{(x,y)\to(0,0)} f(x,y) = \frac{3}{4}$ along the path given by $y = x$ (that is, $\lim_{(x,x)\to(0,0)} f(x,x) = \frac{3}{4}$) whereas limit $= 0$ along path given by $x = 0$.

SECTION 3

(1) $f_x = y^2 + yz$
$f_y = 2xy + z^3 + xz$
$f_z = 3yz^2 + xy$

(3) $f_x = \dfrac{1}{z}$

$f_y = \dfrac{2y}{z}$

$f_z = \dfrac{-(x + y^2)}{z^2}$

(5) $f_x = \dfrac{1}{x + y^2 + z^3}$

$f_y = \dfrac{2y}{x + y^2 + z^3}$

$f_z = \dfrac{3z^2}{x + y^2 + z^3}$

(7) $f_x = 2xy^3$
$f_{xx} = 2y^3$
$f_y = 3y^2x^2$
$f_{yy} = 6yx^2$
$f_{xy} = 6xy^2$

(9) $f_x = \dfrac{1}{2} y(xy)^{-1/2}$

$f_y = \dfrac{1}{2} x(xy)^{-1/2}$

$f_{xx} = -\dfrac{1}{4} y^2(xy)^{-3/2}$

$f_{xy} = \dfrac{1}{4} \dfrac{1}{\sqrt{xy}}$

$f_{yy} = -\dfrac{1}{4}x^{1/2}(y)^{-3/2}$

(11) $f_x = 2e^{2x-3y}$
$f_y = -3e^{2x-3y}$
$f_{xx} = 4e^{2x-3y}$
$f_{xy} = -6e^{2x-3y}$
$f_{yy} = 9e^{2x-3y}$

(13) $f_x = \dfrac{2}{2x + 3y}$

$f_y = \dfrac{3}{2x + 3y}$

$$f_{xx} = \frac{-4}{(2x + 3y)^2}$$

$$f_{yy} = \frac{-9}{(2x + 3y)^2}$$

$$f_{xy} = \frac{6}{(2x + 3y)^2}$$

⑮ $f_x = \dfrac{1}{2x}$

$$f_y = \frac{1}{2y}$$

$$f_{xx} = -\frac{1}{2x^2}$$

$$f_{yy} = -\frac{1}{2y^2}$$

$$f_{xy} = 0$$

⑰ $f_x = e^{xy} + xye^{xy}$
$f_{xx} = 2ye^{xy} + xy^2e^{xy}$
$f_y = x^2e^{xy} - 2y$
$f_{yy} = x^3e^{xy} - 2$
$f_{xy} = 2xe^{xy} + x^2ye^{xy}$

⑲ $f_x = \dfrac{1}{y} - \dfrac{y}{x^2} = y^{-1} - y(x^{-2})$

$f_{xx} = 2yx^{-3}$
$f_y = -xy^{-2} + x^{-1}$
$f_{yy} = 2xy^{-3}$
$f_{xy} = -y^{-2} - x^{-2}$

㉑ $\dfrac{\partial z}{\partial x} = \dfrac{-x}{z}$ $\dfrac{\partial z}{\partial y} = \dfrac{-y}{z}$

㉓ $\dfrac{\partial z}{\partial x} = \dfrac{-3}{4}\dfrac{x}{z}$ $\dfrac{\partial z}{\partial y} = -3\dfrac{y}{z}$

㉕ $\dfrac{\partial z}{\partial x} = \dfrac{z^2 - 6xy}{y^3 - 2zx}$ $\dfrac{\partial z}{\partial y} = \dfrac{-3(x^2 + y^2z)}{y^3 - 2zx}$

㉗ $\dfrac{\partial z}{\partial x} = -\sqrt{\dfrac{z}{x}}$ $\dfrac{\partial z}{\partial y} = -\sqrt{\dfrac{z}{y}}$

㉙ $\dfrac{\partial z}{\partial x} = \dfrac{2 - (e^{x^2y^3z})(y^3z2x)}{x^2y^3e^{x^2y^3z}}$ $\dfrac{\partial z}{\partial y} = \dfrac{3 - 3y^2x^2ze^{x^2y^3z}}{x^2y^3e^{x^2y^3z}}$

㉛ $\dfrac{\partial z}{\partial x} = \dfrac{-x + (1 + x^2 + y^2 + z^2)}{z}$

$$\frac{\partial z}{\partial y} = \frac{-2y + (1 + x^2 + y^2 + z^2)}{2z}$$

(33) $\frac{1}{4}x + \frac{1}{4}y - z - \frac{5}{8} = 0$

(35) a. $f_x = 6xy$ $f_y = 3x^2 - 3y^2$

         $f_{xx} = 6y$ $f_{yy} = -6y$

         $6x - 6y = 0$ for all x, y

     b. $f_x = (x^2 + y^2)^{-1}2x$

$$f_{xx} = -(x^2 + y^2)^{-2}(2x)^2 + (x^2 + y^2)^{-1}2 = \frac{2y^2 - 2x^2}{(x^2 + y^2)^2}$$

$$f_y = (x^2 + y^2)^{-1}2y$$

$$f_{yy} = -(x^2 + y^2)^{-2}(2y)^2 + (x^2 + y^2)^{-1}2 = \frac{2x^2 - 2y^2}{(x^2 + y)^2} = -f_{xx}$$

Thus, $f_{xx} + f_{yy} = 0$

SECTION 4

(1) a. $G(t) = t^6 - 6t^5 + 2t^5 + 2t^3 - 6t^2 + 1$;

        $G'(t) = 6t^5 - 30t^4 + 6t^2 - 12t$

    b. $G'(t) = (2v)(-6t) + (2u + 2v)(3t^2)$

            $= 6t^5 - 30t^4 + 6t^2 - 12t$

(3) $G(t) = \sqrt{10t^2 + 4t + 3}$

    $G'(t) = \frac{1}{2}(10t^2 + 4t + 3)^{-1/2}(20t + 4)$

(5) $g(x, y) = 2x^2 - 2xy$, $\dfrac{\partial g}{\partial x} = 4x - 2y$, $\dfrac{\partial g}{\partial y} = -2x$

(7) $f(x, y) = 2x^2 + 8y^2$, $\dfrac{\partial f}{\partial x} = 4x$, $\dfrac{\partial f}{\partial y} = 16y$

(9) $\dfrac{\partial f}{\partial u}\dfrac{\partial u}{\partial x} = \dfrac{\partial f}{\partial x} = z_x = \dfrac{\partial f}{\partial u}y^2$

    $\dfrac{\partial f}{\partial y} = \dfrac{\partial f}{\partial u}\dfrac{\partial u}{\partial y} = z_y = \dfrac{\partial f}{\partial u}2xy$

    $2xy^2\dfrac{\partial f}{\partial u} - 2xy^2\dfrac{\partial f}{\partial u} = 0$

(11) $z(x, y) = f(u(x, y), v(x, y))$

    $z_x = \dfrac{\partial f}{\partial u} - \dfrac{\partial f}{\partial v}$

    $z_y = -\dfrac{\partial f}{\partial u} + \dfrac{\partial f}{\partial v}$

    therefore, $z_x + z_y = 0$

SECTION 5

(1) Min at $(2, 4)$

(3) Max at $(\sqrt{3}, 0)$, no extrema at $(-\sqrt{3}, 0)$

(5) No extremum

⑦ Max at $(3,3)$

⑨ No extrema

⑪ $(1,1)$ is min

⑬ Let x, y, and z be the length, width, and height of the box, respectively, and let a^3 be its volume. We obtain a minimum when $x = y$ and $z = a^3/x^2$. Solving for x and y, we obtain $x = a \cdot \sqrt[3]{2} = y$ and $z = a/\sqrt[3]{4}$. Notice that the height z is equal to $\frac{1}{2} \cdot x$ (or, $\frac{1}{2} \cdot y$).

⑮ $\$4,333.33$, $\$46,666.66$, $\$20,333.33$
      x y E

(Note that maximum $E(x, y) = E(\frac{13}{3}, \frac{14}{3}) = \frac{61}{3}$)

SECTION 6

① $x = \dfrac{5}{7} = z$ $y = \dfrac{-5}{7}$

③ We obtain $x = y = z$ and $\lambda = -x^4$. Thus, $x^2 = R^2/3$ and finally $f(x, y, z) = (R^2/3)^3$.

⑤ $x = 4\sqrt{2} = y$ $z = 2\sqrt{2}$ (x, y, z are length, width, and height, respectively)

⑦ $x = \dfrac{2}{21}$ $y = -\dfrac{4}{21}$ $z = \dfrac{8}{21}$ minimum $= \dfrac{2}{\sqrt{21}}$

⑨ The equations we must solve are

$$x\left(1 + \frac{\lambda}{a^2}\right) = 0$$

$$y\left(1 + \frac{\lambda}{b^2}\right) = 0$$

$$z\left(1 + \frac{\lambda}{c^2}\right) = 0$$

$$\frac{x^2}{a^2} + \frac{y^2}{b^2} + \frac{z^2}{c^2} = 1$$

Notice that the possible solutions are $x = 0$, $y = 0$, $\lambda = -c^2$; or $x = 0$, $z = 0$, $\lambda = -b^2$; or $y = 0$, $z = 0$, $\lambda = -a^2$. Substituting $x = 0$, $y = 0$, and then $x = 0$, $z = 0$, and then $y = 0$, $z = 0$ into the equation of the ellipsoid, we see that the maximum distance is a, and the minimum distance is c.

---

⑪ $x = \pm 1 = y$ $x = \pm\frac{1}{2} = -y$

Max distance from origin $= \sqrt{2}$; min distance $= \sqrt{\frac{1}{2}}$

(*Hint:* Find the extrema for the square of the distance. The equations you obtain are

$$x + 5x\lambda + \lambda - 3y\lambda = 0$$
$$y - 3x\lambda + 5y\lambda = 0$$

Subtracting, we obtain

$$(x - y)(1 + 8\lambda) = 0$$

Then, $x = y$ or $\lambda = -\frac{1}{8}$.)

(13) $x = 25, \quad y = 50$

SECTION 7

(1) a. 11
    b. 3

(3) a.

    b. 8

(5) 44

(7) 12

SECTION 8

(1) $\frac{1}{66}$

(3) 0

(5) $\frac{31}{15}$

(7) a^2b^2

(9) $e^2 - 3$

(11) $\frac{1}{9} - \frac{e^6}{9}$

SECTION 9

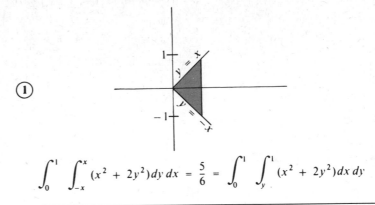

① $$\int_0^1 \int_{-x}^x (x^2 + 2y^2)dy\, dx = \frac{5}{6} = \int_0^1 \int_y^1 (x^2 + 2y^2)dx\, dy$$

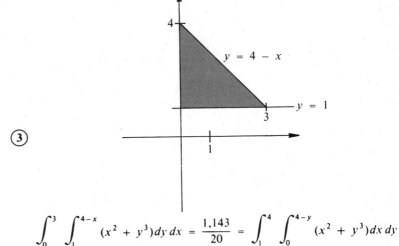

③ $$\int_0^3 \int_1^{4-x} (x^2 + y^3)dy\, dx = \frac{1{,}143}{20} = \int_1^4 \int_0^{4-y} (x^2 + y^3)dx\, dy$$

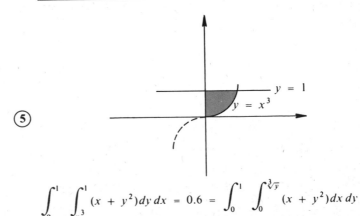

⑤ $$\int_0^1 \int_{x^3}^1 (x + y^2)dy\, dx = 0.6 = \int_0^1 \int_0^{\sqrt[3]{y}} (x + y^2)dx\, dy$$

---

⑦

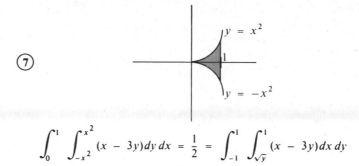

$$\int_0^1 \int_{-x^2}^{x^2} (x - 3y)dy\,dx = \frac{1}{2} = \int_{-1}^1 \int_{\sqrt{y}}^1 (x - 3y)dx\,dy$$

⑨

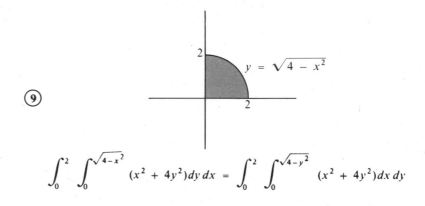

$$\int_0^2 \int_0^{\sqrt{4-x^2}} (x^2 + 4y^2)dy\,dx = \int_0^2 \int_0^{\sqrt{4-y^2}} (x^2 + 4y^2)dx\,dy$$

⑪

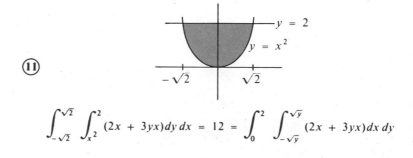

$$\int_{-\sqrt{2}}^{\sqrt{2}} \int_{x^2}^2 (2x + 3yx)dy\,dx = 12 = \int_0^2 \int_{-\sqrt{y}}^{\sqrt{y}} (2x + 3yx)dx\,dy$$

Chapter 9

SECTION 1

(1) $-\pi$

(3) $\dfrac{\pi}{150} \approx 0.021$

(5) $\left(\dfrac{216}{\pi}\right)^{\circ} \approx 68.79^{\circ}$

(7) $\dfrac{\sqrt{3}}{3}$

(9) $\dfrac{\sqrt{2}}{2}$

(11) $\dfrac{\sqrt{3}}{2}$

(13) $\dfrac{-\sqrt{3}}{2}$

(15) -1

(17) 1

(19) $\sqrt{2}$

(21) 2

(23) $\dfrac{2\sqrt{3}}{3}$

(25) 0.2245

(27) 0.1564

(29) 0.8391

(31) **a.** $2 \sin x \cos x$
   b. $\cos^2 x - \sin^2 x = 1 - 2\sin^2 x = 2\cos^2 x - 1$

(33) $-\cos(\pi - x) = -\cos\pi\cos x + \sin\pi\sin x = \cos x$

SECTION 9-2

(1) $\dfrac{\sqrt{2}}{2}$

(3) $\dfrac{\sqrt{2}}{2}$

(5) $0, \dfrac{\pi}{2}, \pi, \dfrac{3\pi}{2}, 2\pi$

(7) $f(x) = -\sin x$

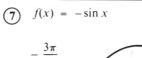

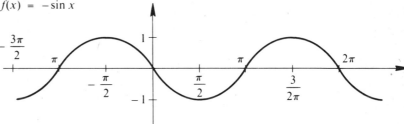

$f(x) = 2 + \sin x$

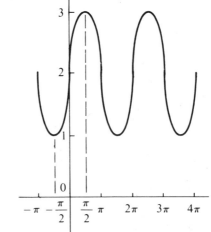

⑪

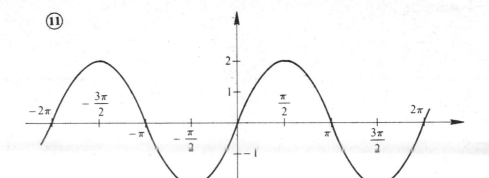

⑬

$f(x) = |\sin x|$

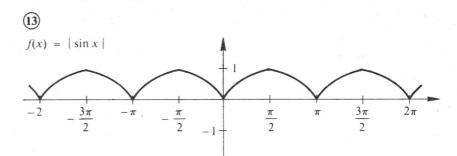

SECTION 9-3

① $\frac{3}{4}$ ③ $\frac{3}{4}$

⑤ 0 ⑦ $\frac{1}{3}$

⑨ 0 ⑪ 1

⑬ 0

SECTION 9-4

① $21 \cos 7x$ ③ $-5 \sin (5x + 2)$

⑤ $8x \sin (4x^2 - 1)$ ⑦ $5x^2 \cos 5x + 2x \sin 5x$

⑨ $2x \sec (x^2) \cdot \tan(x^2)$ ⑪ $\dfrac{5 \cos 5x}{\sin 5x}$

⑬ $2 \tan 2x$ ⑮ $\dfrac{3}{2} \tan^2 \left(\dfrac{x}{2}\right) \sec^2 \left(\dfrac{x}{2}\right)$

(17) $\dfrac{-2 + \cos x}{(1 - 2 \cos x)^2}$

(19) $(\sin^2 x - \cos^2 x) \exp(-\sin x \cdot \cos x)$

(21) $\dfrac{-\cos x}{\sin y}$

(23) $\dfrac{\sin 2y + 2y \sin 2x}{\cos 2x - 2x \cos 2y}$

(25) $(\cos x)^{x^2} (-x^2 \cot x + 2x \ln \cos x)$

(27) $(\sin x)^{\tan x} (\sec^2 x \cdot \ln \sin x + 1)$

(29) (no solution required)

(31) **a.** $2 \cos (2x - 3y)$ **b.** $-3 \cos (2x - 3y)$
      c. $-4 \sin (2x - 3y)$ **d.** $-9 \sin (2x - 3y)$
      e. $6 \sin (2x - 3y)$

SECTION 9-5

(1) $-2 \cos x - 3 \sin x + c$ (3) $\frac{1}{3} \ln |\sec 3x| + c$

(5) $\frac{1}{9} \sin^9 x + c$ (7) $\frac{3}{4} (1 + \cos x)^{4/3} + c$

(9) $2\sqrt{\sin x} + c$ (11) $\sin (\ln x) + c$

(13) $\sin x - x \cos x + c$

(15) $2x \sin x - (x^2 - 2) \cos x + c$

(17) $\frac{1}{2}(x - \sin x \cos x) + c$

(19) $-\cos x + \frac{2}{3}\cos^3 x - \frac{1}{5}\cos^5 x + c$

(21) $\frac{1}{2}\cos x \sin x + \frac{1}{2}x + c$

(23) $\frac{1}{4}\cos^3 x \sin x + \frac{3}{8}\cos x \sin x + \frac{3}{8}x + c$

(25) $\dfrac{1}{2} \ln \left| \dfrac{\sqrt{5} \tan (x/2) + 1}{\sqrt{5} \tan (x/2) + 5} \right|$

(27) 1

SECTION 9-6

(1) $y = -\dfrac{\pi}{2}x + \dfrac{\pi^2}{4}$ (3) (no solution required)

(5) $f\left(\dfrac{\pi}{4}\right) = \sqrt{2}$: max   (7) (no solution required)

$f\left(\dfrac{5\pi}{4}\right) = -\sqrt{2}$: min

(9) $2\pi^2$

(11) a. velocity $= -6\sin 2t$
   b. acceleration $= -12\cos 2t$
   c. at rest: $t = 0, \pi/2, \pi, 3\pi/2, 2\pi$
   d. The object starts (at time 0) 3 units to the right of the reference point, which we shall call equilibrium. From $t = 0$ to $t = \pi/2$, the object moves to the left, passes equilibrium, and ends up 3 units to the left of equilibrium, then stops. The motion then goes to the right, past equilibrium, then 3 units to the right, back to the left and 3 units to the left; this happens from $t = \pi/2$ to $t = 3\pi/2$. From $t = 3\pi/2$ to $t = 2\pi$, the object moves back to the right, passing equilibrium, and ending up where it began at $t = 0$, that is, 3 units to the right of equilibrium.

(13) $\sin x = x - \dfrac{x^3}{3!} + \dfrac{x^5}{5!} - \dfrac{x^7}{7!} + \cdots \pm \dfrac{x^{2n+1}}{(2n+1)!} + R_{2n+1}(x)$

where

$$|R_{2n+1}(x)| = \frac{|\sin w|}{(2n+2)!} x^{2n+2} \qquad w \text{ between 0 and } x$$

Since $|\sin w| \leq 1$ for any w,

$$|R_{2n+1}(x)| \leq \frac{x^{2n+2}}{(2n+2)!} \to 0 \text{ as } n \to \infty \qquad \text{for any } x$$

Hence, for *any x*,

$$\sin x = x - \frac{x^3}{3!} + \frac{x^5}{5!} - \frac{x^7}{7!} + \frac{x^9}{9!} - \cdots$$

Appendix, Part 1

SECTION A-1

(1) {1,2,3}, {1,2,4}, {1,2,5}, {2,3,4}, {3,4,5}, {2,3,5}, {1,4,5}, {1,3,5}, {2,4,5}, {1,3,4}

(3) 35

(5) 252

(7) 2^n

(9) 252

SECTION A-2

Distributive law

(1)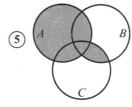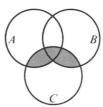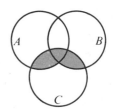

$A \cup B$ $C \cap (A \cup B)$ $(C \cap A) \cup (C \cap B)$

(3) No additional comment (it's done in the book)

(5) True

$$A \cup (B \cap C) = (A \cup B) \cap (A \cup C)$$

(5)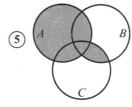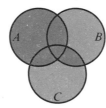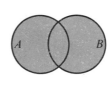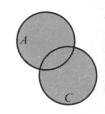

$A \cup (B \cap C)$ $(A \cup B) \cap (A \cup C)$ $A \cup B$ $A \cup C$

Appendix, Part 2

SECTION A-1

① False   ③ False

⑤ False   ⑦ False

⑨ False   ⑪ False

⑬ Yes; yes; yes. Any addition, subtraction, multiplication, or division involving (with nonzero denominator) two rational numbers results in a rational number.

⑮ $x = 2$ ⑰ $x = 0$

⑲ $x = 7$ ㉑ $x \leq 3$

㉓ $x > -1$ ㉕ $x \leq -4$

㉗ $-5 < x < 8$

SECTION A-2

① 22   ③ 6

⑤ $x = -5, 15$ ⑦ $x = -2, 6$

⑨ $x = -2, 3$

⑪ a. When $5 - 2x \geq 0$, that is, $x \leq \frac{5}{2}$

     b. When $5 - 2x \leq 0$, that is, $x \geq \frac{5}{2}$

SECTION A-3

① $2x^2 - 14x$ ③ $x^2 + 6x + 8$

⑤ $x^2 - 9x + 20$ ⑦ $4x^2 - 28x + 49$

⑨ $-6x^2 + 17x - 12$ ⑪ $x^2 - 3$

⑬ $4x^2 - 36$ ⑮ $x^3 + 1$

⑰ $4x^2 + 12xy + 9y^2$ ⑲ $16x^2 - 40xy + 25y^2$

㉑ $9x^2 - 25y^2$ ㉓ $15x^2 + 13xy + 2y^2$

㉕ $6x^2 - 13xy + 6y^2$ ㉗ $x = -\frac{7}{2}, 0$

㉙ $x = 0, \frac{5}{7}$ ㉛ $x = 0, \pm 3$

㉝ $x = -3, 4$ ㉟ $x = -3, -1$

㊲ $x = -\frac{3}{2}, 1$ ㊴ $x = -\frac{3}{2}, -\frac{1}{2}$

SECTION A-4

① $\frac{5}{6}$ ③ $\frac{32}{27}$

⑤ $\frac{8}{5}$ ⑦ Answer appears in text.

⑨ $\dfrac{2a^2 + 7c^2}{abc}$ ⑪ 1

⑬ $\dfrac{2a^2 + ac + c^2}{a^2 - c^2}$ ⑮ $\dfrac{-3a^3 - 12a^2 + a - 4}{a^2 - 16}$

⑰ $\dfrac{a^2}{a^2 + 2a - 3}$ ⑲ $\dfrac{3a^2 + 5a + 7}{a^2 + 3a - 4}$

(21) $\dfrac{-a - 3b}{2a + b}$

(23) $\dfrac{x^3 - 2x^2}{x + 2}$,   or $\dfrac{x(x - 2)}{x + 2}$

(25) $\dfrac{x - 3a}{a - x}$

(27) $\dfrac{1}{x + 1}$

(29) $x = \dfrac{-7}{12}$

(31) $x = \dfrac{-1}{7}$

(33) $x = \dfrac{7}{6}$

(35) $x = \dfrac{-11}{3}$

(37) $x = -11$

(39) False

(41) True

SECTION A-5

(1) 3^{11}

(3) 2^{15}

(5) 5^5

(7) 3

(9) 4

(11) $\dfrac{a}{b^7}$, or ab^{-7}

(13) $a^{10}b^{15}c^5$

(15) $\dfrac{a^4 b^8}{c^{12}}$

(17) $\dfrac{b^8 c^4}{a^{12}}$

(19) $a^4 c^9$

(21) $\dfrac{a^2 c^3 + ab^2 c - 1}{ab}$

(23) $ab^2 c^{1/2}$

(25) $\dfrac{ab^2}{c^2}$

(27) $a^{7/12} bc^{3/2}$

(29) $a\sqrt[3]{a}\, b^3\, \sqrt[3]{b}\, c^{12}\, \sqrt[5]{c}$

(31) $ab^4 c^2 \sqrt{ac}$

(33) $\dfrac{2a + 3b}{\sqrt{ab}}$

(35) $\dfrac{b(a + b)}{\sqrt{a}}$

(37) False

(39) False

(41) False

SECTION A-6

(1) $\dfrac{2}{\sqrt{5} + \sqrt{3}}$

(3) $\dfrac{2}{\sqrt{5} - \sqrt{3}}$

(5) $\dfrac{\sqrt{7} - \sqrt{2}}{5}$

(7) $3\sqrt{5} + 6$

(9) $\dfrac{h}{\sqrt{x + h + 1} + \sqrt{x + 1}}$

(11) $\dfrac{2h}{\sqrt{2(x + h)} + \sqrt{2x}}$

(13) $\dfrac{\sqrt{x + h + 1}}{x + h + 1} - \dfrac{\sqrt{x + h}}{x + h}$

(15) $\dfrac{1}{\sqrt{n^2 + 1} + n}$

(17) $\dfrac{n}{\sqrt{n^2 + 1} + n}$

SECTION A-7

(1) $x = 2 \pm \sqrt{6}$

(3) $x = \dfrac{-3}{2} \pm \dfrac{\sqrt{17}}{2}$

(5) No real solution

(7) $x = \dfrac{-1}{2} \pm \dfrac{\sqrt{5}}{2}$

(9) $x = 2 \pm \sqrt{2}$

(11) $x = \dfrac{3}{4} \pm \dfrac{\sqrt{37}}{4}$

(13) $x = 1, 2$

(15) $x = \dfrac{-3}{4} \pm \dfrac{\sqrt{41}}{4}$

(17) $a = 2, b = -3$

SECTION A-8

(3) $(-10, \infty)$

(5) $[-100, 0)$

(7) $[\sqrt{2}, \pi + 1]$

(9) Open interval: 6
Closed interval: 7
Open half-lines: 3, 4
Closed half-lines: 1, 2

(11) True

(13) False

(15) True

(17) False

(19) $\left(-\infty, \dfrac{-3}{2}\right] \quad \cup \quad \left[\dfrac{3}{2}, \infty\right)$

(21) $[-1, 1]$

(23) $[-5, 1]$

(25) $(-\infty, 0) \ \cup \ (0, 1) \ \cup \ (2, \infty)$

(27) $(-4, -2) \ $ or $ \ (3, \infty)$

(29) $(-\infty, -5) \ \cup \ (-3, -1) \ \cup \ (2, \infty)$

SECTION A-9

(1) $a^4 + 4a^3b + 6a^2b^2 + 4ab^3 + b^4$
$a^4 - 4a^3b + 6a^2b^2 - 4ab^3 + b^4$
$a^6 + 6a^4b + 12a^2b^2 + 8b^3$

(3) $32a^5 - 240a^4b + 720a^3b^2 - 1080a^2b^3 + 810ab^4 - 243b^5$

$\dfrac{8}{27}a^6 - \dfrac{4}{3}a^4\sqrt{b} + 2a^2b - b\sqrt{b}$

$\dfrac{16}{a^4} + \dfrac{32}{a^3b} + \dfrac{24}{a^2b^2} + \dfrac{8}{ab^3} + \dfrac{1}{b^4}$

INDEX

Glossary

$|x|$ — absolute value of x

$\lim\limits_{x \to a} f(x)$ — limit of $f(x)$ as x approaches (or tends to) a

$f'(x)$ or $Df(x)$ or $\dfrac{df(x)}{dx}$ — derivative of f evaluated at x

$f''(x)$ or $D^2f(x)$ or $\dfrac{d^2f(x)}{dx^2}$ — second derivative of f evaluated at x

$f^{(n)}(x)$ or $D^nf(x)$ or $\dfrac{d^nf(x)}{dx^n}$ — nth derivative of f evaluated at x

S.I. — strictly increasing

S.D. — strictly decreasing

$\displaystyle\int f(x)\,dx$ — antiderivative of f

$\left.\begin{array}{c}\displaystyle\int_a^b f(x)\,dx \\[2em] \displaystyle\int_a^b f\end{array}\right\}$ definite integral of f from a to b

$<A_n>$ — sequence $A_1, A_2, A_3, \ldots$

$\displaystyle\sum_{k=1}^{n} a_k$ — sum of the numbers a_k from $k = 1$ to $k = n$

exp — exponential function

ln — natural logarithm function

$\dfrac{\partial f}{\partial x}$ or f_x — partial derivative of f (with respect to x)

$\dfrac{\partial^2 f}{\partial x^2}$ or f_{xx} — second partial derivative of f (with respect to x)